Lecture Notes in Logistics

More information about this series at http://www.springer.com/series/11220

Henk Zijm · Matthias Klumpp
Uwe Clausen · Michael ten Hompel
Editors

Logistics and Supply Chain Innovation

Bridging the Gap between Theory and Practice

Springer

Editors
Henk Zijm
Faculty of Electrical Engineering
University of Twente
Enschede, Overijssel
The Netherlands

Uwe Clausen
Institute of Transport Logistics
TU Dortmund University
Dortmund
Germany

Matthias Klumpp
Institute for Logistics and Service Management
FOM University of Applied Sciences Essen
Essen
Germany

Michael ten Hompel
Dortmund
Germany

ISSN 2194-8917 ISSN 2194-8925 (electronic)
Lecture Notes in Logistics
ISBN 978-3-319-22287-5 ISBN 978-3-319-22288-2 (eBook)
DOI 10.1007/978-3-319-22288-2

Library of Congress Control Number: 2015946105

Springer Cham Heidelberg New York Dordrecht London

Printed on acid-free paper

Springer International Publishing AG Switzerland is part of Springer Science+Business Media (www.springer.com)

Foreword by Minister Melanie Schultz van Haegen

The Netherlands: gateway to Europe. It is this short phrase that underlines the unique position of our country. Our maritime ports in Rotterdam and Amsterdam, for instance, import 536 million tons of goods every year. Most of these are shipped to destinations further inland, traveling by road, rail, or water. Germany as our most important trade partner is a vital destination. Both countries realize that a strong industry cannot survive without a healthy logistics and supply chain sector. Apart from investments in infrastructure as taking place in both countries, innovation in logistics also needs a strong knowledge base.

The fact that Germany and the Netherlands already rank positions 1 and 2 on the World Bank's Global Logistics Performance Index 2014 is no reason to lay back. Faced with a fierce competition, among others from emerging economies, smart innovation has been made a cornerstone of our economic policy, including innovation in logistics and supply chain control.

For that reason, I welcome the initiatives of two of the most important logistics knowledge clusters, the Dutch Institute for Advanced Logistics (Dinalog) and the Effizienzcluster LogistikRuhr in Nordrhein-Westfalen, to join hands and start a long-lasting cooperation—a cooperation which is based on public–private partnership programs in which industry and academic institutes closely work together to the benefit of our industrial and logistics sectors.

In June 2014, a cooperation agreement was signed at the Fraunhofer Institut für Materialfluss und Logistik, attended by King Willem Alexander and Queen Maxima of the Netherlands, as well as the Ministerpräsidentin Hannelore Kraft of Nordrhein-Westfalen. This book is a first proof of the fact that such a cooperation bears fruit. It shows the richness and versatility of the logistics domain but above all it demonstrates that real logistics innovation can be achieved, when all partners are willing to join hands. I welcome the publication of this book and I sincerely hope that it may serve as a source of inspiration to both students and logistics practitioners.

Melanie Schultz van Haegen
Minister of Infrastructure and the Environment
The Netherlands

Foreword by Minister Michael Groschek

This book contains exemplary and innovative contributions regarding the state of the art in logistics research as well as practice, all aiming at improving the contemporary processes in supply chains. This is very important for economic development in general—and especially so for the two involved regions, home of the two leading research clusters DINALOG and EffizienzCluster LogistikRuhr, as the Netherlands and especially North Rhine-Westphalia play a crucial role in the overall setup of global and European logistics concepts.

As incorporated with this book, the two regions—awarded the leading ranks also in the World Bank global Logistics Performance Index—are complementary regarding global seaports as well as excellent Hinterland connections and value-added services in integrated and resilient logistics concepts.

The state of North Rhine-Westphalia has supported the development of the logistics sector on a concept level as well as at important locations such as Duisburg, Cologne, or Dortmund in the best possible manner, as it features Europe's largest inland port, the third largest German cargo airport as well as leading logistics research facilities as the University of Duisburg-Essen and the Fraunhofer Institute for Material Flow and Logistics in Dortmund.

In order to further strengthen the existing logistics excellence, our state is prepared and motivated to provide further support. This is implemented for example with the new ERDF program with the specific cluster support for "mobility and logistics." This competition for the funding of innovative ideas in logistics was started in April 2015 and has submission options in 2015 and 2016.

Logistics researchers, entrepreneurs, and their partners in industry and commerce within our state and beyond proceed to strive for exceptional results in building the supply chains of the future and providing sustainable as well as resilient and responsible services to our society.

Düsseldorf
July 2015

Michael Groschek
Minister for Building, Housing
Urban Development and Transport
of North Rhine-Westphalia

Preface

In today's global economies, logistics has been recognized as one of the key factors that determine the competitive position of both individual corporations and industry-based networks. At the same time, the very nature of supply chains and supply networks is changing rapidly, as a result of both technological and social developments. These developments include advances in ICT and industrial automation (sensors, robotics, 3D printing, and smart mobility) but also environmental concerns (scarcity of natural resources, carbon emission, and congestion) and finally new business models (e.g., e-commerce). The incorporation of these new technologies' potential in modern supply chain operations, while at the same time addressing environmental and societal concerns, is a formidable challenge for companies, economic clusters, and nations. But a challenge that has to be met: The importance of logistics as an indispensable factor of economic development is undisputed, as evidenced also by the annual publication of the World Bank Logistics Performance Index.

Innovation in logistics and supply chain management is a key to respond to the challenges outlined above. Such innovation requires intensive collaboration of industry and research and education institutes, to translate technological developments into sound business models and to train tomorrow's logistic engineer. For Germany and the Netherlands, the two focal innovation clusters are the Dinalog cluster and the EffizienzCluster LogistikRuhr. The editors of this book have ample experience in conducting projects that aim at the implementation of concepts and ideas in the day-to-day business environment (practicality gap). They also concluded there is a strong need in industry to understand the fundamentals of topics such as sustainable logistics, ICT integration, and Web-based businesses (theory gap).

From these experiences, the basic idea for this edited volume was born: to present state-of-the-art advances in logistics theory in different fields as well as to provide case studies for successful and promising logistics applications within important innovation areas in modern logistics management as best practice. This book reports on a number of studies carried out (and still ongoing) in the Dinalog cluster and the EffizienzCluster LogistikRuhr, bringing together different perspectives of basic and applied research. Above all, it should serve to inform the broader logistics and supply chain sector on what can be achieved by implementing novel and smart innovative ideas and what is needed to make these implementations successful.

In order to support this approach of bridging theory and practice in modern logistics, a selected portfolio of theory outlines, practical examples and case studies and in particular project reports or knowledge management documentations within different areas of logistics and supply chain planning is presented in this volume. The editors have selected contributions from a wide variety of projects carried out in the Dinalog cluster and the Effizienzcluster LogistikRuhr. Contributions are grouped into five main parts, each representing key domains in the evolution of logistics and supply chain management:

(A) Logistics innovation and sustainability;
(B) Urban logistics;
(C) Value chain management;
(D) IT-based innovation; and
(E) Logistics training and knowledge management.

Within each part, important topics are outlined and demonstrated through their application in a variety of case studies. This book is intended for both researchers and practitioners in the field of logistics and supply chain management, to serve as an important source of information for further research as well as implementation in practice and hence to stimulate further innovation.

The five parts are preceded by an introductory chapter by Henk Zijm and Matthias Klumpp. After a brief historical overview and a discussion of the need to design more sustainable supply chains, they list chances and opportunities and also discuss an approach advocated by the European Technology Platform for Logistics ALICE. The paper is completed with a discussion on training and competence management in logistics, including a preview on what may be expected.

Subsequently, Part I outlines *basic concepts and strategies* for sustainable and green logistics based on research and the implementation of new developments. Martijn Mes and Maria Iacob outline an approach of synchromodal transport planning in order to optimize transportation in light of greening the supply chain ("Synchromodal Transport Planning at a Logistics Service Provider"). In "DAVINC^{3}I: Towards Collaborative Responsive Logistics Networks in Floriculture", Jack van der Vorst, Robert Ossevoort, Marlies de Keizer, Tom van Woensel, Cor Verdouw, Edwin Wenink, Rob Koppes, and Robbert van Willegen describe research the results of a large research project on the development of a collaborative logistics network in the floriculture industry as a very high-value as well as high-quality example in terms of innovative logistics. In a larger perspective, sustainable multimodal hinterland networks, including the concept of extended gates, are discussed as a major approach toward green and cost-effective logistics by Albert Veenstra and Rob Zuidwijk ("Towards Efficient Multimodal Hinterland Networks"). Thomas Kjaergaard, Martin Schleper, and Christoph Schmidt suggest in "Current Deficiencies and Paths for Future Improvement in Corporate

Sustainability Reporting" that corporate sustainability reporting should be in the center of attention and management action in order to really achieve sustainable logistics. In an operational perspective, Simon Thunnissen, Luke van de Bunt, and Iris Vis are outlining the logistics impediments and chances of the use of LNG as a fuel for both the transport and maritime sector ("Sustainable Fuels for the Transport and Maritime Sector: A Blueprint of the LNG Distribution Network"). The final contribution in Part I comes from Raphael Heereman von Zuydtwyck and Holger Beckmann in "Efficiency Optimization for Cold Store Warehouses Through an Electronic Cooperation Platform", in which they discuss a specialized but promising approach regarding the use of online cooperation in cold store warehousing in order to reduce the environmental impact of this important section of transportation.

Parts II and III outline different levels of the logistics chain and optimization perspective. Whereas Part II deals with the local level in *urban logistics* concepts, Part III addresses the global level of *value chain design* and optimization. In Part II, challenges, failures, and successes of urban freight transportation are discussed by Goos Kant, Hans Quak, René Peeters, and Tom Van Woensel ("Urban Freight Transportation: Challenges, Failures and Successes"). In "The Role of Fairness in Governing Supply Chain Collaborations—A Case-Study in the Dutch Floriculture Industry", Robbert Janssen, Ard-Pieter de Man, and Hans Quak provide an insight into the impact of fairness considerations on local transport regimes in the floriculture industry. A further important aspect of urban logistics is last-mile parcel distribution, increasing steadily with e-commerce—and therefore the contribution of Theodoros Athanassopoulos, Kerstin Dobers, and Uwe Clausen is a welcome contribution that suggests options to reduce its environmental impact ("Reducing the Environmental Impact of Urban Parcel Distribution"). In "Order Fulfillment and Logistics Considerations for Multichannel Retailers" of this part, Kees Jan Roodbergen and Inger Kolman present a framework for decision making on order fulfillment and logistics in multichannel retail distribution.

In Part III, attention is paid to maintenance and service logistics. Maarten Driessen, Jan Willem Rustenburg, Geert-Jan van Houtum, and Vincent Wiers develop control structures for integrating decision making on inventory control and repair shop control for rotable spare parts ("Connecting Inventory and Repair Shop Control for Repairable Items"). In "Knowledge Lost in Data: Organizational Impediments to Condition-Based Maintenance in the Process Industry", Ronald van de Kerkhof, Henk Akkermans, and Nils Noorderhaven present a pilot study on the introduction of condition-based maintenance in the process industry, as a tool to increase asset availability. Jan Willem Rustenburg discusses the merits of a control tower approach for spare parts management as a radical new business model in "Planning Services: A Control Tower Solution for Managing Spare Parts". Finally, in "Impediments to the Adoption of Reverse Factoring for Logistics Service Providers", Christiaan de Goeij, Alexander Onstein, and Michiel Steeman focus on the adoption of supply chain finance methods, in particular reverse factoring, by suppliers in the logistics service businesses, as a tool to enhance chain liquidity.

Part IV is dedicated to the *information technology* enhancements driving many innovations in logistics and supply chain management. In "Towards an Approach for Long Term AIS-Based Prediction of Vessel Arrival Times", Alexander Dobrkovic, Maria Iacob, Jos van Hillegersberg, Martijn Mes, and Maurice Glandrup address how automatic information system data can be used to accurately predict vessel arrival times and thereby optimize logistics. More generally, the use of information technology as a tool for supply chain design, integrating various formerly isolated modules, is discussed and illustrated with case examples by Matthias Parlings, Tobias Hegmanns, Philipp Sprenger, and Daniel Kossmann in "Modular IT-Support for Integrated Supply Chain Design". Even more into current information technology research is the use of multi-agent systems, i.e., in transport coordination as presented by Frank Arendt, Oliver Klein, and Kai Barwig ("Intelligent Control of Freight Services on the Basis of Autonomous Multi-agent Transport Coordination"). Also the supply chain-wide implementation of RFID is still on the table for logistics innovation and value optimization as Kerem Oflazgil, Christian Hocken, Fabian Schenk, Oliver Teschl, Thorsten Lehr, Mareike de Boer, Christoph Schröder, and Rainer Alt outline in "Smart.NRW—RFID as Enabler for an Intelligent FMCG Supply Chain". The need to improve compliance to external regulations (e.g., customs) in supply chains, without delaying the flow more than necessary, presents a further challenge to smart information system design as Melissa Robles, Juan Diego Serrano, Maria Laura Maragunic, and Bernd Noche argue in "Developing Support Tools for Compliance in Supply Chains". A logistic assistance system to support quality control and quality management for logistic processes is presented by Markus Zajac and Christian Schwede ("Cross-Process Production Control by Camera-Based Quality Management Inside a Logistic Assistance System"). The last contribution of Part IV, "Logistics Mall—A Cloud Platform for Logistics" by Damian Daniluk, Maren Wolf, Oliver Wolf, and Michael ten Hompel, discusses the logistics mall, an approach for a domain-specific cloud platform for the trading and usage of logistics IT services and logistics processes.

Last but not least, the final Part V highlights the importance of *competencies and knowledge management for logistics* in bringing most innovation and technology approaches to full fruit. Therein, an approach for problem-oriented knowledge management in logistics is discussed by Natalia Straub, Christoph Besenfelder, and Sandra Kaczmarek ("Problem-Oriented Knowledge Management for Efficient Logistics Processes"). In "Logistics Qualification: Best-Practice for a Knowledge-Intensive Service Industry", Matthias Klumpp is providing an overview regarding measurement concepts as well as political initiatives directed toward best-practice approaches in logistics training and education. Finally, in "Serious Games for Improving Situational Awareness in Container Terminals", Alexander Verbraeck, Shalini Kurapati, and Heide Lukosch discuss the concept of situational awareness at container terminals as a basis for online (re)planning; they have developed various serious gaming-based instruments that have proven their value in the training of both students and practitioners in industry.

We would like to stress that many contributions include pilot or case studies at a large spectrum of industrial companies, which therefore essentially contribute to the objective of this volume: to bridge the gap between theory and practice in logistics and supply chain management. At this place, we extend our appreciation to their willingness to share current processes and data and to jointly work with academic partners toward improving business processes. But most of all, we are grateful to all the authors for their highly valued contributions; working with them was a rewarding experience. Finally, we express the hope that the projects discussed in this book may be of interest to practitioners in industry as well as to industrial engineering and logistics students, and that they may serve as a source of inspiration for further research. We look forward to the further application and implementation of the innovative concepts presented in this volume in industry.

June 2015

Henk Zijm
Matthias Klumpp
Uwe Clausen
Michael ten Hompel

Collaboration Agreement: Signing Ceremony

On May 27, 2014, the Effizienzcluster LogistikRuhr and the Dutch Institute for Advanced Logistics (DINALOG) signed an agreement to jointly work on logistics and supply chain innovation, in close collaboration with industries and the logistics sector in Germany and the Netherlands. The signing ceremony was attended by their Majesties King Willem Alexander and Queen Maxima of the Netherlands, the Prime Minister of North-Rhine Westphalia, Mrs. Hannelore Kraft, The Minister of Economic Affairs of the Netherlands, Mr. Henk Kamp, and the Mayor of Dortmund, Mr. Ulrich Sierau.

At the table (from left to right):

Mr. Willem Heeren, Chairman of the Board of the Dutch Institute for Advanced Logistics
Prof. Dr. Michael ten Hompel, Chairman of the Board of the Effizienzcluster LogistikRuhr

Standing behind the table (from left to right):

Mr. Henk Kamp, Minister of Economic Affairs, the Netherlands
Dr. Thorsten Hülsmann, Director of the Effizienzcluster LogistikRuhr
Mr. Ulrich Sierau, Mayor of Dortmund
Mrs. Hannelore Kraft, Prime Minister of the State of North-Rhine Westphalia
His Majesty King Willem-Alexander of the Netherlands
Her Majesty Queen Maxima of the Netherland
Prof. Dr. Henk Zijm, Scientific Director of the Dutch Institute for Advanced Logistics

Acknowledgments

The editors had to review a large number of abstracts and papers in different stages during 2014 and 2015. They gratefully acknowledge the help of the following referees:

- Dr. Albert Douma (Amersfoort),
- Prof. Hans-Otto Günther (Berlin),
- Prof. Willibald Günthner (Munich),
- Prof. Michael Henke (Dortmund),
- Prof. Kathrin Hesse (Köln),
- Prof. Jos van Hillegersberg (Enschede),
- Prof. Carlos Jahn (Hamburg),
- Prof. Goos Kant (Tilburg),
- Prof. Bert Leerkamp (Wuppertal),
- Dr. Alan Lewis (Nottingham),
- Prof. Torsten Marner (Essen),
- Dr. Martijn Mes (Enschede),
- Dipl.-Logist. Stephanie Möde (Dortmund),
- Prof. Roberto Montemanni (Lugano),
- Prof. Boris Otto (Dortmund),
- Prof. Jürgen Pannek (Bremen),
- Prof. Gregor Sandhaus (Düsseldorf),
- Prof. Thorsten Schmidt (Dresden),
- Prof. Iris Vis (Groningen),
- Prof. Nina Voijdani (Rostock),
- Dr. Gerwin Zomer (Delft).

Contents

Part II Urban Logistics

Part III Value Chain Management

Part IV IT-Based Innovation

Part V Logistics Training and Knowledge Management

Logistics and Supply Chain Management: Developments and Trends

Henk Zijm and Matthias Klumpp

Abstract The demand for sustainable logistic and supply chain processes poses enormous challenges in terms of technology integration, the development of new business models, cultural change and job qualification, and as such requires a real paradigm shift. In this paper, we start with a brief sketch of how modern logistics and supply chains emerged as a result of diversification and specialization of industrial production, globally scattered availability of resources and more demanding consumer markets. Jointly with advances in freight transport and communication technologies, these developments have led to the global economy we face today. The strong growth of trade and consumption however also revealed some essential weaknesses of the system that renders current practices in the long run unsustainable—in social, environmental and economic terms (people, planet, profit). Future supply chains should no longer deplete scarce natural resources or contribute to climate change, should avoid environmental pollution and withstand safety and security threats, while at the same time remaining competitive and satisfying high labor quality standards. This requires not only the application of advanced technologies to mitigate or even neutralize these negative effects, but also the development of smart business models, new job qualification standards and corresponding (lifelong) training and education programs at all levels, including artificial intelligence based learning.

Keywords Logistics · Global supply chains · Sustainability · Circular economy · Physical internet · Logistics education · Logistics trends · Artificial intelligence

H. Zijm (✉)
Dutch Institute for Advanced Logistics, Breda, The Netherlands
e-mail: w.h.m.zijm@utwente.nl

H. Zijm
University of Twente, Enschede, The Netherlands

M. Klumpp
FOM University of Applied Sciences, Essen, Germany

M. Klumpp
University of Duisburg-Essen, Essen, Germany

H. Zijm et al. (eds.), *Logistics and Supply Chain Innovation*,
Lecture Notes in Logistics, DOI 10.1007/978-3-319-22288-2_1

1 Supply Chains: Definition and a Short History

A supply chain concerns the entire production and distribution chain from raw materials to final customers and on top of that the return flow of products and possible re-use of materials or components (closed loop supply chain). Almost always, such a production and distribution chain is not executed by one industry but instead encompasses a number of companies and organizations jointly operating in a chain or network. This so-called end-to-end supply chain is represented in the Supply Chain Operations Reference (SCOR) model, see Fig. 1 (Poluha 2007; Simchi-Levi et al. 2008).

Current supply chains often span the entire globe and involve production, trade and logistics organisations around the world. For instance, in many European countries, most solid materials products and a significant percentage of food products are not produced in the region or country of use or consumption but elsewhere, not seldom even at other continents. In this paper, we investigate why worldwide production and—as a consequence—worldwide logistics flows have become so dominant, what their merits are but also why current practices in the long run are not sustainable, either in social, environmental or economical terms. To turn the tide, a fundamental rethinking of the way we organise production and logistics as well as logistics information management and education is needed. To understand this paradigm change, we first briefly review the way current production systems evolved, see also Hopp and Spearman (2008).

1.1 The First Industrial Revolution: The Principle of Labor Division

In 1776, the English economist Adam Smith published his "*An inquiry into the nature and causes of the wealth of nations*", in which he explained in detail the merits of what has become known as the "principle of labor division" and clearly demonstrated the productivity gains that could be achieved by systematically exploiting the learning curve (Smith 1776). That publication marked the start of the dominant philosophy of efficiency through specialization, worked out towards a first theory on production organisations by Charles Babbage (who later became known as the father of the digital computer) in his "*On the Economy of Machinery and Manufactures*" (Babbage 1835). The first industrial revolution, initiated by the invention of the steam engine, or rather its application in industrial production as engineered by James Watt, meant the definitive change from the classical domestic system and the craft guilds to mass production and mechanization, which has dominated production ever since. Mass production requires physical concentration and so the massive factories were born that colored the industrial landscape in the 19th and large parts of the 20th century. Ideas of mass production were governing the development of the steel industry by Andrew Carnegie, the large food

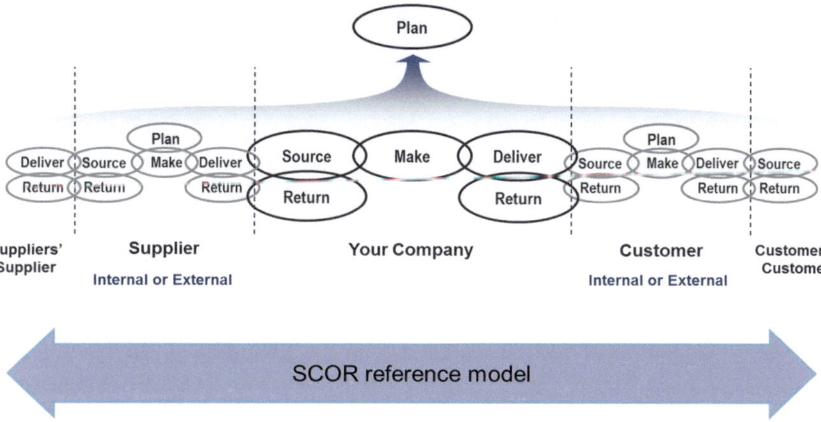

Fig. 1 SCOR model of end-to-end supply chains (*Source* Supply chain council)

conglomerates, the automotive sector and so on: scale, not scope was the leading paradigm. Famous became the reply of Henry Ford to the question in what color the T-Ford was going to be produced and customers might choose from: "Any color, as long as it's black". It was also the time in which the first scientific management theories were developed, with Frederick Winslow Taylor, now often viewed as the first industrial engineer, as its most famous representative. His work on time and motion studies, best working practices and in particular the differential piece rate system, was followed by pioneers such as Frank and Lilian Gilbreth, and Henry Gantt, who introduced the Gantt Chart in project management, while a first attempt to systemize quality management was developed by Walter Shewhart through his work at Bell Labs on Statistical Quality Control methods.

Mass production and limited product diversity continued to be the name of the game, also during the first decades after the Second World War. There was a shortage of the most basic goods; everything that could be made could be sold. However, starting with the sixties, as prosperity grew, consumers began to demand larger variety, leading to more complex products. In response, manufacturing industries introduced more versatile machines that could produce a variety of products, albeit at the cost of large setup or changeover times. The result was still production in large batches: economy of scale remained the leading philosophy. Also, the transfer of production to low wage countries in the Far East and Southern America was an attempt to sustain mass production at affordable costs. Efficiency also marked modern logistics: the introduction of the container and modern materials handling systems meant a big step forward in processing the ever growing logistics and material flows. Functional specialization, concentration of mass production in large factories, the shift of routine production to low wage countries, and as a result supply chains that tend to grow longer and longer, marked the industrial landscape still in the seventies and eighties of the preceding century.

1.2 From Mass Production to Flexible Manufacturing and Logistics

The two oil crises of 1973 and 1979 for the first time revealed also the weaknesses of the prevalent production philosophy. Raw material prices and interest rates raised sharply and industrial companies started to realize that long supply chains represented large amounts of stock and hence capital invested; besides, long supply chains make it hard to quickly adapt to changing market demand. Companies were inert, and not prepared for the flexibility that a changing society required. In addition, publications such as "Limits to Growth" from the Club of Rome (1972) stressed the depletion of natural resources and the pollution of our natural environment at an exponentially increasing rate (Meadows et al. 1972). For the first time industry and the public started to realize that current supply chains were to become economically prohibitive, and socially unacceptable.

And so, large scale batch production once applauded as the most efficient production philosophy now became the cause of all evil. Fortunately, new technologies proved to be at least a partial remedy. The introduction of flexible manufacturing systems, often based on computerized (CNC) machining and robotized assembly, helped to balance efficiency and flexibility, not only in production but also in the nodes of logistics networks, i.e. the material handling and distribution centers. In addition, attempts were made to synchronize and integrate supply chains by means of administrative information systems such as MRP and ERP, or by introducing new production philosophies such as Just-in-Time, or lean and agile manufacturing that focus on rigidly removing any buffer stocks as these were primarily seen as indications of waste or slack that characterize non-synchronized production. These were the heydays of the Toyota Production System and the SMED (Single Minute Exchange of Die) system, an engineering philosophy advocated by Shigeo Shingo, who systematically sought to reduce machine setup or changeover times, thereby again aiming at flexible, synchronized manufacturing and logistics (Shingo 1985).

Hence, although factories became more flexible, long and expensive supply chains due to functional specialization and dispersed production of parts and components continued to be the overarching story. At the same time, these supply chains contribute significantly to the BNP of those countries for which logistics is a strong economic sector, including Germany and The Netherlands (which ranked positions 1 and 2 on the World Logistics Performance Indicator 2014, published by the World Bank). The unprecedented growth of production and logistics and its far reaching rationalization as a result of modern manufacturing methods, the introduction of the container and above all the penetration of automation and computing technologies is definitely one of the sources of prosperity in most developed countries.

But this growth comes with a price and more and more it is realized that current supply chains are fundamentally unsustainable. This will be outlined in the next section.

2 Unsustainability of Today's Supply Chains

Current production and logistics systems cause serious and in the long run unacceptable environmental damage, due to for instance the emission of hazardous materials (CO_2, NO_x, particulate matter), congestion, stench, noise and more general the high price that has to be paid in terms of infrastructural load. While the European Committee has set clear targets to reduce Greenhouse Gas Emissions (GGE) in 2015 to 60 % as compared to 1990, the percentage of transport related GGE has increased from 25 % in 1990 to 36 % today (ALICE 2014). Besides, the pressure of the infrastructure needed on land use gives rise to additional social and environmental problems which hit urban areas in particular. Below, we first describe various phenomena which sometimes represent threats but in all cases pose at least important challenges to future supply chain management. In Sect. 3, we list ways and developments that may help to address these challenges.

2.1 *Scarcity and Sustainability*

Natural resources are scarce and not evenly distributed in terms of type and geographical location in the world. Logistic chains enable the distribution of materials, food and products from the locations where they are extracted, harvested or produced to people's homes and nearby stores. Current supply chains and logistics systems are global, partly due to natural conditions but certainly also because of labor rate differences between emerging and mature economies. First indications of reshoring production however become visible, not only because wage rates are moving upwards also in a number of Far-Eastern countries, but also since the amount of manual labor needed in high tech products continues to diminish, while logistic costs are increasing. As a result, future supply chains are believed to be *"glocal": global when needed, local when possible*. On the other hand, global supply chains will remain inevitable in cases where conditions for growing food ingredients are only satisfied in some regions in the world, or when minerals are only locally available. They will also continue to exist in cases where material processing consumes such an immense amount of energy that this is only sustainable at places where energy is abundantly and sustainably available, such as locations with geothermic energy, locations with water-powered energy generation, and locations with long periods of sunshine.

2.2 *Demographic Trends*

The current world population of 7.2 billion is projected to increase by 1 billion over the next 12 years and reach 9.6 billion by 2050, according to a recent United

Nations report (LOG2020 2013). Within Europe, population size is predicted to be stable—but a severe shift in population movements is expected from Eastern to Western Europe. Ageing continues, meaning that people in general will work longer in order to maintain a reasonable standard of living. Europe-based companies should be prepared for scarcity of human resources and should be able to provide working conditions that extend the working life of employees. The need to further increase productivity while at the same time diminishing the ecological and social footprint, requires a quality upgrade of the human resource pool, e.g. by better education and training, including lifelong learning programs. In parallel productivity can be improved by better support tools, easier access to relevant information, and finally further automation of both technical processes (i.e. robotics) and decision making (i.e. artificial intelligence).

2.3 *Urbanization*

As urbanization continues[1] it becomes an unprecedented challenge to keep cities livable, which includes a sustainable logistics planning and execution. The development of wealth in Asia and Latin America has resulted in a huge shift from agricultural and nomadic forms of living to urban life. More and more cities with over 10 million inhabitants will emerge requiring different modes of transport and logistics systems than available today. There is an increasing interdependency between supply chain design or management and urban planning or land-use management. It is not yet clear whether mega cities are sustainable when wealth increases to the levels currently accessible for the population in developed countries. Innovative sustainable, safe and secure logistics might inspire agencies and institutions towards new patterns of sustainable urbanization.

2.4 *Supply Chain Safety and Security*

Border-crossing supply chains and logistics systems often concern high-value goods, and therefore are vulnerable to crime and illicit acts. Within the European research programs, various projects have developed roadmaps to enhance supply chain safety and security. Regarding safety, extensive attention has been paid to safe working conditions (and for instance driving hour regulations) but the fight for supply chain security, abandoning crimes and illegal activities, appears to be a harder one. Economic crimes for example include: theft (robbery, larceny,

[1]In 2007 the world passed the point in which more than half of its population is living in urbanized areas, in some developed countries the urban population percentage is well above 70 %, and continues to rise.

hijacking, looting, etc.), organized immigration crime (human trafficking, illegal immigration), IPR violations and counterfeiting and customs law violations (tax fraud, prohibited goods). Alternatively, ideologically or politically motivated crimes occur, next to obvious vandalism (Hintsa 2011). A legislative framework may in principle safeguard society against these unwanted practices and provide a mandate for government authorities to act. However, the challenge often is to find a balance between required inspections and interventions, and the economic interests of shippers and logistic service providers who wish to minimize delays, inefficiencies and additional costs.

Another aspect of supply chain security is supply chain resilience, which can be defined as the ability to maintain, resume, and restore operations after a major disruption (Gaonkar and Viswanadham 2007). This is a critical aspect of supply chain risk management and is generally seen as one of the major future challenges. Disruptions to supply chains can prove costly, as highlighted by a variety of natural disasters. According to research conducted by Accenture, significant supply chain disruptions have been found to cut the share price of impacted companies by 7 % on average (WEF/Accenture 2013).

2.5 *Changing Consumer Markets*

Commercial product life cycles tend to become still shorter. At the same time we observe an increased re-use of products, components and materials, both via (electronic) second markets and in so-called closed-loop supply chains (cradle-to-cradle, circular economy). Mass customization is an important aspect of current consumer markets, enabled by fast technological developments (to be discussed below). The rapid advance of e-commerce is another characteristic of today's markets: on the one hand it reduces the number of links in the supply chain, but without adequate regulation of both forward and reverse flows of packages it often leads to a rapid additional increase of urban congestion and pollution. Finally, we note that in some sectors customers no longer buy an actual product but only the service the product represents (e.g. cloud computing, music streaming, car sharing). These phenomena will have a profound impact on the ecological footprint of mankind and as such also on the design and planning of future supply chains.

All phenomena sketched above pose important challenges to future supply chain design, planning and control. Fortunately, technological innovations are extremely helpful to at least partially address some of these challenges. But technological innovation alone is only a part of the story; at least equally important are the development of smart and fair business models based on joint responsibilities and fair allocation of revenues instead of on individual profit maximization, and the mind shift needed for all stakeholders concerned, which in turn requires high investments in (lifelong) training and education programs (cf. Sect. 4).

3 Chances and Opportunities for Future Supply Chains

The observations outlined in the preceding section call for a fundamental paradigm change when redesigning future-proof supply chains, i.e. supply chains that are able to efficiently deliver goods and services when and where needed, while respecting social and environmental constraints. Fortunately, both technological and social-economic innovations provide adequate tools that may help to address that challenge.

3.1 New Materials and Manufacturing Technologies, Design for Logistics

The design of new and lightweight (bio-)materials and their application in a wide variety of products poses exciting new possibilities to diminish both the costs and ecological footprint of these products. Rapid advances in such fields as polymer technology, bio-engineering and nanotechnology already lead to products that could not have been imagined only 10 years ago. Technologies like 3D-printing and micro-machining are also a step forward towards mass-customization but in addition have a profound logistic impact, for instance in stimulating "local for local" production. In addition, 3D printing which is believed to have a future in particular in small batch and one-of-a-kind manufacturing, may lead to far shorter lead times and hence a reduction of so-called anticipation (safety) stocks, because it allows production at the place and time needed. Another manifestation of improvement through technology is the continuous development of cleaner engines and non-fossil fuel based engines (e.g. electric, hybrid or LNG-powered vehicles for city distribution and local passenger transport, but also for both inland and sea vessels as an attempt to diminish the environmental footprint. It is important to realize the importance of an integrated supply chain view when focusing on reducing their negative impacts. As an example, consider product design. Modular product design allows the transport of components instead of full products which not only results in a higher package density but in addition again allows customization closer to the end-user. Also note that that 3D-printing and additive manufacturing in general is based on material addition, instead of material removal as in classical machining, hence in principle has a waste avoidance potential. Smart packaging logistics also may help to reduce volumes and to avoid waste, in particular in the case of bio-degradable package materials.

3.2 Automation and Robotics, Internet of Things

The impact of robotics has already been visible for a long time e.g. in automotive assembly lines but also in warehouses and distribution centers, in so-called ASRS

(Automatic Storage and Retrieval Systems), often consisting of high bay storage racks which are served by fully automated cranes, and equipped with automatic identification, i.e. RFID technology. Apart from the visible hardware, innovative warehouse management systems help to coordinate and synchronize activities, in close communication with information systems covering both suppliers and customers. Similar developments can be found at container terminal sites in both seaports and inland harbors. Without exception, all such systems rely heavily on smart sensor and actuator systems, recently evolving towards the so-called Internet of Things, where devices are equipped with sensors that automatically signal when actions such as ordering or replenishment have to be initiated. Additionally, materials and machinery themselves are able to communicate with each other and find solutions based on decentralized and autonomous decision making using state-of-the-art algorithms. In particular the world of both passenger and freight transport is currently innovating rapidly, as demonstrated for instance by various experiments with freight vehicle platooning, in which a convoy of freight trucks is controlled by a single driver. Vehicle transportation in 2050 is foreseen to be largely unmanned transportation.

3.3 Business Information Systems, New Business Models

Although many scholars view business information systems and architectures as belonging to the field of technology development, it is essentially more than that. Complex modern supply chains are first and foremost characterized by the fact that many stakeholders are involved in shaping its ultimate manifestation. Direct stakeholders are of course suppliers of raw materials, product designers, manufacturing and trading companies, logistics service providers, forwarders and transport companies, and ultimately the customer. Indirect stakeholders are supply chain financiers, ICT consultants, local and regional governments in their role as infrastructure providers but also as representatives of societal interests, customs authorities and indeed the public at large. The multi-stakeholder and multi-decision maker environments we deal with require adequate mechanisms to respond to their requirements, including distributed architectures, cloud computing solutions, cognitive computing and agent-based decision support systems. Organizational innovations are indispensable to fully exploit the potential of advanced information and decision support architectures. The recent attention for data driven models (big data analytics) marks an important further step towards full-blown automated decision architectures.

The design and acceptance of decision models based on both horizontal and vertical cooperation in supply networks however proves to be one of the most difficult steps to make. Although many stakeholders quickly recognize the potential win-win situation arising from collaboration they find it in general extremely hard to give up their autonomy. Mathematically, game-theoretical approaches have proven to provide adequate tools to handle such multi-stakeholder games, for instance the Shapley value calculation defines a "fair" allocation of cooperation gains to individual actors. But the key idea—established in the Nash equilibrium theory—that players may give up their individual optimal solution in order to achieve an overall stable equilibrium solution is still hard to accept in particular for private companies that were used to concentrate on their individual profits. This is perhaps the biggest hurdle to be taken to arrive at sustainable logistics; it involves not only smart business solutions but more important a change of mindset and indeed trust in the value of collaboration.

3.4 Circular and Sharing Economy, Servitization

The key idea behind servitization is the realization that both private consumers and industrial asset owners basically need the functionality of assets and products, rather than the products itself (Cohen et al. 2006; Neely 2008). Initially, this idea has led to the establishment of after sales service models that aim to deliver improved availability and system performance based on smart service level agreements. One step further is not to sell products anymore but to lease them, or to provide "power by the hour" support as some industrial equipment suppliers already do. Apart from the long-term relationship between supplier and customer and the emphasis on lifecycle support it also enables a planned take-back and renewal of systems at the end of their functional lifetime. The circular economy is based on the idea that products and systems that are disposed of can be either restored and reused or disassembled after which components and parts are given a second life in next generation equipment. Another option is to jointly use equipment in a predefined group of people. Those products or systems are either owned by individual group members or remain property of the supplier and can be leased or hired at moderate costs. It may be clear that such developments may have important consequences for supply chain design, planning and control in that the focus may at least partially switch from delivering products to delivering services. An example can be recognized in the trend towards car sharing, implying for logistics that cars may not have be to delivered to the individual end customer (via the usual dealers) but more centralized towards several car sharing operators, changing the setup of the distribution logistics concept entirely.

Other important developments concern e.g. the design of new supply chain finance models, such as reverse factoring, and developments in marketing and sales. Figure 2 displays a key summary of the main developments that may serve as drivers for innovation in supply chain management, regarding the physical, information and financial flows (Zijm and Douma 2012).

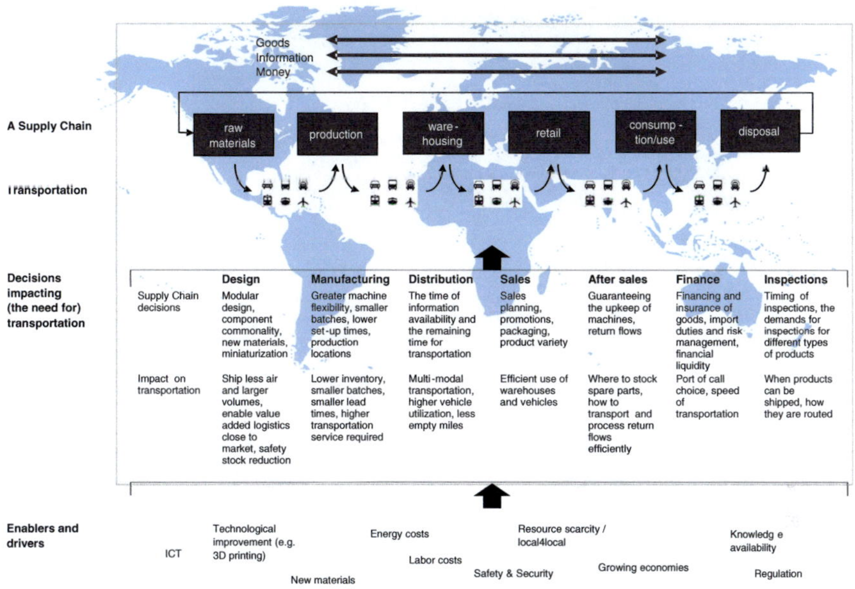

Fig. 2 Decisions impacting supply chain sustainability (including transportation)

4 Roadmaps Towards Sustainable Logistics and Supply Chain Management

In this section we briefly sketch some developments at a European level, more precisely the foundation of the European Technology Platform (ETP) for Logistics ALICE (Alliance for Logistics Innovation and Cooperation in Europe). Within ALICE, five working groups have developed roadmaps that address the issues sketched in the preceding sections (ALICE 2014), converging to the Physical Internet paradigm. The five roadmap topics are listed below and visually integrated in Fig. 3):

- *Sustainable, safe and secure supply chains*: this roadmap aims at a thorough rethinking of the contents of the goods flow. How to design products and processes such that efficient logistics is enhanced at minimal social and environmental costs, and how to enhance safety and security (protection against theft and other illicit actions, but also supply chain resilience in case of natural disasters,
- *Global supply network coordination and collaboration*: to further enhance logistics efficiency, cooperation is needed, both along the supply chain and across various heterogeneous supply chains (i.e. horizontal and vertical synchronization). Combining flows and integrating forward and reverse flows may prove to provide adequate remedies against too low transport loads and empty vehicles drives.

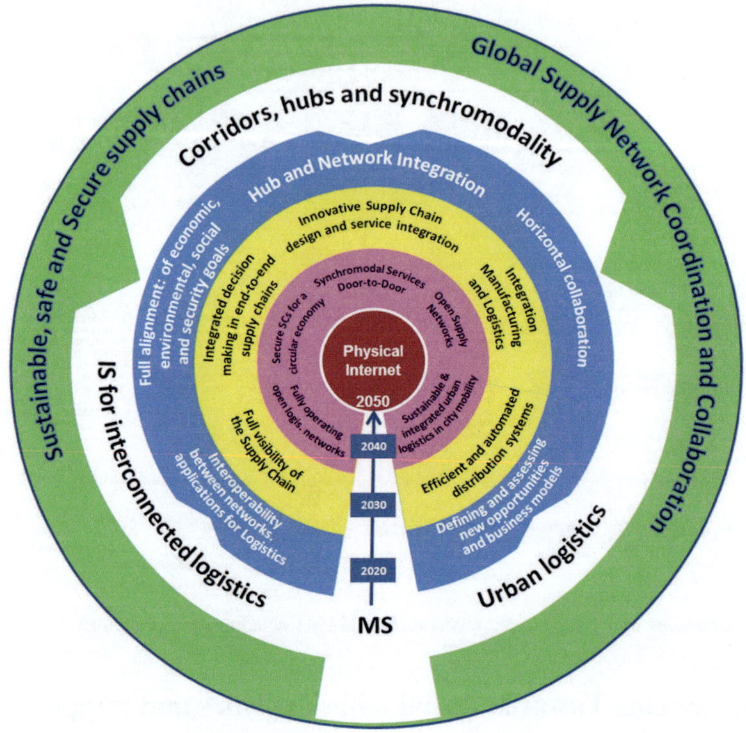

Fig. 3 ALICE roadmaps and their interactions: towards the Physical Internet

- *Corridors, hubs and synchromodality*: hubs and corridors are key infrastructural elements to increase efficiency of transport. Synchromodality is the concept to exploit multiple modes in a flexible, dynamic way, and in this way to shift loads from road to e.g. rail or water, based on a reconciliation of economic and environmental goals (people, planet, profit).
- *Information Systems for interconnected logistics*: there is no doubt that a sound and, more important, uniform (single window) and transparent information infrastructure is essential to achieve a tight interplay between the many stakeholders involved, to really achieve economic and ecological sustainable solutions that integrate decision making along the supply chain.
- *Urban logistics*: the trend towards further urbanisation still continues which calls for drastic policies to preserve or enhance quality of life. The rapid advance of *e-commerce* causes a sharp growth of small load (forward and reverse) freight flows which further increases pollution, noise and congestion. Sound infrastructural policies are needed to address these problems in order to keep cities attractive as centres of economic activities.

The roadmaps and instruments outlined above all form important stepping stones towards more sustainable logistics and supply chains. In the long run, however, the concept of a "Physical Internet" as proposed by Montreuil (2011) may serve to integrate all these elements in a radical new logistics framework. The Physical Internet is defined as a logistics system in which modular packages are automatically routed from source to destination through a network of hubs and spokes. Major elements of such a network more or less exist for parcels, pallets, containers and "swap bodies". Carriers of these types of loading units do optimize between various alternative routes in their networks, e.g. by bypassing hubs, either in advance through offering more time definite services, or real time during the actual transport. A full-fledged physical internet may be built upon all these elements with the holistic integration of existing elements and concepts as the main challenge.

The Physical Internet should not be confused with the Internet of Things; the latter refers to the possibility of communicating devices, often followed by local actions initiated by software agents. Internet of Things technology may be an important building block of the Physical Internet, e.g. in determining alternative routes in case of congestion on the preferred route, or in signalling a potential quality loss in case of delays (e.g. in food and flower transport). But the Physical Internet is a full-fledged alternative to a classical, manually operated logistics network, with important consequences for all stakeholders involved but first and foremost transport companies and service providers.

Why is the Physical Internet a radical solution? Basically, because decentralized market economy mechanisms usually prevent holistic optimization as many providers of transport and logistics services are "locked-in" in their current ways of working and acting. This happens for instance by fixed or sunk costs in specific equipment, preventing the individual actor from agreeing to large-scale collaboration and optimization. To arrive at such enhanced cooperation levels, shippers, manufacturers, retailers, carriers and other providers of logistics services should take the broader sustainability goals into the economic equation. This requires new ways regarding decision making in the system on financial and market criteria but also on safety, security and environmental aspects. In particular, transnational governance and regulation is needed to achieve such a cultural shift, and to encourage collaboration, coordination and horizontal partnerships.

A major challenge is to design a multifaceted decision support system for the Physical Internet, with partly automated execution via intelligent agents. Radical new business models based on openness and sharing or resources are required, as opposed to the current local ownership and control of resources. Note that this notion of openness is almost in contrast with the core of e.g. supply chain security. Therefore, the adoption of a Physical Internet will require radical changes with respect to the roles and responsibilities of many stakeholders. Achieving such a combination of physical and electronic infrastructure is just one step, stimulating shippers and logistics operators to connect to it, is an even bigger challenge.

5 Logistics Training and Competence Management

So far, we only obliquely touched the topic of human skills and competences in logistics. However, it goes without saying that a fruitful contribution of technology innovation and smart information systems to sustainable supply chains critically depends on the presence of a competent workforce at all levels in both private organisations and government (LOG2020 2013). Training is needed to adopt and adequately apply new technologies while the design of smart business models requires academic skills to understand the increasing complexities of modern global supply networks. Figure 4 displays the essential three elements and their mutual relations that are essential in shaping future supply chains. Indeed, apart from natural resources and despite the growing world population, the quality and availability of human competencies appears to be the most important limiting factor (Wu 2007). That recognition was an important argument for several countries to invest not just in physical but above all in logistics competence clusters, of which the DINALOG cluster in the Netherlands and the Effizienzcluster LogistikRuhr in Germany are two notable examples. Such clusters not only govern applied research but moreover play an important role, together with the available knowledge infrastructure, in training and upgrading logistics operators and management alike. In this section, we take a closer look on logistics competence management.

The first scientific description of sustainability as a general concept is due to the 18th century German agriculture and forest academic *Hans Carl von Carlowitz*. In 1713 he defined sustainability in relation to wood cutting, as follows: "Wird derhalben die größte Wissenschaft und Einrichtung hiesiger Lande darinnen beruhen, wie eine sothane Conservation und Anbau des Holtzes anzustellen, daß es eine

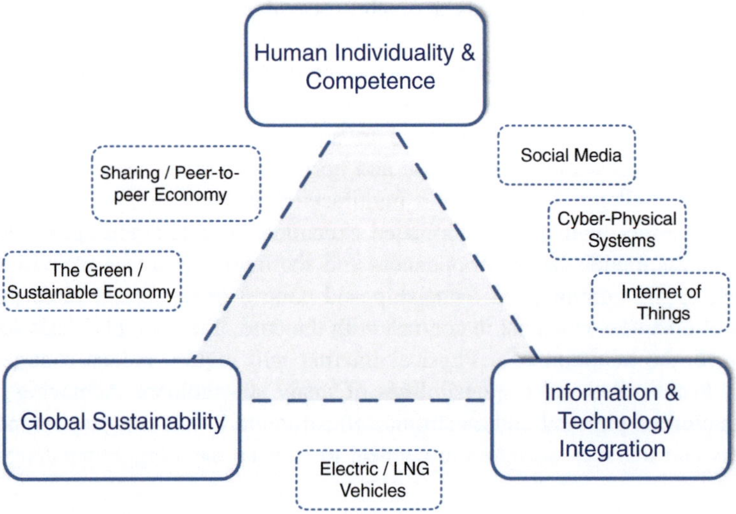

Fig. 4 Essential elements in shaping future global supply chains

continuierliche beständige und *nachhaltende* ["sustainable"] Nutzung gebe, weil es eine unentberliche Sache ist ohne welche das Land in seinem Esse nicht bleiben mag"[2] (*Sylvicultura Oeconomica*, pp. 105–106).

If transferred to the question of available qualification and competencies of the human workforce, it is striking that the amount of research that has studied the development of human knowledge and competence in a general sense is fairly limited, in spite of the recognition that innovation essentially requires a close match between technology, policy and business support, and human skills development (i.e. Aghion et al. 2009). This is further outlined in Fig. 5, depicting an average competence level of the human workforce in a very simplified manner, inspired by an "application of knowledge paradigm".[3]

It can be stated that, starting from the industrial revolution ("A"), the necessary or expected competence level of the workforce has increased on average. For logistics processes especially it can be argued that this still ongoing process has a "double nature". First, existing activities such as truck driving, warehouse processes or production processes increasingly demand higher level competences—as demonstrated for example by the new vocational training for truck drivers in Germany or by EU regulations that request further training of drivers regarding safety regulations, sustainability and hazardous goods as well as technology usage. Whereas only thirty years ago truck driving was a typical "unskilled" profession without any necessary training to do the job properly (apart from a truck driver's licence), today no individual could just "start driving" a truck in complex transport processes as a multitude of systems (toll systems, routing systems, communication systems, auto ID systems etc.) have to be mastered. This development can be named "knowledge and competence *enrichment*" of existing processes. Second, new activities arise in logistics and global supply chains, typically with a very high knowledge and competence requirement, such as IT systems management, logistics consulting, logistics and supply chain finance, logistics tender management, logistics controlling and so on. This trend can be titled "knowledge and competence *extension*" as new processes increasingly demand new and higher knowledge and competence levels (see e.g. Klumpp 2013).

Between the ever-increasing expectations and requirements regarding human competence levels (the upper black line in the figure), a distinctive "gap" is opening up over time as the required training for humans has for each and every person to start "anew": learning cannot be "inherited" or automated. Longer schooling and training programs are needed in order to arrive at the required competence levels of a modern logistics and business environment. This can be termed a "*knowledge accumulation gap*" (grey field in the figure) that arises due to the fact that humans

[2] "The first and foremost objective of research and practical application shall therefore be a method securing the establishment and preservation of tree cultivation, enabling a continuous and sustainable use as this is an essential prerequisite for the well-being of the land."

[3] Competence is here understood as the *application* of knowledge and information for a given (exemplified) *real-world* problem, e.g. the transportation and placement of a container in a seaport terminal or the production of a car specified by a customer.

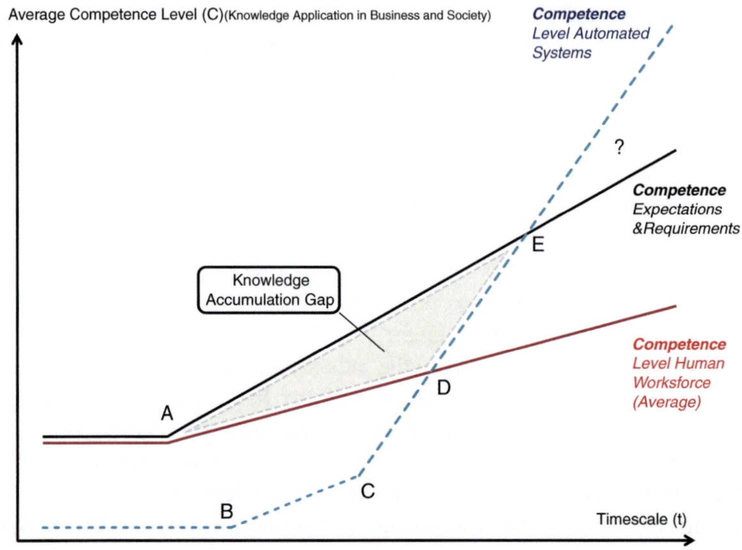

Fig. 5 Competence development for humans and artificial intelligence

are not able to accumulate knowledge over generations, as opposed to machines and computers that are increasingly able to do exactly that. Artificial intelligence based machines may in the future mitigate or close this gap.

Besides human knowledge and competence levels, automated or artificial intelligence (non-human) competence levels (dotted line in the figure) are expected to impact technology and business development. Though artificial intelligence had a somewhat "slow start" during the sixties, seventies and eighties of the preceding century (Newell 1982), it has significantly accelerated in solution contribution width and depth recently. This is connected to the trend of "deep learning", allowing computers autonomously to acquire new knowledge and to find links as well as directions of further learning and meaning *themselves* (Erhan et al. 2010). As outlined in Fig. 5 (point "B"), automated systems were initially very slowly adopted;, examples in logistics include the *automated gearbox* in trucks, partly automated cranes and warehouse equipment as well as automated communication and transmission devices in logistics management (EDI systems, automated decision protocols). These "separate" and limited systems never really "matched" human competence levels, which is why the dotted line traverses significantly below average human competence levels between B and C. This was further demonstrated by introducing only slightly different context variables or *external changes* that proved to be too hard for these early automated systems, making human intervention necessary in every case.

In recent years however—symbolized by point "C" in Fig. 5—automated systems have undergone a drastic change, only partly described by a "merging" or an "integration point" as formerly separate systems now are increasingly coupled and are beginning to *interact*. This is on a rudimentary level the idea of the "Internet of Things" as outlined above. For example, current automated warehouses are integrated systems of software (warehouse management systems), hardware (moving goods) and even optimization (error analysis, automated storage optimization). This integration tremendously increases the capability of such systems and accelerates their "innovation speed" significantly.

In some cases artificial intelligence and automated systems are already "overtaking" human competence levels (point "D"). For truck driving for example, the "intelligent" combination of the "old" automated gearbox with GPS-based navigation systems allows a state-of-the-art truck to actually downshift before the steep slope of an oncoming mountain street is even visible to the human driver. This form of "foresight and decision as well as action" is a new capability of automated systems, which has recently passed new levels in freight platooning and automated passenger car driving experiments. The general prediction is that automatically driven vehicles are in the long run safer than those controlled by humans; the main obstacles to their further introduction are of a legal rather than a technical nature. More and more we see machines taking over increasingly complex tasks—not incidentally the year in which machines started to play continuously better chess than the best human being was in 2005.[4]

What comes after point "D" can only be speculated but it is not unlikely that in the future a point "E" is reached, where automated systems even *exceed* the expectations (set by humans) of society and business. This may sound risky, as "unintended and unforeseen behavior" of automated systems may rightly worry humans—not unintentionally artificial intelligence is listed as a risk by the "Centre for the Study of Existential Risk" in Cambridge. But as most technologies, it can easily be argued that risks and opportunities are usually embedded in any development, from the taming of fire to the atomic chain reaction and similar regarding artificial intelligence applications. Just for a short insight, some of an unknown multitude of applications and developments can be listed for the area beyond "E" signified with a question mark:

- Automated trucks may for example leave the motorway having information about a jam ahead or even a severe accident in order to make way for emergency operations.
- Automated production systems may increase output on specific workdays (Thursday say) due to an identified structure and repeating sequence for customer demand.
- Automated logistics systems may decide to switch to a different supplier in another country having analysed reports about imminent hostilities or fraud in the current supplier's country.

---

[4]The chess computer IBM "Deep Blue"; the first victory was already in 1997 against Garry Kasparov, but until 2005 human players still were able to score in some cases against computers. Today in chess ranking "ELO points" computers lead unchallenged with 3304 against the best human with 2.882 (Magnus Carlsen in 2014).

In any case, the future regarding logistics education and training (human as well a non-human) will be increasingly interesting and important to innovation processes in logistics.

In the light of the described qualification and training developments, the general model of innovation in a partnership of technological development and human qualification and training in implementation has to be scrutinized. In the past, the sequential model "A" in Fig. 6 was mostly implemented. This model of technology development, followed by implementation and finally training has a clear structure and also a very distinctive risk mechanism—workers were only trained for technologies already developed and implemented, "lost training" was therefore seldom. But in the modern world, this model is outdated as (i) the total time for change and innovation is shrinking, leaving no time for the "luxury" of sequential approaches and (ii) the volume and depth of training has severely increased, but is at the same time essential to complete a successful timely innovation.

The current model "B" is using a parallel approach for at least a part of the timeline, regarding implementation and training experiences as essential input for further technology development ("user involvement" in research and development). In some cases trainings even start before the actual implementation, i.e. before new machinery or software arrives at the manufacturing or office floor. A future outlook is suggested in model "C", where in an environment of largely automated blue and white collar work the innovation process may even be devoid of any large-scale human training. In such systems, human roles are limited to technology development as well as general oversight. Artificial intelligence and robot appliances may take over the innovation process completely by introducing new manufacturing as well as management decision concepts without necessary detailed human training.

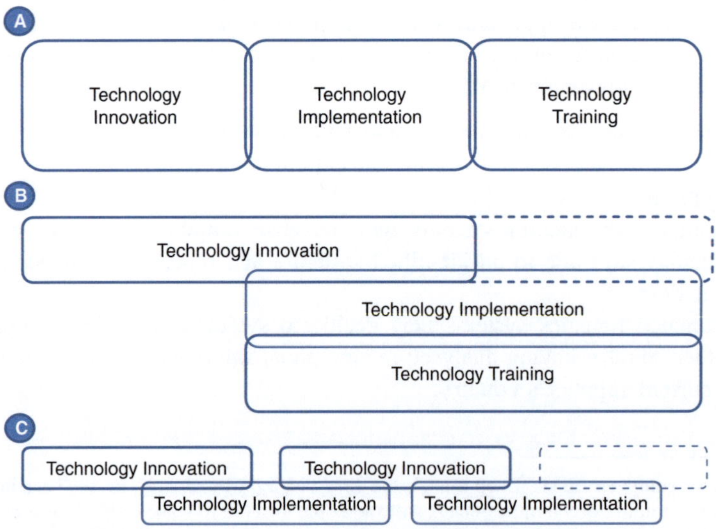

Fig. 6 Innovation and training regimes in the past, present and future

This may also enable the phases of technology development and implementation to go largely "hand in hand" as innovation cycles grow shorter and shorter. Finally, such a foresight implies that technology development and implementation are two intertwined parallel infinite processes—as can be observed already in the smartphone app market today.

6 Outlook

As argued in this chapter we expect the landscape of modern logistics and supply chain management to change significantly in the next two decades, a shift unprecedented in its magnitude and impact. A comparative figure may be the development of the global *smartphone* market in the consumer arena: in 2007 there were only some "smart phones" as Apple introduced the iPhone on January 9. In 2015, only eight years later, there are more than 2 billion smartphone *users* on the planet, each on average with a computing power larger than the NASA Apollo 11 lunar module used in 1969 to put the first human being on the moon. In 2013, the annual sales volume exceeded one billion smartphones alone, bringing the average user lifetime per device to some one and a half years—a very fast and flexible market indeed. The smartphone device has already revolutionized private life, from buying to dating, communicating and learning to even selling goods and services, increasingly in a sharing economy—and all in less than a decade. An increasing number of applications and services—from travel booking to health analysis—are based on the smartphone device and are evolving faster every year. This comparative picture may indicate what the Internet of Things as well as the full use of automated systems and artificial intelligence may imply for the business context and global supply chains, as well as for its associated learning processes: nothing less than a revolution on how we do business and logistics lies ahead.

References

ALICE (2014) European technology platform for logistics. http://www.etp-logistics.eu/

Aghion P, David PA, Foray D (2009) Science, technology and innovation for economic growth: linking policy research and practice in 'STIG Systems'. Res Policy 38:681–693

Babbage Ch (1835) On the economy of machinery and manufactures, 4th edn. Charles Knight, London

Cohen M, Agrawal N, Agrawal V (2006) Winning in the aftermarket. Harvard Bus Rev 84 (5):129–138

Erhan D, Bengio Y, Courville A, Manzagol P-A, Vinvent P, Bengio A (2010) Why does unsupervised pre-training help deep learning? J Mach Learn Res 11(1):625–660

Gaonkar RS, Viswanadham N (2007) Analytical framework for the management of risk in supply chains. IEEE Trans Autom Sci Eng 4(2):265–273

Hintsa J (2011) Post-2001 supply chain security—impacts on the private sector. Doctoral Thesis, HEC University of Lausanne (Chapter 7)

Hopp W, Spearman M (2008) Factory physics: foundations of manufacturing management, 3rd edn. Waveland Press Inc, Long Grove
Klumpp M (2013) How to structure logistics education: industry qualifications framework or topical structure? In: Pawar KS, Rogers H (eds) Resilient supply chains in an uncertain environment, ISL 2013 Proceedings, Vienna/Nottingham, pp 895–902
LOG2020 (2013) Logistics and supply chain management 2020: training for the future. WP2: positioning paper. ERASMUS LLL project ID: 527700-LLP-1-2012-1-NL-ERASMUS-EMCR
Meadows DH, Meadows DL, Randers J, Behrens WW III (1972) The limits to growth. Universe Books, New York
Montreuil B (2011) Towards a physical internet: meeting the global logistics sustainability grand challenge. Logistics Res 3(2–3):71–87
Neely A (2008) Exploring the financial consequences of the servitization of manufacturing. Oper Manag Res 1(2):103–119
Newell A (1982) The knowledge level. Artif Intell 18(1):87–127
Poluha RG (2007) Application of the SCOR model in supply chain management. Youngstown, New York
Shingo S (1985) A revolution in manufacturing: the SMED system. Productivity Press
Simchi-Levi D, Kaminski P, Simchi-Levi E (2008) Designing and managing the supply chain: concepts, strategies and case studies. McGraw-Hill International, Boston
Smith A (1776) An inquiry into the nature and causes of the wealth of nations. W. Strahan, London
von Carlowitz HC (1713) Sylvicultura Oeconomica oder haußwirthliche Nachricht und Naturgemäßige Anweisung zur Wilden Baum-Zucht, (edited by Johann Friedrich Braun), Leipzig
World Economic Forum/Accenture (2013) Building resilience in supply chains. An initiative of the risk response network in collaboration with Accenture, Geneva
Wu YC-J (2007) Contemporary logistics education: an international perspective. Int J Phys Distrib Logistics Manag 37(7):504–528
Zijm H, Douma A (2012) Logistics: more than transport. In: Weijers S, Dullaert W (eds) Proceedings of the 2012 freight logistics seminar. Venlo, The Netherlands, pp 395–404

Part I
Logistics Innovation and Sustainability

Synchromodal Transport Planning at a Logistics Service Provider

Martijn R.K. Mes and Maria-Eugenia Iacob

Abstract In this chapter, we consider synchromodal planning of transport orders with the objective to minimize costs, delays, and CO_2 emissions. Synchromodal planning is a form of multimodal planning in which the best possible combination of transport modes is selected for every transport order. The underlying problem is known as the multi-objective k-shortest path problem, in which we search for the k-shortest paths through a multimodal network, taking into account time-windows of orders, schedules for trains and barges, and closing times of hubs. We present a synchromodal planning algorithm that is implemented at a 4PL service provider located in the Netherlands. We illustrate our approach using simulation with order and network data from this logistics service provider. On the corridor from the Netherlands to Italy, an average cost reduction of 10.1 % and a CO_2 reduction of 14.2 % can be achieved with synchromodal planning.

Keywords Synchromodal transport · Intermodal transport · Transportation planning · Shortest path problem · Decision support systems

1 Introduction

Growing freight flows, increasing road congestion, and increasing pressure to lower emissions all stimulate the use of intermodal transport. Commonly, in intermodal transport different modes of transport are decided upfront for certain shipment types, such as shipments on a particular corridor. Since the actual transport orders and network characteristics may vary (e.g., rush orders and low water levels), a safe choice is often to rely on road transport. A recent development to cope with the inflexibility of pre-determined modes is the use of synchromodal transport. Here,

M.R.K. Mes (✉) · M.-E. Iacob
Department of Industrial Engineering and Business Information Systems,
University of Twente, Enschede, The Netherlands
e-mail: m.r.k.mes@utwente.nl

H. Zijm et al. (eds.), *Logistics and Supply Chain Innovation*,
Lecture Notes in Logistics, DOI 10.1007/978-3-319-22288-2_2

the best possible combination of transport modes is selected dynamically for each incoming order, based on aspects like costs, duration, reliability, and sustainability. A shipper agrees with a logistics service provider (LSP) on the delivery of products at specified costs, duration, and sustainability but gives the LSP the freedom to decide on how to organize the transportation according to these specifications. The LSP is able to deploy different transport modes flexibly depending on real-time information, traffic conditions, and resource availability. Synchromodal planning offers the opportunity to improve transportation services in terms of sustainability, costs, and quality.

In this study, we investigate the applicability of synchromodal transport planning at a Dutch 4PL service provider active in the European transport market. An exploration of the concept of synchromodal planning reveals its high complexity compared to truck planning. The option to select from more alternatives also increases the demand for computer aided planning. Furthermore, a potential increase of stakeholders, like multiple carriers and terminal operators, leads to higher planning complexity and also requires more coordination tasks for the LSP.

We propose a planning algorithm with the objective to minimize costs and CO_2 emissions, while maintaining delivery reliability. Our goal is to provide the human planner with a list of the top k multimodal routes. The complexity of our algorithm depends on the number of legs available for transport, where a leg is defined as a combination of a direct connection between two hubs (points where a modal switch can take place), a carrier operating on this connection, and—possibly—a pre-defined departure and arrival time.

The goal of this chapter is twofold: First, we propose a synchromodal planning algorithm that is easy to work with, easy to implement, and able to cope with many practical challenges. Second, we illustrate and test our approach using a real case study, thereby showing the potential benefits of synchromodal planning.

The remainder of this chapter is structured as follows. In Sect. 2 we provide a literature review on planning methods for synchromodal transport. After introducing the problem in Sect. 3, the synchromodal planning approach is proposed in Sect. 4. In Sect. 5, we shortly describe the implemented system and how it is used in practice. Our analysis and the results of the case study can be found in Sect. 6. We end with conclusions and recommendations (Sect. 7).

2 Literature

Synchromodal transport is closely related to intermodal transport. Literature reviews on intermodal transport are given by Macharis and Bontekoning (2004), Caris et al. (2008), and SteadieSeifi et al. (2014). They all conclude that the majority of research on intermodal transport focuses on strategic decisions, such as network design and location of terminals. Operational aspects, like mode choice and routing decisions for incoming shipments, have received limited attention

(Caris et al. 2008). We mention some related studies on operational aspects in intermodal transport.

Boardman et al. (1997) describe a decision support system to determine the best combination of transport modes to use for intermodal shipments. Their system evaluates multiple routes based on costs, service level, and type of commodity being shipped. Ziliaskopoulos and Wardell (2000) consider a similar situation, but explicitly take into account transfer times and time-dependent arc travel times; the latter being derived from transit time tables. Also in the area of public transport, some related work can be found. For example, Horn (2004) considers a traveller information system that provides a sequence of journey-legs (combination of walking, taxi, bus, and train) that meet the traveller's requirements against lowest possible costs. The proposed planning procedure first generates routes consisting of only a single leg, and gradually increases the number of legs in a route while simultaneously establishing upper bounds. Horn (2004) states that this procedure generally leads to an optimal schedule quickly, because good routes tend to contain only few legs. We also utilize this idea, since in intermodal transport, each transhipment results in additional handling costs.

The planning problem we consider is related to the *k*-shortest paths problem (Eppstein 1998). Our goal is to provide the human planner with a list of the top *k* multimodal routes taking into account costs, CO_2 emissions, and delivery reliability. Many of the *k*-shortest path algorithms, such as those proposed by Eppstein (1998) and Yen (1971), provide good results, but their running time complexity is quite large, which is inconvenient in practical situations when the available time for calculating a new set of options might be very short. Vanhove and Fack (2012) propose a heuristic that is capable of doing this, although there is no absolute guarantee that the *k* shortest paths are always found.

Shortest path problems typically have one objective, namely minimizing the travel distance. Typically, for barges and trains, the costs and CO_2 emission per kilometre would be lower compared to trucks; however, the total transport time is usually larger. Therefore, our problem can be best characterized as a multi-objective *k*-shortest paths problem, because we have to balance costs, emissions, and delivery punctuality. A few existing approaches are relevant in this context. Martins (1984) presents two algorithms, tested on a shortest path problem, in which multiple objectives can be included. Climaco et al. (1982) presents an algorithm that returns the set of Pareto optimal paths. An algorithm to solve bi-criterion shortest path problems is developed by Mote et al. (1991); this algorithm also returns the set of Pareto optimal paths. Drawback of this method is that only two parameters can be taken into account for each leg. Mote et al. (1991) have tested their method against the *k*-shortest path approach of Yen (1971); they conclude that their parametric approach is significantly faster.

Besides the requirement to provide the planner multiple options and select these options based on multiple criteria, additional complexity is introduced in our problem due to (i) the decision which carrier to use on which leg at what time, (ii) soft time windows, (iii) pre-defined timetables for rail and barge transport causing variable waiting times, (iv) contractual agreements, (v) complex costs functions depending on time, weight and volume, and (vi) the possibility of freight consolidation. The contribution of this chapter is that we provide a multi-objective *k*-shortest path algorithm that is able, in combination with the human planner, to handle the afore-mentioned practical complexities. In addition, we describe the planning system that has been implemented at the LSP, and provide numerical results quantifying the benefits of synchromodal planning at this LSP.

3 Problem Description

Our problem concerns an LSP that provides logistics services (i.e., transportation and warehousing) to its customers. These customers (i.e., the shippers) have freight that needs to be transported from a pickup location to a delivery location. The pickup is only possible after an earliest pickup time, and delivery after a latest delivery time is penalized. A lane is defined as the combination of a pickup and delivery point. The LSP has contracts with multiple carriers that have one or more modalities available to transport the goods.

The network where carriers operate consists of hubs and legs. Hubs are points where containers can be transferred from one modality to another. Legs are route segments between two hubs. Since multiple carriers might operate on the same leg, we introduce the term carrier-leg. This can be defined as the leg a carrier is assigned to with the related values of the criteria (e.g., costs). We extend this concept by adding another level by connecting the resource (e.g., the train leaving at 10 a.m.) to it: resource-carrier-leg. We make a distinction between main legs and free legs. Main legs have a fixed schedule (in location and time), whereas free legs offer more flexibility. A route consists of a sequence of legs connecting the pick-up and delivery point of a transport order. In case of 'traditional' road transport, we only have a free leg connecting the origin with the destination (left part of Fig. 1),

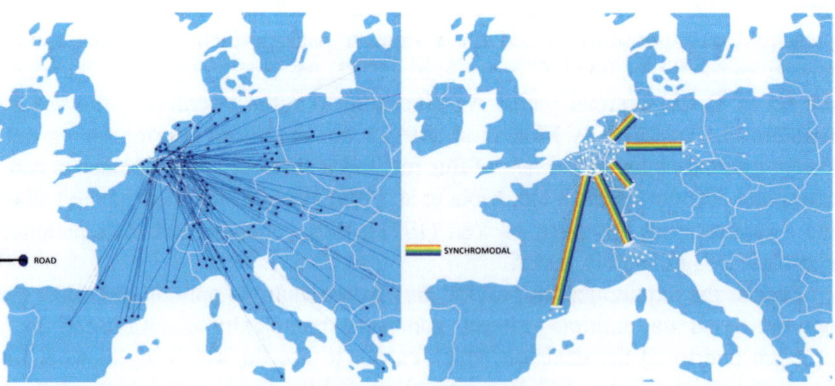

Fig. 1 Example of road transport versus intermodal transport

whereas in intermodal transport, the route typically consists of a few main legs connected by free legs (right part of Fig. 1). The challenge is to find, for each incoming order, k routes that score well on the criteria costs, duration, and emission.

To simplify the presentation of the planning algorithm, we make some assumptions. Later on, in Sect. 5, we describe how to generalize these assumptions in practice. First, we assume that all costs, times, and emissions are known and deterministic. This holds for the legs, but also for hubs that have closing times and transhipment costs. Second, we ignore contractual restrictions, such as volume agreements and preferred carriers. Finally, orders are planned one by one, without considering consolidation opportunities.

Our proposed algorithm will be used within a control tower. A control tower is software that helps a planner to direct the transport through the network. It takes restrictions into account and tries to find the best way (based on user preferences) to schedule all transports. Next to this, the control tower provides all users with real-time data about all ongoing transport orders.

4 Approach

In our intended implementation environment, we face several challenges. First, the optimization criteria and constraints are 'negotiable', which means that the exact objective coefficients (weights for various optimization criteria) are not given up front, and constraints on duration (delivery windows) and contracts (volume agreements) are soft (flexible). Second, the human planner does not want a single 'optimal' recommendation for multi-modal transport, but wants to choose from a top k of potentially good options, such that the planner can balance various optimization criteria as well as constraint violations (through sorting and filtering). Given these characteristics, we decide to use a sequential heuristic approach.

Our approach consists of 2 stages: (i) offline steps in which no order-specific characteristics are taken into account, and (ii) online steps in which order-specific characteristics are taken into account, to make sure that the generated options are suitable for a given order at a given time.

In the offline steps, also called pre-processing, we attempt to reduce the network for each lane (combination of origin hub and destination hub), which will speed-up the process of determining the top k options when an order arrives. Pre-processing starts from a full network including all legs and hubs, an origin hub and a destination hub. By removing hubs (and their associated legs) from this network, which are geographically too far from the route between the origin and destination hub, a reduced network for a specific lane can be created. We use a parameter α to indicate a threshold on the maximum increase in distance when inserting a hub in a route from the origin to the destination hub (i.e., with $\alpha = 1.5$, we allow a 50 % increase in total distance when adding a hub). The resulting reduced network is a subset of the complete transport network, and legs that are likely not to be used for this order are not part of this subnetwork. The trigger for executing these offline steps does not

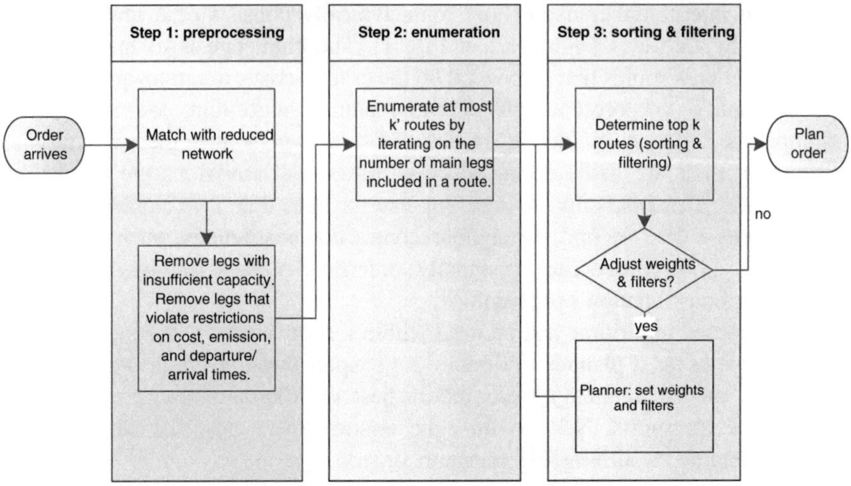

Fig. 2 General overview of the approach

have to be an incoming order, but may also result from a change of the network. The smaller the reduced network is, the faster the online steps can be carried out.

The online steps are triggered when an order arrives. We distinguish between three steps, as shown in Fig. 2. In the remainder of this section, we explain the algorithm in more detail.

In the first step, the incoming order will be matched with a certain lane (or a few lanes) so that a reduced network is selected (the selected lane has a pickup hub closest to the pickup location of the order and a destination hub closest to the delivery location of the order). This network can be reduced even further at this step, because order characteristics are known. Legs with insufficient capacity are removed. For all remaining main legs, resource-carrier-legs will be generated (see the definition in Sect. 3) for each possible departure time within a given planning horizon (e.g., the period between the earliest pickup time of the order and a few days after latest delivery time). In the remainder, we simply refer to these time-resource-carrier-legs as *main legs*.

The next step is to enumerate all possible combinations of main legs (which result in routes after adding free legs where necessary) to determine, for given weights, a relative large number k' of 'shortest' routes, with $k' \gg k$. In the third step, the human planner can apply filters (filtering out less desirable results taking into account criteria not explicitly included in the planning algorithm) and change the weights attached to the three criteria. After changing the filters and weights, the results are sorted and at most k routes are shown.

The first and third step can be performed very fast (in milliseconds). Note that even for a large value of k', the third step only requires (i) selecting the k-shortest paths and (ii) sorting these k paths, i.e., making sure that the remaining paths are worse than the *k*th path. However, the enumeration step (second step) might cause computational complexities.

In the enumeration step, we iterate on the number *m* of main legs included in a route. We do this because (i) good routes are likely to have only a few main legs and (ii) during iteration we can check whether it is efficient to extend existing routes with additional main legs. An enumerated route is characterized by its sequence of main legs. Free legs are added to these routes to make sure they are connected (connected path from pickup to delivery location through all hubs). In case there are multiple free legs available, we choose the one with lowest costs (alternatively, the cheapest one based on default settings of the weights).

In iteration *m*, we take all routes from iteration $m - 1$ (routes with $m - 1$ main legs), and for each of these routes we remove the last free leg connecting to the delivery location of the order. A route up to the last hub (i.e., a route excluding the last free leg connecting to the delivery location) is called a partial route. In an iteration, we loop over all partial routes and loop over all possible main legs to be added at the end of these partial routes. Possible main legs to add are those that have (i) a destination hub not yet present in the route and (ii) a closing time later than the earliest arrival time at the origin hub. In case there are multiple resource-carrier-legs of the same carrier-leg (different departure times within our planning horizon), we only consider the first suitable departure time. For each of the extended partial routes, we add free legs where necessary (to create a connected path from pickup to delivery location).

At the end of each iteration, we check which partial routes are eligible for consideration in the next iteration. For this, we introduce two thresholds: β and γ. These thresholds represent the maximum allowable increase in costs (β) and delay (γ) of the partial route compared to respectively the cheapest route and the route with the least delay found so far. After updating the lowest costs or lowest delay, we will be more critical in adding new routes. However, we might also come back to routes generated earlier that now might become infeasible regarding the maximum deviation to the lowest costs (β) and to the lowest delay (γ). In addition to these two thresholds that limit the number of routes enumerated, enumeration is also limited to k' routes at most.

Obviously the number of routes explodes due to the number of main legs. To illustrate this, suppose we have *M* main legs in a reduced network; then we already have *M!* possible routes containing all of the *M* legs (we have a full connected graph since trucks can drive between all hubs). The total number of routes *R* of all possible lengths *m* is be given by:

$$R = \sum_{m=0}^{M} \frac{M!}{(M-m)!}$$

For a network with $M = 35$ carrier-legs, a typical amount faced by the LSP under study, we have 2.8×10^{40} different routes (note that our algorithm uses the resource-carrier-legs as input, which automatically results in even more routes).

Obviously, the majority of these routes will contain many main legs, whereas in practice one would expect the better routes to contain only a few main legs (less transhipment costs). For the example with 35 legs, we have approximately 1.2 million routes containing at most 4 main legs. In addition, the number of routes is reduced using the thresholds β and γ.

The values of the thresholds have a large impact on the output of the algorithm. Setting appropriate vales for them is therefore important. We use three tuneable parameters: α (used in the offline steps), β (threshold on the maximum costs of a partial route), and γ (threshold on the maximum delay of a partial route). In the remainder of this section, we discuss how to find appropriate values.

It is difficult for the human planner to determine good values for the thresholds as it is hard to estimate the impact of certain values, e.g., on the number of routes R that will be generated. Setting the thresholds too high results in long computation times, while setting the thresholds too tight results in limited options to choose from. Also note that the human planner needs enough options because (i) he has to balance the three criteria (changing the weights in step 3 of our planning approach) and (ii) has to apply other criteria not explicitly included in the enumeration step (using the filters to, e.g., to filter out results that violate contractual agreements). Obviously, when no additional criteria play a role, β can be set to 1, i.e., only accepting the 'shortest' route for the given weights.

With respect to α, it is useful to have an α that is lane-specific; lanes with an origin and/or destination in densely populated areas will have a lower α than when they connect sparsely populated areas. This way we can make sure that for each lane a comparable number of hubs remains in the reduced network, which is likely to result in a comparable number of routes to enumerate. We can assist the human planner by showing a map that displays the result for a given α. In addition, we can use a feedback method: when the number of enumerated routes exceeds k' for a certain lane, we advise the planner to reduce α for this lane. A possible extension is to use an automated approach, where all possible realizations of the number of enumerated routes and used threshold per lane/corridor are stored, and used to assist the human planner in the best choice of parameter settings. Whenever a planner (or system administrator) changes the values of a threshold, the user is provided with feedback on the expected number of routes that will be enumerated. If this exceeds the limit k', visual warnings can be given. Moreover, the system may learn to propose the 'right' set of k routes by memorizing the routes most often selected by the human planner. If the human planner always selects a route consisting of hubs close to the direct route (hence which would have been selected with a low value of α), the value of α could be decreased. This will make future calculations faster and more efficient.

The same line of reasoning for setting α can also be followed for setting β and γ. The right choice for these parameters depends on k'. Ideally, β and γ are set in such a way that a number close to k' routes are generated. To give an indication of reasonable parameter settings, we used the following values in our case study: a value of $k' = 100{,}000$ (a lower value $k' = 10{,}000$ did not provide a sufficiently large number of results to handle all additional constraints taken care of in the filtering step), $\alpha = 2$ (accepting hubs that result in routes twice as long as a direct connection), $\beta = 2$ (accepting partial routes that are twice as expensive as the cheapest route found so far), and $\gamma = 2$ (accepting routes that have 1 additional day of lateness compared to the route with the least amount of lateness).

5 Relaxation of Assumptions

In Sect. 3, we made several assumptions for our algorithm. Specifically, we assumed that (i) input data is known and deterministic, (ii) no contractual restrictions have to be taken into account within the algorithm, and (iii) only one job at a time is considered. In this section, we describe how to relax these restrictions in practice.

The proposed algorithm treats the current status of the network as static and deterministic input. However, execution of plans might work out differently than planned. We might face unexpected delays, e.g., caused by traffic jams, missed connections, and breakdowns. In these situations, we have to perform re-planning. The system can be used by modifying the changed network status (e.g., removing legs or changing the duration of legs) and planning the order from its current location onwards. The main challenge here is to adjust, preferably automatically, the input data used by the algorithm.

We previously mentioned that the existence of contracts and governmental regulations is not considered within the algorithm. However, contractual restrictions can be taken into account in the sorting and filtering step. The benefit of this approach is that the algorithm itself (i.e., the enumeration step) remains fully independent of contractual changes. Restrictions like required legs and carriers can be incorporated both in the pre-processing stage (by changing the input data) and the sorting and filtering phase. Regarding the latter, if a leg or carrier is required, we can simply filter out all routes that violate this requirement. If we have volume contracts (e.g., on a certain corridor, 50 % of the loads should be rewarded to carrier A and 50 % to carrier B), we have another performance criterion that states for each route the deviation from contractual agreements. We can then filter out large deviations, or we can weigh this additional criterion against the others.

Finally, our assumption to schedule one job at a time makes it hard to consolidate jobs. Even though we typically deal with containers in intermodal transport, consolidation might be possible. Again, by changing the input data, we can support consolidation decisions with the implemented planning system. In the time between planning and execution of an order, we can create 'copies' of the legs planned for this order. These copies have reduced capacity and possibly a reduced price. In the planning of a subsequent order, we also consider these copies, thus enabling consolidated shipments on certain legs.

6 Implementation

In this section, we shortly describe the Synchromodal Control Tower (SCT) system and its relationship with other surrounding systems from the LSP and with external actors. The structure of the business network served by the SCT is depicted in Fig. 3. The system needs inputs from the shippers and the warehouses where the cargo will be loaded or dropped. The planning is communicated with the carriers handling the cargo on the different legs along the chosen lanes. The system is also connected with the customs for the automated handling of paper work.

The functional components of the SCT's application architecture are shown in Fig. 4. In the old situation, the LSP's architecture consisted only of the software components shown in Fig. 4 in blue or green. More precisely, the LSP had an enterprise resource planning system in place supporting all operational tasks (e.g., the static planning of shipments and invoicing), a Geographical Information System for the mapping and routing of shipments, and a Track and Trace module for the monitoring of ongoing transports. In the new situation, a synchromodal control tower application (shown in red) has been added. Due to this major change and the need to establish connections between the new tower and the existing systems, some of the existing application components (those coloured green) required some modifications in order to make the exchange information with the SCT possible. Furthermore, the architecture also shows the external parties that have access, or provide information to the new SCT.

The SCT consists of several components, the most critical one being "Top k lanes calculation". The main task of this component is to carry out the calculation of the *k*-shortest routes, based on the algorithm explained in Sect. 4, and using a

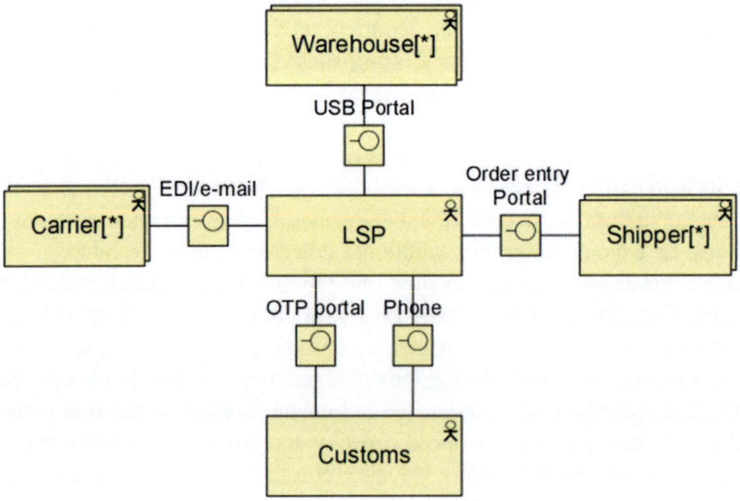

Fig. 3 Actor model

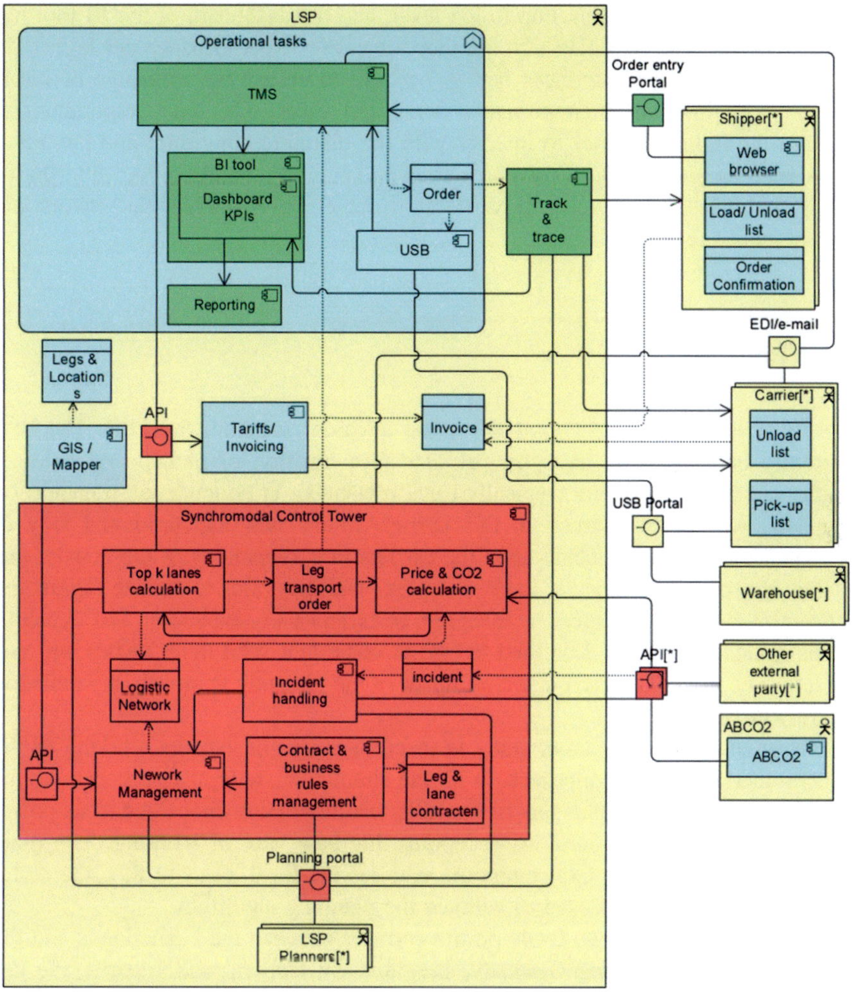

Fig. 4 Application architecture

network of legs and hubs as input (the management of which is done by the "Network management" component).

Another important function of the SCT concerns the price and CO_2 emissions calculation for the chosen lane, which is dealt with through the "Price & CO_2 calculation" component. For the CO_2 emission calculation, we use an external, cloud based, carbon management system ($ABCO_2$) that has been designed on the basis of the reference architecture proposed by Iacob et al. (2013).

The implementation of the SCT has also lead to better "Track and Trace" and business performance monitoring (i.e., the "BI tool") functions. Orders can be traced now (nearly real time) up to leg-level, and the dashboard of the BI tool has been enriched with new KPI's to monitor the carrier's performance per leg.

A number of new interfaces had to be built to ensure the exchange of information with internal and external systems and users. The most important one enables the human planner to interact with the planning algorithm, and has functions such as setting values for parameters and running the algorithm, filtering of orders and lanes, management of contracts, management of the logistic network and of incidents, etc.

7 Validation

We illustrate and test our approach using an order set from the LSP. The LSP currently outsources its transport per lane (combination of pickup and delivery point) to a preferred carrier (typically truck transport). In principle, all transport on this lane will be outsourced to this carrier. In the new synchromodal way of working, this LSP will decide upon the carriers in a dynamic way. Obviously, this makes the planning process more complex, but they aim to realize significant efficiency gains with respect to transport costs and CO_2 reduction. The synchromodal planning system described before is consistent with their wishes and can therefore be used. To determine the impact of the planning approach, a simulation study will be carried out.

The goal of the simulation study is to provide insight into the working of the synchromodal planning approach. With the simulation study, the approach can be clearly explained to the planners of the LSP, which will convince them of its power and will help in the process of accepting this new way of working. The main indicators are the costs, CO_2 emission, and service level. Next to this, the simulation study will also be used to validate the planning algorithm.

For our experiment, we focus on the corridor between the Netherlands and the middle of Italy, including Germany, Belgium, Switzerland and Austria. We use orders on this corridor during one quarter of 2013, originating in The Netherlands (1728 in total). We use the historical data set to determine the following characteristics. For the orders, we use the pickup and delivery location, earliest pickup time, latest delivery time, and volume. For the network information, we use 110 available barge and rail connections (legs) on this corridor. For each leg, we define the carrier, modality, distance, CO_2 emission, schedule of departure and closing times, and duration. For each hub, we use the coordinates, opening times, possible modality switches, and handling and transfer times. As costs components we use a price per kilometre for each carrier-leg, and for the handling- and transhipment costs per hub.

The results showed that for 39 % of the orders it is beneficial to make a modal shift (for at least part of the route) from road to rail (30 %), barge (5 %) or a combination of both (4 %). Mainly orders with destination in Italy, Switzerland and Austria showed to be suitable for synchromodal planning because of their relative long transport distance and good connections over rail and water (we observe a minimum distance of approximately 400 km before intermodal transport is preferred over direct truck transport). On average, a cost reduction of 10.1 % can be achieved with synchromodal planning, while CO_2 emissions reduce with 14.2 %, which comes down to a yearly reduction of 712 tons of CO_2 on this single corridor. Additional savings can be achieved by extending the deadlines with 1 day (1.5 % cost reduction) and by limiting the FTL size of orders to the capacity of 40 ft containers (4.0 % cost reduction).

The results so far are based on the assumption that the LSP does not consolidate shipments, which in practice does not hold. For a fair comparison, we extend the synchromodal planning algorithm with a consolidation option (see Sect. 5). In this case, the cost reduction of synchromodal transport with consolidation compared to truck transport with consolidation is 3.9 %. This value is lower than the original reduction of 10.1 %, because consolidation opportunities arise when there is sufficient slack in the delivery windows. Since intermodal transport typically takes longer than truck transport on this corridor, the cost reduction of synchromodal transport gets smaller.

8 Conclusions

We developed an algorithm to support synchromodal transport planning. The underlying problem is known as the multi-objective k-shortest path problem, in which we search for the k-shortest paths through a multimodal network taking into account time windows of orders, schedules for trains and barges, and closing times of hubs. We provide the human planner with a list of the top k multimodal routes, with the objective to minimize costs, delays, and CO_2 emission.

Our approach consists of an offline phase and an online phase. In the offline phase, also called pre-processing, we attempt to reduce the network for each lane (combination of origin hub and destination hub). The online phase is triggered when an order arrives. We distinguish between three steps. In the first step, the incoming order will be matched with a certain lane so that a reduced network is selected. The next step is to enumerate all possible routes from origin to destination within the reduced network. In this step, we iterate on the number of main legs included in a route, since routes with a limited number of transfers tend to be better. In the third step, the human planner can apply filters, change the objective weights, and make the final choice.

We illustrate our approach using simulation with order and network data from a Dutch 4PL service provider. On the corridor from the Netherlands to Italy, an average cost reduction of 10.1 % and a reduction in CO_2 emission of 14.2 % can be achieved with synchromodal planning.

The proposed synchromodal algorithm has been implemented at the LSP within their supply chain control tower. This system contains all required information on the LSP's main transport routes (time tables, costs, duration, emission, etc.), provides real-time decision support on the best combination of transport modes, and monitors the actual execution of transport. The synchromodal control tower was put into use in 2013, and already during the second half of 2013 a modal shift of 1500 TEU with a CO_2 reduction of 3230 ton has been reported.

References

Boardman BS, Malstrom EM, Butler DP, Cole MH (1997) Computer assisted routing of intermodal shipments. Comput Ind Eng 33(1):311–314

Caris A, Macharis C, Janssens GK (2008) Planning problems in Intermodal freight transport: accomplishments and prospects. Transp Planning Technol 31(3):277–302

Climaco J, Martins E (1982) A bicriterion shortest path algorithm. Eur J Oper Res 11(4):399–404

Eppstein D (1998) Finding the k shortest paths. SIAM J Comput 28(2):652–673

Horn ME (2004) Procedures for planning multi-leg journeys with fixed-route and demand responsive passenger transport services. Transp Res Part C Emerg Technol 12(1):33–55

Iacob ME, van Sinderen MJ, Steenwijk M, Verkroost P (2013) Towards a reference architecture for fuel-based carbon management systems in the logistics industry. Inf Syst Front 15(5): 725–745

Macharis C, Bontekoning YM (2004) Opportunities for OR in intermodal freight transport research: a review. Eur J Oper Res 153(2):400–416

Martins E (1984) On a multicriteria shortest path problem. Eur J Oper Res 16(2):236–245

Mote J, Murthy I, Olson D (1991) A parametric approach to solving bicriterion shortest path problems. Eur J Oper Res 53(1):81–92

SteadieSeifi M, Dellaert NP, Nuijten W, Van Woensel T, Raoufi R (2014) Multimodal freight transportation planning: a literature review. Eur J Oper Res 233(1):1–15

Vanhove S, Fack V (2012) An effective heuristic for computing many shortest path alternatives in road networks. Int J Geogr Inf Sci 26(6):1031–1050

Yen J (1971) Finding the K shortest loopless paths in a network. Manage Sci 17(11):712–716

Ziliaskopoulos A, Wardell W (2000) An intermodal optimum path algorithm for multimodal networks with dynamic arc travel times and switching delays. Eur J Oper Res 125(3):486–502

DAVINC^{3}I: Towards Collaborative Responsive Logistics Networks in Floriculture

Jack G.A.J. van der Vorst, Robert Ossevoort, Marlies de Keizer, Tom van Woensel, Cor N. Verdouw, Edwin Wenink, Rob Koppes and Robbert van Willegen

Abstract Today most flowers physically pass through the Dutch auction houses on their fixed routes from growers located all over the world to European customers. Physical presence is necessary to allow for inspection, quality control and break-bulk activities. Several developments, such as increased internationalization and virtualization, stimulate the sector to develop an efficient European hub distribution network, in which cut flowers, plants and other products are delivered to detail, retail and e-tail (i.e. webshop) customers using different logistics concepts. The DaVinc^{3}i project has had the objective to strengthen the international leading competitive position of the Dutch horticulture sector in a global, virtualized trade network by researching the opportunities for new logistics coordination, consolidation and collaboration concepts in extended international trade park networks. In the last years over 30 case studies with business partners have been conducted within the DaVinc^{3}i project. In this chapter we highlight the main research advances and lessons learned when moving towards collaborative responsive logistics network designs for perishables, illustrated with five case studies.

Keywords Virtualisation · Collaboration · Perishables · Logistics network design

J.G.A.J. van der Vorst (✉) · R. Ossevoort · M. de Keizer
Operations Research and Logistics Group, Wageningen University, PO Box 8130
6700 EW Wageningen, The Netherlands
e-mail: Jack.vanderVorst@wur.nl

E. Wenink · R. Koppes
FloraHolland, Aalsmeer, The Netherlands

T. van Woensel
Industrial Engineering and Innovation Sciences, Technical University of Eindhoven, Eindhoven, The Netherlands

C.N. Verdouw
LEI Wageningen UR, Wageningen, The Netherlands

R. van Willegen
Association of Wholesale Trade in Horticultural Products (VGB), Aalsmeer, The Netherlands

H. Zijm et al. (eds.), *Logistics and Supply Chain Innovation*,
Lecture Notes in Logistics, DOI 10.1007/978-3-319-22288-2_3

1 Introduction

The floriculture sector in the Netherlands is of world-class quality, and serves as main trading hub for Europe. The sector as a whole has a huge impact on the Dutch economy, being the largest exporter of fresh products in Europe, the top-3 largest exporter in the world with still significant opportunities for further growth. Most flowers physically pass through the Dutch auction houses on their fixed routes from international growers to European customers to allow for physical inspection, quality control and break-bulk activities. However several developments, such as the upcoming of new markets in Eastern Europe and increased virtualization, will move the centre of gravity eastwards and stimulate the chain to become an efficient floriculture hub-network, in which cut flowers, plants and other products are delivered to international customers taking different (direct) routes and using different logistics concepts. Cross-dock centres and hubs (trade parks) are being set-up in Europe (linking local with global flows) and the sector is searching for efficient coordination and control mechanisms for the complete logistics network to consolidate flows and fulfil market demands. However, this is not easy as the sector is characterized by a large number of independent SMEs (many growers, traders, and small LSPs) and a large auction on a cooperative basis, each with their own objectives and views on roles and functions of parties in the supply chain network.

The Dutch sector wants to continue being the (virtual) floriculture-trading hub of Europe, and has therefore initiated the DaVinc^{3}i project. DaVinc^{3}i is an acronym for *D*utch *A*gricultural *V*irtualized *I*nternational *N*etwork with *C*oordination, *C*onsolidation, *C*ollaboration and *I*nformation availability (see www.davinc3i.com). The project aims to develop innovative logistics concepts supported by an information platform and collaborative business models in support of the Dutch competitive strength. More specifically, we study:

- the functional specifications for potential logistics coordination, consolidation and collaboration concepts, with particular attention for responsive quality driven logistics networks and synchromodal transport management;
- opportunities for advanced information exchanges and architectures to facilitate the advanced planning and control concepts developed in the project;
- relevant collaborative business models that work for specific settings.

The next sections, successively, present sector specific characteristics and developments resulting in industry needs, key issues in the design of a responsive synchromodal logistics network for perishables, and five illustrative case studies resulting in determining key factors for successful and sustainable collaborative logistics.

2 Sector Characteristics and Industry Needs

The Netherlands is at the heart of the international floriculture sector. It has an intricate and high-quality network of companies, ranging from breeders and growers to sales experts and export firms, representing every aspect of the business. The supply chain network consists of the following links: growers, auctions, traders, logistics service providers and outlets (Fig. 1).

The FloraHolland flower auction consists of five auction centres for trading in cut flowers (about 70 % of turnover) and ornamental plants (about 30 %), a national intermediary organization (facilitating direct trade between growers and traders) and an internationally active import department. FloraHolland is a primary cooperative: its roughly 5000 members, especially *growers* in the Netherlands, own the business. *The traders* can be split up in three groups: wholesalers, exporters and importers. Sometimes this overlaps, when a Dutch wholesaler also acts as an exporter. There are about 1200 Dutch traders, dealing with many (inter)national customers. Most important import countries are Kenya, Ethiopia, Israel, Ecuador and Germany. Key export countries are Germany, United Kingdom, France, Italy, Belgium and Russia. In many cases, the transport between two chain stages is outsourced to one of over 70 *logistics service providers*. In some cases, the providers execute extra activities like quality control, handling and packaging. On average there are 20,000 truck movements per day and about 1800 truck movements per day between the five marketplaces.

Different *sales channels* can be identified in the market. The DaVinc^{3}i project distinguishes between three types of sales channels: the traditional “detail”

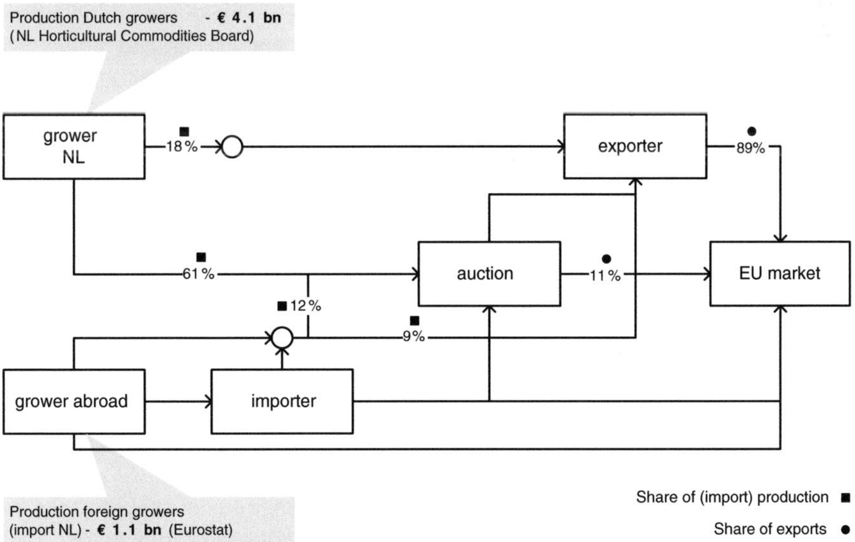

Fig. 1 The floriculture supply chain network (de Keizer et al. 2014, p. 162)

(specialist shops) and "retail" (including supermarkets, garden centres and construction outlets) channels, and the expanding "e-tail" channel (online shops). Retail industry has seen significant consolidation and concentration, which led to dominance of the market by large retailers. Retail sells flowers and plants as by-product and aims for large volumes of specific products guaranteed via long-term (preferred supplier relationship) contracts and fixed prices. Specialist shops often gain their competitive advantage due to a deep product assortment (and hence small volumes per individual product) and a focus on high-quality products. They market value-added products via small-scale shops using day-to-day prices and volumes available. Online shops are relatively new to the sector and the result of increased digitalisation and virtualisation. It is not yet clear how this channel will further develop and what kind of assortment will be offered as well as which order fulfilment strategies will be applied.

The *cut flower detail chain* is a supply driven chain with small but very perishable product flows to, and main inventory placed at, geographically dispersed detail shops. The cut flower retail chain is a growing demand driven chain with large product flows and customizing processes like bouquet-making, labelling and packaging, at locations at a one to three days distance from customers. The *potted plant detail chain* is a relatively small chain which resembles the flower detail chain but has more slack in lead times in production and distribution due to the less perishable nature of the product. The *potted plant retail chain* is a rather predictable pull chain with large product flows and customizing processes at locations at a one to three days distance from customers in which products are sold piece by piece. In the world of the ornamental plants the role of the garden centres and lumber yards is stronger than in cut flower chains. This leads to direct deals between retailers and growers with a much higher volume.

Using expert interviews and chain process analyses, we identified multiple types of supply chains in the complete network, specific bottlenecks and floriculture industry dynamics and needs. Important examples are the following:

- Product quality deteriorates during distribution, whilst there is an increasing demand for guaranteed vase life.
- Traders increasingly order in higher frequencies and in smaller quantities. Current logistics does not fit this need.
- Very limited supply chain visibility causes suboptimal internal processes at all chain actors. There is a need for sharing real time information and supply chain transparency.
- The transportation market is fragmented with few possibilities for consolidation of volume. However, faster transport at lowest possible cost between market places is needed with guaranteed product quality.
- High dynamics and uncertainty in supply and demand result in many last-minute changes, rush orders and inefficient transportation.
- There are many possibilities for information exchange in the supply chain, but they are inefficiently used; a lot of manual/phone communication occurs.

In summary and simply put, the sector is characterized by a diversity of products, actors, and market outlets (each with their own specific demands), and these characterizations require a redesign of the current logistics concepts.

3 Logistics Research Challenges

The specific developments and sector characteristics result in research challenges on the level of the design and management of logistics processes. Based on desk research and stakeholder interviews we identified the following issues.

3.1 *Need for Differentiated Demand-Driven Logistics Concepts*

As indicated earlier, different market channels can be distinguished: retail, detail and e-tail. There is a trend towards retail and e-tail chains, which are foremost demand driven, whereas traditional detail chains are foremost supply driven (using the auction clock). In all these channels, product quality (vase life) is one of the most important product attributes (for flowers nowadays about 7 days); hence speed and responsiveness are key. As a result, order lead times are continuously being reduced and there is a trend towards smaller order batch sizes. Furthermore, growers are more and more shifting from mass delivery to auctions to direct delivery to traders. As a result they have to change from supply driven to demand driven working practices.

3.2 *Need for Collaborative Logistics Hub-Network Design*

The floriculture sector is characterised by intensive cooperation between all actors in the network. However, from a supply chain perspective still many logistics flows from source to sink are managed independently by many chain actors, resulting in less efficient transport flows. This becomes more and more difficult due to increasing end-customer demands and a growing political pressure to reduce logistics movements. Flowers and plants are sourced internationally and might in the future, instead of being transported via the market places in the Netherlands, be directly distributed via a logistics hub network in Europe to regional customers. These customers require value-added products packed and delivered within a complete assortment with specific logistics service constraints. More logistics collaboration between different actors in the chain, vertical as well as horizontal, may improve the efficiency of processes like harvesting and transport, and reduce product waste. Key issue is that in a virtualised network, opportunities arise for different trade park network configurations, for different route and process (e.g. where to assemble and pack) configurations of supply chains through the network, and for transport consolidation practices of actors in the network. Sophisticated facility location models are necessary to determine the best supply chain network design (cf. Melo et al. 2009; Klibi et al. 2010).

3.3 Need for Robust and Flexible Quality Controlled Logistics Concepts

One of the main logistics challenges for the sector is to deal with strong dynamics and uncertainty in supply and demand, regarding fresh product quality as well as the available volume in time at a specific place. The sector is characterized by last-minute changes and rush-orders. A particular characteristic is the difficulty to predict the exact quality of fresh produce before it has been harvested. The prediction of these quality changes is even more difficult during the trade, transport and storage processes (resulting in potentially large product losses if logistics is not organized adequately). At the same time there is a trade-off between expensive measures that could prolong the vase life of the flowers and the use of slower and cheaper transport modalities with often less carbon emissions. Typically, next to biological variations, the quality of flowers and plants is determined by time and environmental conditions (such as temperature and humidity during transport). Environmental conditions may be influenced by, for example, the type of packaging, way of loading and the availability of temperature conditioned transportation means and warehouses. Customers demand guarantees on quality specifications leading to strict requirements on the logistics network concepts used in the sector. As a consequence, the required prediction and planning concept and accompanying logistics system need to be very flexible, enabling last minute changes and reallocations, but also to provide a robust planning (compared to the many rush orders and transports at the moment). More specific, it should allow for advanced logistics decision making taking real-time information on product quality behaviour into account, resulting in the delivery of the right product to the right outlet in time; a concept called “Quality Controlled Logistics” by Van der Vorst et al. (2011).

3.4 Need for Innovative Collaborative Distribution Strategies

The horticulture sector is confronted with many emergency orders requiring more efficient and responsive logistics processes. In order to be responsive, a supply chain can make use of multiple delivery modes in which the slower and cheaper modes are employed for shipments under usual planning (push process) to enjoy the economies of scale and contribute to a cleaner way of transportation by emitting less carbon emissions (lean and green), while the faster and more expensive delivery modes are used for speedy and emergency replenishments by market demand (pull process) (Chan and Chan 2010). Synchromodal transport receives increased attention in this sector. Rail and sea transport using conditioned containers instead of speedy air transport has already proven to be a successful technology. This holds true especially for import flows as these containers usually contain large volumes of the same flower or plant species. In export flows multiple types of flowers or plants have to be distributed together (often using a diversity of load carriers), but they each respond differently to specific temperatures and humidity. If these new conditioning technologies can be used to transport products over long distances, it could also provide us with opportunities to hold inventories at strategic locations within the network, i.e. at international distribution hubs. This shows that it is relevant to research the optimal temperature when facing different quality decay profiles for different products; as well as, given the demand for multiple products, which products could be combined in a common (flexible) container transported by rail, road, water or air.

4 Case Studies

In the course of the project over 30 case studies have been investigated in close collaboration with the projects' business partners, dispersed over the flower and plant supply chain as well as over the different market outlets as research focus areas. It is of course impossible to discuss all case findings in depth, hence we take a few illustrative cases to show our findings related to the defined logistics research challenges in the previous section. Afterwards, we will present the overall findings and lessons learned from all cases.

4.1 Case 1: From Supply to Demand Driven Grower Practices

One of the practical implications of virtualisation and increasing direct trade between growers and traders is the increase of customer ordering frequencies and decrease in ordering volumes. Multiple growers in the DaVinc^{3}i project took up a research project how to address this challenge in their logistics processes.

We now focus on a grower of gerberas, that are mostly sold through specialised shops, such as florists. Over time, the grower has seen a significant increase in the share of flowers sold directly to wholesalers/exporters instead of via the traditional auction clock. Auction flows need to be ready for shipment by around 6 pm and can be considered a push flow: a grower offers a bulk volume of products assuming it will be sold to traders. On the contrary, shipments for direct sales are ordered even before harvesting takes place and can thus be considered a pull flow: they are picked and shipped only after the customer orders have been received. Direct sales orders, for the large part, have to be processed between 7 and 11 am. Practically, the increasing share of direct sales means that an increasing number of orders arrive up to two hours before they have to be shipped, creating significant logistics challenges.

Given existing constraints in processing capacity, the grower faced increasing difficulty in preparing batches within the given timeframes. Our research has shown that some changes to the classical way of working enable the grower to deal with these practical implications of virtualisation. The basic principle is in line with a shift of the customer order decoupling point (cf. Olhager 2010); instead of harvesting the products bulk wise in the most efficient way (i.e. using the shortest route through the nursery to harvest flowers), the grower changed his practices to first harvest those products that need to leave the building first. As a result the efficiency of the harvesting process reduces (due to smaller lots and more order picking time), however, modelling results showed the total system performance increases (i.e. reduced costs, reduced throughput time and, most important, increased fill-rate) due to much more efficient demand-driven processing and handling of customer orders. In the new design, all customer orders could be harvested and handled in time resulting in higher service levels. The change in working practice turned out to be relatively easy to implement. Goods flows were actually rescheduled, order handling was changed and a new layout for the processing area was implemented, requiring only a relatively small investment.

4.2 Case 2: Design of a Consolidated European Hub-Network

In this study, we explore the potential of an advanced logistics concept, i.e. logistics orchestration, that aims collaboration between actors in the European potted plant supply chain network. We optimized the design of a global supply chain network, thereby incorporating routing in supplier and customer regions and analysing the trade-offs between multiple objectives. One of the key aspects in our study is effective and efficient consolidated distribution of potted plants. Consolidated distribution could improve logistics performance, e.g. increasing capacity utilization by combining less than full truck loads when the volume of the goods to be distributed is smaller than the transport unit size; or decreasing total travelled distance by splitting and re-combining full truck loads based on sources and destinations of the goods to be distributed.

A multi-layer, multi-product mixed integer linear programming model was developed to investigate the benefits of logistics orchestration in three scenarios, i.e.

a scenario with no collaboration, a scenario with consolidation in collection of products at supplier regions, and (on top of the previous one) a scenario with consolidation in distribution to points of sale in customer regions (see Fig. 2). For these scenarios the effects of enhanced hub-network designs and logistics consolidation on logistics costs, working times and CO_2 emissions were quantified. An enhanced network design implies a shift in the number of hubs, the hub locations, and the degree of consolidation. Modelling assumptions, data and results were

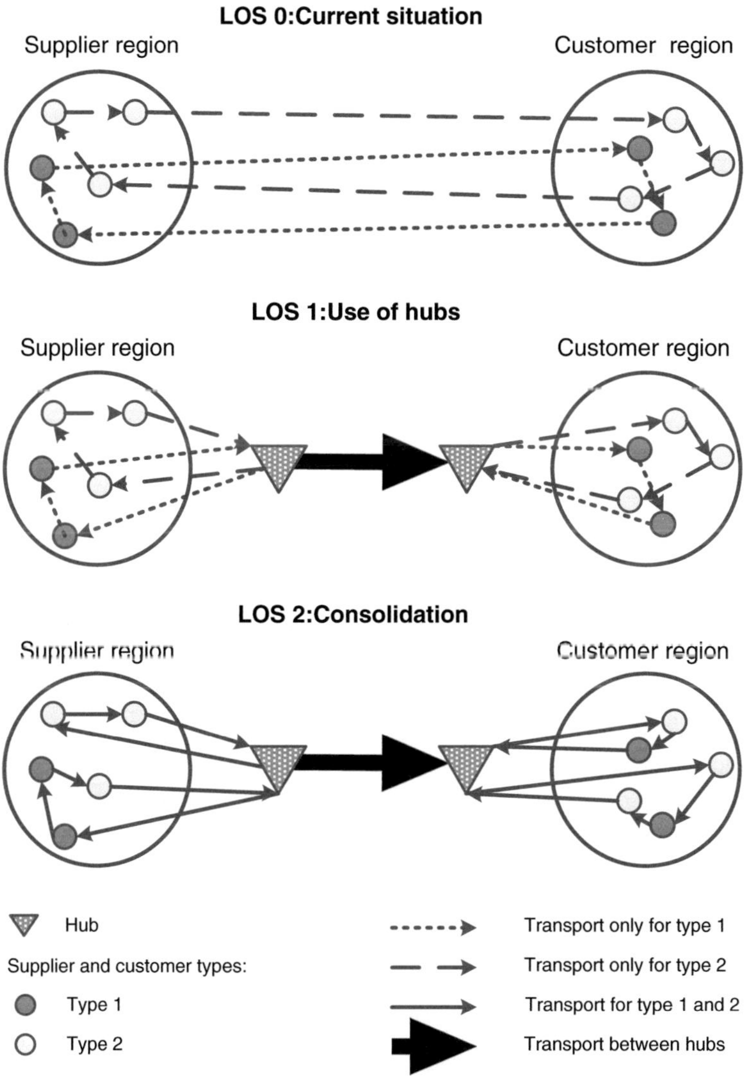

Fig. 2 Logistics orchestration scenarios (de Keizer et al. 2014, p. 164)

Table 1 Normalized performances for different scenarios

	Hubs	Normalized KPIs		
		Costs	Time	Emissions
LOS 0	–	100	100	100
LOS 1	15	81	73	91
LOS 2	15	72	62	80

validated in collaboration with business partners. More information can be found in de Keizer et al. (2014).

The results of enhanced network designs and consolidation can be seen in Table 1. Opening hubs across Europe (LOS1) decreases logistics costs by 19 %, distribution time including work shift durations by 27 %, and kilometres travelled and CO_2 emissions by 9 %. This is mainly due to time savings because long transportation routes in the current situation are split into routes to and from a hub, which gives less violations of work shift durations. If one also consolidates the flows to, between and from these hubs (LOS2), costs, time and CO_2 emissions decrease even further by 9, 11 and 11 % respectively. The optimal hub network for scenario 1 and 2 (equal in both scenarios) has three hubs in the Netherlands and 12 across Europe (excluding the Netherlands). When we look at the locations of the hubs, there is a gradual path of development. At first hubs are opened in the centre of Europe, and after these centres are covered hubs in the east and south of Europe are suggested to be opened.

4.3 Case 3: Quality Controlled Logistics Concepts

The management of product quality is of vital importance in supply chains of fresh produce such as flowers and plants. Until now, tracking and tracing of quality conditions in the logistical process from production to market was mainly carried out with the help of data-loggers. These are capable of recording trajectories of the environmental conditions like temperature and relative humidity, but the data can only be accessed afterwards (Verdouw et al. 2013). In this case study, innovative technologies are introduced to realise real-time management of plant and flower quality all through the supply chain. It involves technologies in the field of tracking & tracing (like RFID), quality monitoring (like wireless sensor networks) and internet (like cloud computing and web services). The case study aimed to design and demonstrate an innovative system for real-time virtualisation of product quality in floricultural supply chains. This system enables that logistic processes throughout the supply chain can be controlled, planned and optimized continuously based on real-time information of the relevant quality parameters.

Focal company is a Dutch trader with the role of supply chain orchestrator. Via this trader, also a grower, transporter and auction are incorporated. The study is leveraging the trader's current logistics tracking system, which is based on the ultrahigh frequency RFID tags that are attached to the complete pool of plant trolleys. In the study we: (i) defined the user requirements, (ii) designed an information systems architecture, and (iii) developed a prototype system. The requirements analysis and system design has addressed three use case scenarios:

- *Quality Monitoring*: real-time access to quality information including ambient conditions (e.g. temperature), prediction of remaining shelf life using prediction models, and early warning in case of deviations of plans;
- *Quality Controlled Distribution*: flexible (re)planning and (re)scheduling of distribution based on real-time quality information;
- *Quality Controlled Vendor Managed Inventory*: intelligent replenishment of retail stores by proactively balancing retailer demand and grower supply, concerning the availability and quality of flowers and plants.

The prototype system focuses on the Quality Monitoring scenario. During the development, mock-ups were used to define the detailed specification in interaction with the stakeholders (see for example Fig. 3).

The designed system provides practical functionalities for (i) sophisticated quality monitoring from producer to the market; (ii) dynamic and intelligent logistics management based on up-to-date quality information (including early warning, rapid scheduling, remote control and shelf life simulation); and (iii) seamless and secure interoperability. The involved end users expect that implementation of these functionalities will especially result in a significant reduction of product waste throughout the supply chain and in improvements of product quality for end-consumers.

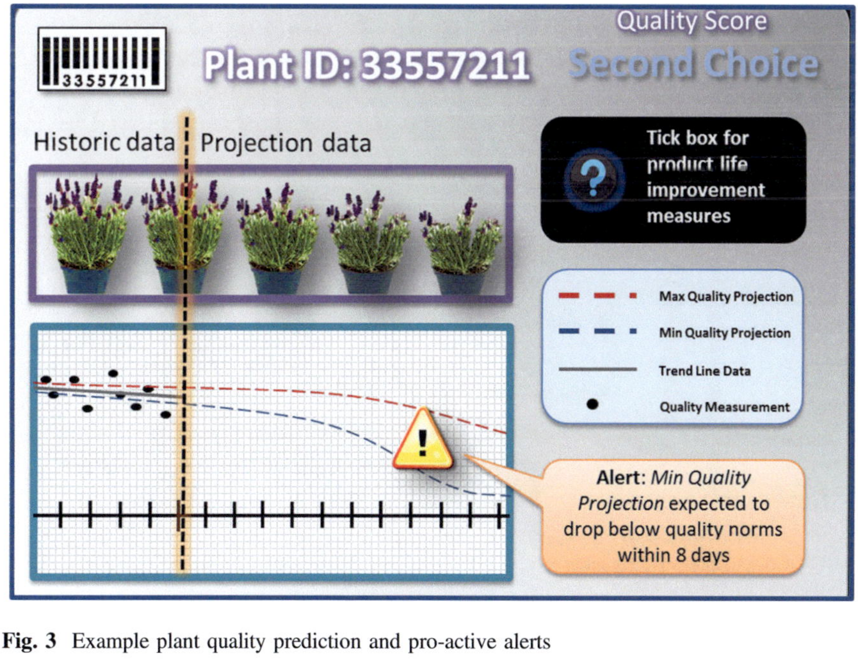

Fig. 3 Example plant quality prediction and pro-active alerts

4.4 *Case 4: Synchromodal Logistics Network Design*

Whereas business case 2 focussed more on the location and number of hubs in the network, DaVinc^{3}i also investigated the use of multiple transport modalities and route options in the network (e.g. SteadieSeifi et al. 2014). This is done by introducing a synchromodal *metro model*. This metro model consists of multiple inbound hubs (points to enter the metro model after leaving the grower), break-bulk and consolidation hubs (where value adding activities take place), and exit points (where products leave the network). After leaving the network the products are ready for the last mile distribution to the customer. In the metro model it is possible to travel from point A to point B through different routes using different modalities (truck, ship, train, airplane); either directly or through one or multiple transhipment points. The flowers and plants are transported according to a high frequency fixed time schedule. The metro model is designed to achieve more flexible routes between growers and different kind of customers. Figure 4 gives a schematic overview of the metro model.

One of the studies conducted focused on the effect of multimodal transportation and the choice of the greenest, fastest, or cheapest transportation routes. The first part of the research consisted of long-haul transportation, which is described by a service network design (SND) model. The second part of the research is a short-haul last mile distribution problem, which is described by a vehicle routing problem (VRP). Results showed that the use of a metro model in the floriculture

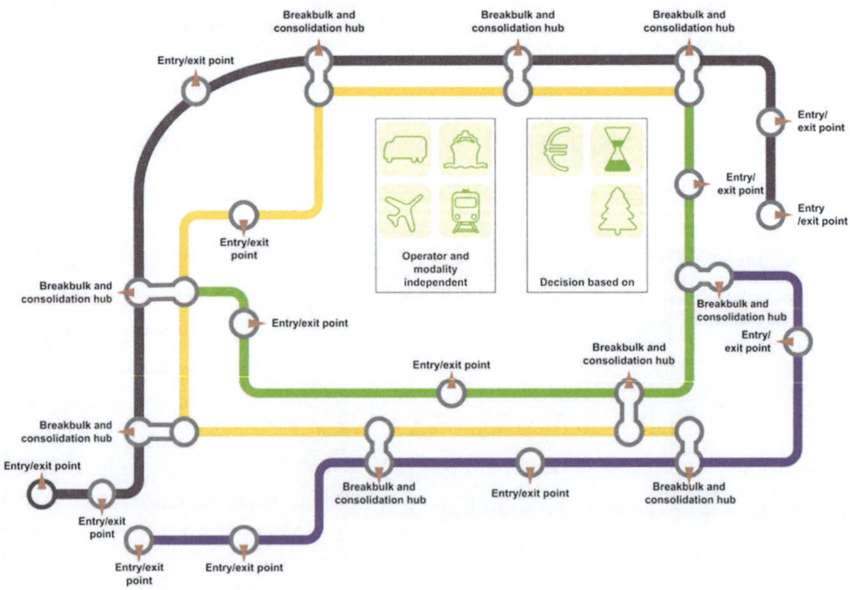

Fig. 4 Overview of the metro model

sector can result in a cost reduction of 5.1 % (from 2.85 to 2.70 euro per trolley)—which is quite interesting in this low margin business. By contrast, the average lead-time will increase from 0.44 to 0.74 days. This automatically indicates the need for forward inventory positions close to the market, reducing the need for high speed logistics networks.

4.5 *Case 5: Collaboration on Returnable Transport Items*

Internationalisation in floriculture takes place across the supply chain. The growth of distant demand regions and distant supply regions not only increases distances, but also creates new challenges for Returnable Transport Items (RTIs). RTIs used for distant markets have longer turn-around times decreasing the return on investment. Most Dutch exporters currently use their own type of RTI to deliver products to their customers abroad. The question is whether this is the most efficient solution. Together with exporters and logistics service providers, the DaVinc^{3}i project has therefore researched what opportunities exist to realise gains in this area. These studies indicate significant savings are possible while maintaining the service levels and flexibility that would be required.

In our research we distinguish between six possible options for export RTI cooperation: (1) a decentralised structure where each exporter manages its own (type of) RTIs, or (2) a centralised depot. In both cases there may be (a) the current RTIs, (b) a uniform export container used by all exporters (UEC) and (c) a uniform supply chain container (UCC) used throughout the complete chain. The expected benefits include time savings and error reductions associated with sorting and distributing returned RTI. The impact for different scenarios is shown in Table 2.

This case clearly shows the benefits of cooperation. However, the distribution of the benefits of cooperation between groups of stakeholders turns out to be uneven. Modelling results show exporters would see their costs decrease by 20–40 % (respectively for a centralised deport with the current containers and a centralised depot with a uniform container). For logistics service providers the costs would decrease by 6–17 %. It is clear that more insights on pain and gain sharing are needed to facilitate implementation of identified favourable scenarios.

Table 2 Impact indicators for new scenarios of RTI management

	Decentralised structure			Centralised depot		
	Current container[a]	UEC	UCC	Current container	UEC	UCC
Operational cost	0.00 %	−10.29 %	−15.58 %	−9.52 %	−16.69 %	−22.80 %
Average trip cycle time (days)	17.90	16.27	15.78	17.93	16.13	15.66

[a]This scenario describes the current situation

5 Lessons Learned for the Design of a Responsive Collaborative Logistics Network for Perishables

Next to the five business cases presented above, more than 25 other projects were conducted within the DaVinc^{3}i project with actors in the supply chain. Many of these projects entailed chain or business process analyses of specific business cases and quantitative modelling studies to analyse the advantages and disadvantages of new logistics concepts for the businesses involved. Other studies focussed more on extended literature reviews on the related topic to formulate an analytical research framework (comprising all relevant factors that influence decision making, e.g. on pooling inventories or transport flows for perishables) and test it, successively, using expert interviews. Outcomes of the last studies resulted in the formulation of hypotheses, hence theory building. The modelling studies focussed on theory testing, hence indicating the gap between theory (what is possible in the ideal case?) and practice (what can be achieved in this particular practice?).

The nice thing about the DaVinc^{3}i project is that it brings together theory and practice. Using real life data, we have conducted strategic research at the sector level into logistics network design, consolidation strategies, alternative transport modalities, information systems and collaboration models. These projects resulted in theoretical insights that help businesses prepare for the future. But we also conducted very practical studies with companies via master students internships resulting in concrete improvement options and actual implementations.

The project team has made an attempt to abstract the lessons learned from all projects and to categorise them. Table 3 presents findings according to the design elements of the logistics concept (e.g. Ribbers and Verstegen 1992), i.e. configuration, planning and control system, information system and organisation. Furthermore, we defined a number of eye openers and more generic lessons learned when designing a responsive collaborative logistics network. The main are as follows:

- Most chain actors in the floricultural sector are aware of current developments, but are waiting to take action until it is urgent;
- Virtualisation requires collaboration and synchronization of processes and information in the complete chain network;
- Responsive, high frequent delivery of high quality cut flowers to the international market requires an international hub network with quality controlled logistics principles;
- The added value of such an international hub network depends on collaboration: it requires high volumes and frequent flows to be cost efficient;
- Due to the advantages of consolidation, supplying to the nearest location is not always the cheapest;
- Advanced conditioning technologies and reefer containers enable the use of more sustainable transport modalities;
- Supply chain redesign is very complex comprising many different elements and disciplines;

Table 3 Lessons learned in the DaVinc^{3}i project

Element	Lesson learned
Network configuration	• Build a common synchromodal logistics infrastructure that allows for consolidation and economies of scale • Create a logistics network with hubs close to the markets to be responsive at low cost; especially when high product quality is demanded • Create and manage a closed cold chain from grower to consumer • Create inventories close to market using conditioning technologies • Postpone value adding logistics (VAL) activities as much as possible; this increases flexibility and reduces cost and lead times • Use standardized (conditioned reefer) containers
Planning and control system	• Build a central planning, dynamic and pro-active control system for collaboration, coordination and consolidation • Be responsive to changes in supply and demand by establishing flexible order fulfilment processes, and predict where needed • Re-allocate inventory positions closer to the market and pool inventory (this can reduce inventory costs whilst the same service levels are obtained) • Align planning processes with supply chain partners and pool inventory • Integrate quality management systems with supply chain partners • Quality controlled logistics enables dynamic pricing • Adapt and synchronise business processes to customer demand to create responsiveness (establish demand driven chains)
Information system	• Use one primary information and communication network/platform • Build an open trading platform • Know your customers; get information on their specific demands • Create transparency: exchange POS, product quality information • Harmonise product coding and create integrated business information systems (create interoperability between systems) at sector level • Manage data ownership, data reliability and security
Organisation	• You do not have to be the best in everything; collaborate with partners at the vertical as well as horizontal level • Collaborate with local partners if one enters a new international market; they understand the culture and successful practices • Have common logistics means (and outsource the management of these means to dedicated partner) • Collaboration will not work without a good governance structure in which pain and gain sharing is defined • Collaborate out of strength and not out of weakness • The strength of a business model is increasingly dependent on that of the supply chain: have a shared approach to maximise consumer value

- Information also has a best before date;
- "One size fits all solutions" for logistics concepts, IT solutions or business models will not work;
- "Trust is nice, control is better": contracts are increasingly required to support cooperation between supply chain partners.

6 Conclusions

This chapter discussed innovative logistics concepts in the floriculture industry. In collaboration with business partners, the DaVinc^{3}i project has developed many new insights on responsive collaborative logistics networks for perishables, and in most cases the business projects resulted in concrete changes of daily practices. Overall we see the sector needs to move—and is moving—towards:

- demand driven logistics concepts, linking growers in different international sourcing areas directly to customers, thereby enabling new collaborative supply and logistics management concepts—while considering the continuous need for supply driven concepts (using the (virtual) auction clock);
- coordinated logistics control concepts with emphasis on responsiveness and guaranteeing product availability and product quality to customers;
- new (synchromodal) network designs in which hubs are established in the European market that allow for value added activities and sustainable responsive supply to (web based) customer orders;
- dynamic configurations of information systems, advanced information exchanges and transparency to facilitate virtual trade and advanced coordination and collaboration concepts.

This development will strengthen the international leading competitive position of the Dutch horticulture sector in a global, virtualized trade network. Of course, more research and foremost business case studies are needed to translate the developed theoretical concepts and business opportunities into successful business practices.

Acknowledgments DaVinc^{3}i is co-financed by Dinalog and the Horticultural Commodities Board of the Netherlands. We would like to thank the participating companies for their continued support: FloraHolland, VGB, Bunnik Plants, Gro4U, Holstein Flowers, Kwekerij A. Baas, Marjoland, Oriental Group, Muscari Magic, Hamifleurs, Hamiplant, DGI, OZ Import, Noviflora, Vida Verde, KiPlant Concept, Floréac, Greenport Logistics, Stet Heemskerk, Nover Logistics, Frugicom, Rijnplant, RoseLife, Florpartners, Baas Plantenservice, FloraCadena. We would also like to thank all involved students of Wageningen University, Eindhoven University of Technology and Free University of Amsterdam for their contributions.

References

Chan HK, Chan FTS (2010) A review of coordination studies in the context of supply chain dynamics. Int J Prod Res 48(10):2793–2819

de Keizer M, Groot JJ, Bloemhof J, van der Vorst JGAJ (2014) Logistics orchestration scenarios in a potted plant supply chain network. Int J Logistics Res Appl 17(2):156–177

Klibi W, Martel A, Guitouni A (2010) The design of robust value-creating supply chain networks: a critical review. Eur J Oper Res 203(2):283–293

Melo MT, Nickel S, da Gama FS (2009) Facility location and supply chain management: a review. Eur J Oper Res 196(2):401–412

Olhager J (2010) The role of the customer order decoupling point in production and supply chain management. Comput Ind 61(9):863–868

Ribbers AMA, Verstegen MFGM (eds) (1992) Toegepaste logistiek. Deventer, Kluwer Bedrijfswetenschappen (in Dutch)

SteadieSeifi M, Dellaert NP, Nuijten W, Van Woensel T, Raoufi R (2014) Multimodal freight transportation planning: a literature review. Eur J Oper Res 233(1):1–15

Verdouw CN, Beulens AJM, van der Vorst JGAJ (2013) Virtualisation of floricultural supply chains: a review from an Internet of Things perspective. Comput Electron Agric 99(1):160–175

Van der Vorst JGAJ, van Kooten O, Luning P (2011) Towards a diagnostic instrument to identify improvement opportunities for quality controlled logistics in agrifood supply chain networks. Int J Food Sys Dyn 2(1):94–105

Towards Efficient Multimodal Hinterland Networks

Albert W. Veenstra and Rob A. Zuidwijk

Abstract Terminal operator ECT in the port of Rotterdam sought to establish a multidisciplinary research group that would support the development of a strong multimodal hinterland network concept. This was the start of the ULTIMATE project. ULTIMATE aimed to support the development "towards efficient multimodal hinterland networks". The Ultimate research agenda consisted of four elements: (1) The operational consequences of integrating transport and cargo handling activities for supply chains. This has resulted in efficient container stacking and barge routing algorithms. (2) Incorporating new business models in the design of hinterland networks. This delivered new insights in the role of information and pricing in hinterland networks. (3) The legal consequences of mixing transport and storage activities by container terminals. The contribution of this element is a more fundamental understanding of the importance of legal concepts in multimodal transport for different actors. (4) The role and position of the port authority vis-a-vis the activities of container terminals. This led to new insights on the role and contribution of a port authority in port and hinterland network development. This chapter discusses some of the main research outcomes of the work for these four research problems.

1 Developments in Multimodal Hinterland Networks

The research into seaports underwent a transition in the last fifteen years or so. While ports have always been considered as special locations in spatial structures [see, for instance (Hoyle and Hilling 1984)], the links between ports and their

A.W. Veenstra (✉)
TU/e, PO Box 513, 5600 MB Eindhoven, The Netherlands
e-mail: a.w.veenstra@tue.nl

R.A. Zuidwijk
EUR-RSM, PO Box 1738, 3000 DR Rotterdam, The Netherlands
e-mail: rzuidwijk@rsm.nl

H. Zijm et al. (eds.), *Logistics and Supply Chain Innovation*,
Lecture Notes in Logistics, DOI 10.1007/978-3-319-22288-2_4

hinterland have not been studied much. This is surprising, since, in the field of port economics, it is commonly stated that ports compete not only on their sea access, but also on the quality of their hinterland access, see, for instance (Roso et al. 2009). Such links consist of, among others, a (not necessarily dedicated) collection of inland nodes and hubs, the mix of transport modes used to connect these locations to the seaport, the coverage of origins and destinations in the hinterland from the seaport and the organizational mechanisms that drive the flow of cargo to and from the hinterland.

In 1998, van Klink et al. (1998) suggested that ports might compete on the quality of their intermodal transport connections. Somewhat later, Notteboom and Rodrigue (2005) developed a framework for a more structured analysis of port–hinterland relationships. They position the port as a pivot linking two networks: the ocean shipping connections on the seaside, and the hinterland connections to various destinations and intermediate hubs on the landside. Their framework suggests that integrated transport structures such as corridors will develop to connect regional clusters in the hinterland to the seaport.

Around the years 2004/2005, ports in Europe were confronted with a sudden surge of containerized cargo from China. In the Port of Rotterdam, the fully automated terminals from ECT had great difficulty handling this strong sudden increase of incoming container flows, and in a few cases, when the stacks ran full, the AGVs became gridlocked, and the terminal systems came to a complete standstill. The solution for this seemed to be a mechanism that allowed the terminals to actively manage the levels of containers in their stacks, by pro-actively pushing batches of containers into the hinterland. Most multimodal hinterland transport systems were, and still are, pull systems: the receivers of containers call for transport. As a result, ECT began to take an active interest in strategically placed hinterland nodes, such as Duisburg in Germany, Venlo in the Netherlands, and Willebroek in Belgium. In addition, ECT began to search for transport partners to offer frequent services from Rotterdam to these hinterland nodes. In some cases (in particular Venlo), the network would offer several transport services per day. A final element of this concept was paperless transport, and extended releases for the container. This would facilitate speedy transport out of the seaport, while maintaining a high level of security and control over the containers in the network. While the container flow from China became more manageable and predictable, another development in the Port of Rotterdam gave new importance to the activities of ECT in the hinterland: the concessions for new terminals at the Second Maasvlakte area that were awarded to two competing terminal groups: APM Terminals and Rotterdam World Gateway. Suddenly ECT was faced with the prospect of intense competition on the seaside. This could perhaps partly be compensated with a strong competitive position in the hinterland.

In 2009, ECT sought to establish a multidisciplinary research group that would support the development of a strong multimodal hinterland network concept. This was the start of the ULTIMATE project, funded by the Dutch Institute for Advanced Logistics (DINALOG). ULTIMATE is an acronym that stands for "towards efficient multimodal hinterland networks". This was at the same time the title and the aim of the project. Ultimate was a four year projects that funded or partly funded six PhD projects. A research agenda for the project consisted of four questions:

1. what are the operational consequences of integrating transport and cargo handling activities for supply chains?
2. how can new business models be incorporated in the design of hinterland networks?
3. what are the legal consequences of mixing transport and storage activities by container terminals?
4. what is the role and position of the port authority vis-a-vis the activities of container terminals?

This chapter discusses some of the main research outcomes of the work for these four research questions. First we will elaborate on the state of the art research in multimodal transport. Four subsequent sections will each deal with one of the research questions. This chapter ends with some concluding remarks and suggestions for further research.

2 Multimodal Transport Research

There are many terms for a transport solution that consists of more than one mode: intermodality, multimodality, combined transport, and more recent inventions, such as co-modal, exomodal and synchromodal transport. In this chapter, we consistently use multimodal transport, which we take to mean a transport solution, in which several modes of transport are used simultaneously and interchangeably. This is different from intermodal transport that usually means transport of standard cargo units by more than one mode of transport in a transport chain (which presumes the cargo unit is transferred from one mode to another, somewhere underway). Almost always, and especially with container transport, rail or barge transport is intermodal transport. This is because pre- and end-haulage for these modes is usually done by truck. Integrated multimodal transport is much less well developed.

There is a lot of research on intermodal transport. We briefly describe the transition of this research from the recognition of intermodality as an economics subject to a more elaborate analysis of intermodality as a competitive factor and a platform for advanced operational research. As a basic introduction into intermodal transport, we recommend (Vrenken et al. 2005). Much of this research is also relevant for the understanding of multimodal transport.

In the early 1990s, intermodal transport, and intermodal terminals began to be noticed by researchers as a phenomenon worthy of investigation. The European Commission had, at that time, adopted a transport system approach to its policy. Giorgi and Schmidt (2002) observe that after the adoption of the common market policy in the mid-1980s, the European Commission shifted from a strongly infrastructure based transport vision towards a vision on the economic structure of transportation. The European transport system was there to support the common market. Many elements in this transport system then received attention: railways, ports, pan-European motorways, short sea shipping, inland shipping, and, the integration of these modes in intermodal transport. Academic research followed this development, first with exploratory research, such as Konings (1996), on what intermodal terminals are, and Wiegmans et al. (1999), on the economics of intermodal terminals. Internationally, McCalla et al. (2001), among others, introduced a strongly geography oriented economic view on intermodal nodes, focusing on the relationship between these nodes and their surroundings. This type of research developed over the years in extensive transport network modelling efforts [see for instance Janic (2007), and more recently, Zhang (2013)].

Another, more strategically oriented research stream emerged from an early contribution of van Klink et al. (1998), who point out that intermodalism could and should be a competitive factor for seaports. This research is followed, among others, by Notteboom and Rodrigue (2005), who elaborate on the development of hinterland networks and place this development within the framework of general port development models. Strategy making of port authorities should thus include a vision on hinterland connectivity. This line of research was reflected in the ULTIMATE project; see Sect. 6. This line of research also seems to lead to fruitful new insights for management theory proper, as van der Lugt et al. (2013a) note, since port authorities are hybrid organizations of a particular kind.

While the term intermodal transport and intermodal terminal are relatively new, the notion of inland nodes that are connected to seaports is much older. In economic development circles, the term dryport has been used for a long time to indicate hinterland extensions of, often congested and poorly managed, seaports. Roso et al. (2009) revitalize this concept, and place it in the context of freight transport networks and intermodal gateways. Veenstra et al. (2012) contribute to this discussion by pointing out that the developments of extended gate networks are in fact modernized versions of dry ports in the hinterland of the Port of Rotterdam.

In the meantime, little attention was given to operations in intermodal and multimodal transport networks. Much of the research that emerged focused on isolated operational processes, such as crane operations, truck movements in terminals, or terminal equipment routing. In recent years, a more integrated view on multimodal transport operations was kicked off by surveys from Macharis and Bontekoning (2004) and, more recently, SteadieSeifi et al. (2014), who identify many white spots in the operations research literature for intermodal networks: network topologies, focusing on corridor structures in particular, integrating the planning of forward and backward hauls of transport modes, simultaneous planning of modes, distributed planning in complex networks, and solution methods for the more complicated planning problems all deserve attention.

In the ULTIMATE project, this research is extended in several ways. We extend the research on strategy making of port authorities, and consider several integrated multimodal network problems. In addition, we introduce a relatively new field of research in intermodal transport that deals with legal aspects of multimodality.

3 Integrated Planning of Transport and Cargo Handling

Multimodal transport is a complex logistical arrangement, in which the activities of multiple modalities, as well as one or more nodes or terminals need to be coordinated. Fazi (2014) unravels the mode allocation problem in choices for bundling, scheduling and routing. He also investigates the impact of container transport specific constraints that relate to time windows and due dates that are imposed by various stakeholders on the transport problem. Such constraints originate from the transport service itself—cut off times for a departing vessel—or from the use of the container on land. In particular these last constraints, i.e. the time before a full container needs to be collected in a terminal, and the time before the empty container has to be delivered back to the shipping lines—in practice the terms demurrage and detention are used—are particularly binding.

Fazi (2014, Chap. 2) considers the problem of bundling as many containers into one barge as possible. He models this problem as a variable size bin packing problem with time constraints, both in a deterministic and a stochastic variant. The model represents the realistic situation of an inland terminal that want to bring a number of containers from a seaport terminal to its own terminal inland, preferably by barge. The barge has a limited capacity, and the containers have deadlines by which the owners of the cargo need them. When the barge has to sail, remaining containers will be shipped by truck. Containers with a tight deadline will also be transported by truck. These problems are NP-hard, and they are solved with a heuristic. The results underline the negative impact of variability in the container flow, as well as time constraints, on the solutions of the problem. In this problem formulation, the actual impact of transportation was not considered.

The problem of mode allocation and scheduling for import containers is described as a heterogeneous vehicle routing problem (Fazi 2014, Chap. 3). This problem describes the situation in which an inland terminal has a number of barges and trucks, that need to collect containers at various terminals in the seaport. Again, time constraints for the containers are considered. A planner needs to allocate containers to barges and trucks in order to minimize transport time, while making optimal use of capacity and meeting the deadlines. Again a heuristic was developed to solve this problem. Some further analysis on the availability of information showed that more information does not always lead to better results. Sometimes, shorter time spans make better barge schedules. Fazi concludes from this that the choice of the planning horizon is critical for an efficient inland shipping operation.

A further extension of the mode allocation and scheduling problem is obtained by including export containers. Now the problem becomes a vehicle routing problem with pick-up and delivery and due dates for containers. This problem comes close to the problems inland terminal planners solve on a daily basis. Fazi deploys several heuristics and greedy algorithms to compare solution strategies. The greedy algorithms resemble how planners work in practice. The analysis shows that the heuristics perform markedly better than the greedy algorithms.

The practical relevance of Fazi's approach was proven in an experiment where a planner of one inland terminal in the Brabant Intermodal Group of terminals competed against Fazi's pick-up and delivery algorithm that includes import and export flows. This experiment showed that optimization of the operation did deliver better results than the planner, and could potentially save thousands of euros per trip.[1]

Finally, Fazi (2014, Chap. 5) studies the impact of demurrage and detention of containers. The due dates that shipping lines impose on their customers for full container pick up, and for empty container delivery back into an empty container depot are serious barriers to logistics optimization, and add significant costs to shippers' transport bills, as a result of penalties for violation of the due dates. Fazi shows that the due date and fee structure actually extends dwell time of the containers in terminals. He also shows that with the absence of demurrage and detention penalties, transportation costs could go down (because of less last minute trucking), and dwell time in the terminals (and therefore container turnaround time) could be reduced. Fazi also developed a new demurrage and detention cost function, i.e. a daily fee, chargeable from the moment of arrival of the container in the ocean terminal. This results in better transport costs, lower dwell times, and still a reasonable revenue for the shipping lines. This work sheds a whole new and critical light on the impact of the current demurrage and detention policy of shipping lines on multimodal transport in the Port of Rotterdam.

4 Network Design and New Business Models

As explained in the introduction, maritime container terminal operating companies have assumed a new role as multimodal transport network operator. They connect deep sea terminals with inland terminals via high capacity transport means such as barge (river vessel) and train. They have also extended the gate of their deep sea terminal to the gate of the inland terminal by the offering of (sea)port-to-(inland) port services and (sea)port-to-door services. These services compete with truck transport services when service frequencies are high enough. In his PhD research, Ypsilantis has discussed the role of new business models in the design of multimodal network services. In his work for the ULTIMATE project, he has considered (i) joint pricing and design of multimodal networks, (ii) dwell time analysis based on shared information, and (iii) the impact of collaboration on performance of network services.

First, Ypsilantis and Zuidwijk (2013) considered joint pricing and design of network services. In their research, they focused on Extended Gate networks, in which operators face the following three interrelated decisions: (1) determine which inland terminals act as extended gates of the seaport terminal, (2) determine

[1]w.a. Wiskundige module containertransport biedt kansen. Logistiek.nl, November 2014.

capacities of the transport services based on vessel capacities and frequencies of service, and (3) set the prices for the transport services on the network. Based on the work of Brotcorne et al. (2008), a bi-level programming model has been developed which jointly designs and prices extended gate network services. The network operator maximizes profit in anticipation of customers that route their cargo via minimum cost paths. The customers also have the option to purchase direct trucking services offered by the competition. Compared with the existing literature, Ypsilantis and Zuidwijk (2013) in addition incorporate time constraints and economies of scale.

Moreover, a heuristic is developed, similar to the one proposed by Brotcorne et al. (2008), that exploits the specific structure of the problem and that provides near optimal solutions to the problem in substantially less time. The ULTIMATE project provided data that could be used to model realistic scenarios, for which the authors study optimal network designs while comparing the aforementioned port-to-port and port-to-door services. One of the interesting results of the paper relate to the differences between the design of port-to-port services and port-to-door services. In the case of port-to-door services, the prices of services are determined by the competition. This implies that the design of the network is driven by cost minimization, and freight flows are consolidated through a limited number of extended gates to achieve economies of scale. In the case of port-to-port services, extended gates are opened to target particular market segments for which the competition leaves room for higher port-to-port tariffs. As a consequence, in the case of port-to-port services, optimization of network design truly comes down to revenue enhancement, and pricing and designing services cannot be distangled.

In the second contribution by Ypsilantis et al. (2014), joint data from a deep sea terminal operator and hinterland terminal operators allowed for an in-depth analysis of dwell times of import containers at sea ports and dry ports. The statistical analysis performed in the paper demonstrates that a considerable part of the dwell time variance is explained by factors that are under direct influence of the shipper. The analysis is based on real data, in particular milestone timestamps of import containers in the specific region in 2011.

A third contribution by Ypsilantis et al. (2014) relates to the design of barge services between a number of deep sea terminals and a number of inland terminals. Important principles in the design of multimodal transport services are the deployment of barges of certain sizes to reap economies of scale and the routing and circulation of these barges to attain frequencies of services on the network. The aim is to make the barge services on the network competitive with road transport, despite the costs and delays associated with the additional transshipments required. The authors develop a tight MIP formulation for the Fleet Size and Mix Vehicle Routing Problem, which is especially adapted to the mulitmodal barge service network design at hand. Particular attention is given to the benefits of horizontal cooperation between inland container terminals through capacity sharing. Results show that in the case of cooperation, not only costs are saved, but also service levels are enhanced.

5 Legal Consequences of Transport by Terminals

There are generally two standard ways in which hinterland transport can be arranged [see, for instance, Veenstra et al. (2012)]: either the shipping line extends its service to bring cargo from a seaport to a hinterland location, or the receiver of the cargo picks up cargo at the seaport and brings it to its own premises. The former is called carrier haulage, while the latter is called merchant haulage, referring to the buyer of the goods (i.e. the merchant) taking care of haulage. Usually both the merchant and the carrier hire a specialist to take care of the transport operations: a freight forwarder or transport operator.

In the development of multimodal transport networks, a new form of transportation is emerging: terminal haulage. In this transport concept, the terminal is the party that offers inland transportation as part of its service offering. Usually, the transportation offered by the terminal is limited in the sense that the terminal offers transport from its seaport location to one of its inland terminal locations. The final transport leg is then usually performed either by the merchant or by the inland terminal, by order of the merchant.

In transport law, in general, the roles of these parties, as well as the terminals that perform the loading and unloading operations, are carefully defined; see for more details, Haak and Kluwer (2010). In many countries in the world, transport law, as it applies to these different parties is mandatory law. In contracts, parties are not allowed to deviate from the stipulations in the law. This makes the development of new roles, such as a terminal operator that also offers a transportation service as part of its storage service, complicated. If the terminal moves a container from one terminal location to another one, without an order from its customer, it is not considered transportation, but an inter-terminal move, for which the standard limited liability of transport operators does not apply. The terminal is then fully liable for the value of the container. See for a legal discussion, the legal case of the General Vargas (Haak 1997).

As part of the ULTIMATE project, Susan Niessen has investigated several related legal questions. First of all, when is a transport operator responsible for the goods? Legally, this is from the moment of acceptance of the cargo until delivery of the cargo at destination. Delivery is a two-sided transaction, in which both the transport operator and the receiver of the goods have to interact, and the receiver actually takes control over the goods. Second, what is the legal obligation that a terminal undertakes in an extended gate service? In fact, an extended gate service consists of activities by five different legal actors: the stevedore, the storage keeper, carrier, forwarder or customs agent, and the legal treatment of problems such as damage depends strongly on which role was being performed at the time of occurrence of the problem (Niessen 2012).

The newly adopted Rotterdam Rules might provide a solution. Niessen investigated the role the terminals receive under the Rotterdam Rules regime (Niessen 2014). Terminals are so-called Maritime Performing Parties, under the Rotterdam Rules. This stipulation provides a better basis for the legal position of, among others, terminals than the current legal framework offers. At the same time, however, inland transport as an activity is excluded as part of the activities of the maritime performing party. The solution is a mixed contract, in which the terminal performs stevedoring activities as a maritime performing party, and inland transport as a (non-maritime) performing party, which is subject to applicable national law. What is required for the terminal offering a combination of cargo handling and transport services, is therefore a mixed contract that clearly identifies what activities is part of what legal role. Niessen recommends that the boundaries between activities are clearly identified in these contracts.

In a practical sense, Niessen's research has resulted in the formulation of intermodal transport conditions that are an improvement upon the amalgam of mode specific transport conditions that transport operators currently use. For the industry association VITO, representing Dutch inland intermodal container terminals, such a set of intermodal transport conditions were formulated.

6 The Role of Port Authorities

The extended gate concepts, as they are being developed by container terminals, as well as by ocean carriers, seem to by-pass port authorities. At the same time, port authorities should take a keen interest, since success of these concepts will enhance the competitive position of the port as a whole. As a result of this, the ULTIMATE project has addressed the topic through two lines of research. Larissa van der Lugt has concentrated on the shifting role of the port authority in governing a port area. Roy van den Berg has worked from inside the Rotterdam Port Authority to develop the strategy of the port authority towards carriers and terminal operators.

Verhoeven (2010) states that port authorities are publicly owned organizations in a competing world. The role most port authorities perform in this position is that of landlord: they own and control the port area, while mostly private parties invest in superstructure and perform activities such as cargo handling. Van der Lugt et al. (2013b) argue that the landlord model no longer covers the range of activities and roles of a port authority. Among these are activities related to the development of hinterland networks, facilitating private port companies and the creation of incentives for behaviour of port companies. In van der Lugt et al. (2013a), the authors elaborate on the particular nature of strategy making of port authorities. They characterize port authorities as hybrid, shared value organizations that are very interesting research subjects from a management research point of view, since their complex strategy making challenges some of the key concepts in management strategy literature.

Van den Berg provides an inside-out view on this complicated strategy making at a port authority. He investigates the modal split obligations that were included by the Rotterdam Port authority in the concessions on the Maasvlakte area in Rotterdam (van den Berg and De Langen 2014b). He finds that the modal split obligations do focus the attention of the container terminals on the quality of their hinterland networks, and in the case of the new terminals, has led to changes in the terminal design to better meet the obligations.

With respect to shipping lines' strategy, van den Berg and De Langen (2014a, c) find that a new service strategy is being introduced: the port-to-inland terminal (ILT) service, which lies somewhere in between the classic door-to-door and port-to-port services shipping lines have offered. They also find, surprisingly, that the port-to-ILT service actually competes more with the door-to-door service than with the port-to-port service. This is in contrast with earlier research [for instance, Franc and Van der Horst (2010)], that claimed the door-to-door service is not as sensitive to competition as the port-to-port service. Port authorities, for which shipping lines are important clients, need to be aware of these competitive dynamics. Apart from excellent cargo handling facilities, they could and should strive to offer a portfolio of inland terminals that have all the relevant capabilities and facilities to enable the shipping lines' new value proposition.

The results of this research have translated back into the Rotterdam Port Authority's strategy making, in various areas. First of all, the Port of Rotterdam has considered the importance of the quality of hinterland connections a key element in the design of the operational cooperation between the competing terminals at the Second Maasvlakte. Instead of treating these terminals as independent entities, the Port Authority has tried to forge an operational cooperation agreement between these terminals on inter-terminal transport, and on the joint handling of container trains. Second, the better understanding of the strategy of ocean shipping lines has directed the focus on inland terminal development by the Port Authority in especially southern Germany. The Port Authority closely cooperates with the container terminals in these efforts.

7 Concluding Remarks

The extended gate concept is an important development in the Port of Rotterdam. The research in the ULTIMATE project has contributed significantly to the understanding of the potential and the bottlenecks for extended gate services. Both the research of Fazi (Sect. 3) and Ypsilantis (4) showed that collaboration among inland terminal operators and between deep sea and inland terminal operators creates efficiencies and enhanced service levels. Moreover, Panagiotis demonstrated that design and pricing of network services are intertwined. Fazi showed that optimisation of the hinterland container network brings benefits for transport and terminal operators. In addition, Fazi showed that current demurrage and detention policies of ocean carriers are detrimental to this optimization process. Van der Lugt and van den Berg (Sect. 6) showed that port authority strategy making could play an important role in the further development of hinterland networks of terminals and ocean carriers, and this will contribute to the overall competitive position of the seaport. Niessen (Sect. 5), finally, pointed out some serious legal impediments in the development of extended gate concepts for ocean terminals. She also presented a solution, which is a hybrid contract under the new Rotterdam Rules regime.

All of this research also has practical applications, which are either within research on the short term, or are important ambitions for the medium term.

Many research questions remain. Much work needs to be done in the development of heuristics to solve more advanced hinterland transport problems, as well as more elaborate multimodal network design problems. Strategy making of port authorities turns out to be fruitful research area in business strategy, due to the particular nature of port authorities as hybrid organization in the public-private arena. In the legal profession, role changes for transport and logistics businesses are a complicated research area in need of more work. The ULTIMATE project focused on one particular role change: from terminal to transport operator. In addition, the complicated legal matter of a uniform framework for multimodal transport remains to be solved. The Rotterdam Rules provide a complicated solution for only one instance: intercontinental transport with an ocean leg. In addition, the Rotterdam Rules refer to national law for the multimodal transport leg, in which different legal structures exist for different modes of transport. What is required is a new framework that integrated all these frameworks into one.

The definition of new logistics concepts that re-establish new business models and roles in supply chains has impacts beyond the logistics realm. The project ULTIMATE has spurred a new research agenda where logistics, legal, business, and public administration aspects play a role and interact. For instance, both horizontal and vertical integration and collaboration have logistics and legal consequences. The ULTIMATE project has shown the benefit of a multidisciplinary approach, where mutual understanding between researchers strengthens their own effort, and, at the same time, provides more practical outcomes for business parties. As such, the ULTIMATE project has set a significant step in the direction towards integrated multimodal hinterland networks.

Acknowledgments This research was funded by Dinalog, the Dutch Institute for Advanced Logistics. We thank the business partners in the project for their continued support of the PhD projects and a number of master student projects.

References

Brotcorne L, Labbé M, Marcotte P, Savard G (2008) Joint design and pricing on a network. Oper Res 56(5):1104–1115

Fazi S (2014) Mode selection, routing and scheduling for inland container transport. PhD thesis, TU Eindhoven, Beta Research School, 2014

Franc P, Van der Horst M (2010) Understanding hinterland service integration by shipping lines and terminal operators: a theoretical and empirical analysis. J Transp Geogr 18(4):557–566

Giorgi L, Schmidt M (2002) European transport policy—a historical and forward looking perspective. Ger Policy Stud 2(4):1–19

Haak K (1997) Hr 28 november 1997, nj 1998, 706;& 1998, 33 (general vargas)., 1998

Haak KF, Zwitser R (2010) Van haven en handel: hoofdzaken van het handelsverkeersrecht. Kluwer

Hoyle BS, Hilling D (1984) Spatial approaches to port development. In: Hoyle BS, Hilling D (eds) Seaport systems and spatial change. Wiley, Chichester, pp 1–19

Janic M (2007) Modelling the full costs of an intermodal and road freight transport network. Transp Res Part D: Trans Environ 12(1):33–44

Konings JW (1996) Integrated centres for the transshipment, storage, collection and distribution of goods: a survey of the possibilities for a high-quality intermodal transport concept. Transp Policy 3(1):3–11

Macharis C, Bontekoning YM (2004) Opportunities for or in intermodal freight transport research: a review. Eur J Oper Res 153(2):400–416

McCalla RJ, Slack B, Comtois C (2001) Intermodal freight terminals: locality and industrial linkages. Can Geogr/Le Géographe Canadien 45(3):404–413

Niessen S (2012) The legal position of terminal operators within the legal framework of intermodal transportation. Poster, 2012

Niessen S (2014) A new era for terminal operators: the terminal operator as a maritime performing party under the rotterdam rules when concluding mixed contracts for sea port and hinterland activities. European transport law:= Droit européen des transports= Europäisches Transportrecht= Diritto europeo dei trasporti= Derecho europeo de transportes= Europees vervoerrecht 49(1):25–30

Notteboom TE, Rodrigue JP (2005) Port regionalization: towards a new phase in port development. Maritime Policy Manage 32(3):297–313

Roso V, Woxenius J, Lumsden K (2009) The dry port concept: connecting container seaports with the hinterland. J Transp Geogr 17(5):338–345

SteadieSeifi M, Dellaert NP, Nuijten W, Van Woensel T, Raoufi R (2014) Multimodal freight transportation planning: a literature review. Eur J Oper Res 233(1):1–15

van Klink HA, van den Berg GC (1998) Gateways and intermodalism. J Trans Geogr 6(1):1–9

van den Berg R, De Langen PW (2014a) Assessing the intermodal value proposition of shipping lines: attitudes of shippers and forwarders. Marit Econ Logistics 1–20 (ahead-of-print)

van den Berg R, De Langen PW (2014b) An exploratory analysis of the effects of modal split obligations in terminal concessioning contracts. Int J Shipp Trans Logistics 6(6):571–592

van den Berg R, De Langen PW (2014c) Towards an inland terminal centred value proposition. Marit Policy Manage 1–17 (ahead-of-print)

van der Lugt L, Dooms M, Parola F (2013a) Strategy making by hybrid organizations: the case of the port authority. Res Trans Bus Manage 8:103–113

van der Lugt LM, de Langen PW, Hagdorn L (2013b) Beyond the landlord: typologies of port authority strategies. In Cariou P (ed) 2013 IAME Conference, Marseilles, France. Euromed Management, Kedge Business School, Euromed Management, Kedge Business School, 2013

Veenstra AW, Zuidwijk R, van Asperen E (2012) The extended gate concept for container terminals: expanding the notion of dry ports. Marit Econ Logistics 14(1):14–32

Verhoeven P (2010) A review of port authority functions: towards a renaissance? Marit Policy Manag 37(3):247–270

Vrenken H, Macharis C, Wolters P (2005) Intermodal transport in Europe. European Intermodal Association Brussels, Belgium

Wiegmans BW, Masurel E, Nijkamp P (1999) Intermodal freight terminals: an analysis of the terminal market. Trans Plann Technol 23(2):105–128

Ypsilantis P, Zuidwijk RA (2013) Joint design and pricing of intermodal port—hinterland network services: considering economies of scale and service time constraints. Technical Report, 2013

Ypsilantis P, Zuidwijk RA, Pourakbar M (2014) Joint fleet deployment and barge rotation network design: the case of horizontal cooperation of dry ports. manuscript, 2014

Ypsilantis P, Zuidwijk RA, van Dalen J (2014) Dwell times of containers at container terminals: the shippers' effect. manuscript, 2014

Zhang M (2013) A freight transport model for integrated network, service, and policy design. TRAIL Thesis Series, 2013

Current Deficiencies and Paths for Future Improvement in Corporate Sustainability Reporting

Thomas Kjaergaard, Martin C. Schleper and Christoph G. Schmidt

Abstract Current International Accountability Standards for sustainability reporting, such as The United Nations Global Compact and the Global Reporting Initiative are subject to criticism from two sides, researchers and practitioners. Through interviews with key persons from audit firms and a systematic literature review, we identify major deficiencies in current corporate sustainability reporting practices. Based on these findings, we derive five propositions which address the need for future improvements, i.e. we propose that a dynamic standard for corporate sustainability reporting must capture a firm's longitudinal learning and development of intra- and inter-organizational sustainability capabilities by integrating them as leading indicators. We conclude the article with an outlook on future paths for an improved sustainability reporting framework focusing on intra- and inter-organizational capabilities and best practices which are proposed to have an impact on sustainability performance along the entire supply chain.

Keywords Sustainability reporting · Best practices · Sustainability capabilities · Supply chain · International Accountability Standards

T. Kjaergaard
Interdisciplinary Centre for Organizational Architecture (ICOA), Aarhus University, Fuglsangs Allé 20, 8210 Aarhus V, Denmark

M.C. Schleper (✉)
German Graduate School of Management and Law (GGS), Bildungscampus 2, 74076 Heilbronn, Germany
e-mail: Martin.Schleper@ggs.de

C.G. Schmidt
Institute for Supply Chain Management (ISCM), EBS University for Business and Law, Burgstraße 5, 65375 Oestrich-Winkel, Germany

H. Zijm et al. (eds.), *Logistics and Supply Chain Innovation*,
Lecture Notes in Logistics, DOI 10.1007/978-3-319-22288-2_5

1 Introduction

Corporate efforts in assuring sustainability activities have increased continuously during the last decades. In 2013 more than 90 % of the top 250 organizations listed on the Fortune Global 500 ranking used a sustainability report to display their sustainability undertakings (KPMG 2013). At the same time, the field of sustainability reporting has undergone a significant consolidation in which some major International Accountability Standards (IAS) have emerged. Among these reporting frameworks are the United Nations Global Compact (UNGC), the Global Reporting Initiative (GRI), the AccountAbility 1000 (AA1000), as well as the Social Accountability 8000 (SA8000).

Nonetheless, the trend towards increasing reporting practices has not implicitly caused more excellence in every aspect. Despite the positive effect of overall and particularly firms' awareness of the need to strive for sustainability and to assure these activities, there is still a call for improvements of IAS from scholarly research and practice. Particularly two major areas for improvement can be identified.

On the one hand a wider integration of activities and factors which are lying beyond the direct impact of the corporation is needed. In its 2013 annual review, the World Business Council for Sustainable Development (WBCSD) pointed to this improvement gap and called organizations to provide evidence to show they are engaging with suppliers and customers to address material risks and opportunities identified along the supply chain (WBCSD 2013). This is in line with the fourth generation of the GRI guidelines (GRI4) for sustainability reporting which also emphasize the requirement to focus on supply chain aspects and presents an extended set of indicators for supply chain reporting compared to previous GRI guidelines. Since GRI4 has only recently been issued, corporations' adoption is still very low, but these new reporting structures are bound to influence corporate reporting practice and potentially also corporations' management of their sustainability practice in the future.

On the other hand a move away from lagging indicators towards a more capability-based view referring to leading indicators is often called for (e.g. Peloza 2009). Although achieving corporate sustainability is often a time intensive learning process (Kaptein and Wempe 1998) that requires organizational capabilities, this learning perspective is mostly neglected when it comes to sustainability performance and the reporting of its indicators. Consequently, corporations engaged in sustainability reporting primarily collect and disclose data for lagging indicators. The "work related injury rate" from the GRI framework for example shows only factual past performance and does not inform corporate management and stakeholders on the actual capabilities in this regard. On the contrary, practices like a written company policy on labor and risk and impact assessments in the area of labor from the UNGC framework can be perceived as leading indicators, potentially representing capabilities which are essential for corporate decision-making and the advancement of sustainability performance. Furthermore, focusing on leading indicators and capabilities enhances transparency within reporting processes and stakeholders will get an improved understanding of the sources for sustainability performance. Thus, in contrast to less successful attempts to design IAS analogous to financial reporting (Etzion and Ferraro 2010) relying mostly on lagging indicators, this article[1] follows the direction of proposing a learning and capability focused framework for sustainability reporting building on best practices for sustainability performance.

By actively embracing an inter-organizational supply chain perspective within the context of sustainability reporting practices, we not only provide a remedy for the WBCSD gap mentioned above but also pave the way for a reporting framework that accounts for environmental and social performance along the entire supply chain. From this perspective, entangling supply chain management with sustainability reporting allows for a new interpretation of both. Sustainability reporting should be considered as a management instrument of sustainable supply chain management and no longer as a mere reporting of past performance.

The paper is structured as follows. In the next section, we briefly describe the applied research process, followed by our findings which result in the presentation of five propositions. Simultaneously, we focus on identifying best practices for sustainability reporting. The final sections provide a future outlook on the new framework and conclude this research note.

2 Research Process

In order to leverage the purposes described above, a research process with two major steps was applied. The first part of our research focuses on interviews with key informants from auditing firms, while the second part of our study provides a systematic literature review of supply chain related best practices with a positive effect on sustainability performance.

In order to gain a broader insight from corporate practice regarding the identification of deficiencies in current reporting practices, we chose to interview senior consultants specialized in sustainability reporting from three of the "Big Four" auditing firms. This choice was particularly motivated by two reasons. First, senior consultants in the area of sustainability assurance and auditing are by definition well-informed on current developments in the area of sustainability reporting and possess an extensive expertise and experience in that area. Second, these key informants advise their clients in IAS implementation processes and their audit firms are consulted on or directly involved in the development of both multi-stakeholder standards and industry-specific standards.

[1]This article is a scientific excerpt from an ongoing research project. We gratefully acknowledge that this project is funded by the German Federal Ministry of Education and Research, grant no. 01IC10L14A.

The three semi-structured interviews followed the same standardized approach. An interview guide was developed with a series of standard questions based on relevant topics, but the answers in one interview also led to additional questions in the following interview and the interviewers had the freedom to pose ad hoc questions during the interviews.

The key findings framed our approach to focus on sustainability best practices and served as input to the further research process, which was undertaken as a systematic literature review of scholarly articles published in the years 2005–2015. The sampling process was based on a keyword search in the databases ScienceDirect and EBSCO searching for articles that included combinations of the keywords such as "drivers", "antecedents", "practices", or "factors" and variations and synonyms of "supply chain performance". From the initial set of relevant articles we removed those published in journals with a rating of below 0.3 according to the "Handelsblatt Ranking BWL 2012". We then screened the remaining papers and excluded articles of which the titles did not fit our research purposes. Subsequently, we eliminated articles with abstract that did not match our goals. The final set consisted of 38 scientific articles which were analyzed in-depth by two researchers independently with the goal of identifying potential supply chain related best practices. The identified best practices were discussed, clustered, and aggregated.

3 Findings

In this section we describe the results of our research process in detail. First, we provide a summary and overview of findings from the key informant interviews. These new insights are then linked to current IAS and reporting literature. A set of five propositions addresses the identified deficiencies and paves the road for a revised and improved framework for corporate sustainability reporting according to IAS, thereby integrating a broader supply chain context. As we particularly emphasize the capability-oriented logic as the most interesting and important characteristic of the proposed framework, which influences all other design issues, we extract best practices for sustainability performance from the academic literature in the second part of our result section.

3.1 Insights from Auditing Firms on IAS and Clients' Adoption

In the following we outline the results of interviews with key sustainability professionals of three of the "Big Four" auditing firms. Table 1 illustrates the interviewees' quotes for each of the topics raised during the interviews. In the subsequent part these different issues are analyzed in order to provide a basis for developing propositions towards a framework.

As a first perspective, we address the process of IAS adoption, which includes the view on how the interviewees and their clients perceive and use these standards. All sustainability audit professionals (SAPs) stated that most of their major clients use a version (3.0, 3.2 or 4) of the GRI guidelines which they all assess to be the most adopted IAS among their clients. Two of the SAPs stated that the UNGC is also widely followed among their clients, whereas SAP_2 indicated a lower adoption rate for the UNGC.

Although the IAS adoption rate already indicates the extent to which corporate sustainability reporting is present in practice throughout different industries, determination of the quality of the reporting practice requires supplementary information with regard to how corporations perceive, prioritize, and apply IAS. SAP_1 illustrated this aspect by stating that reporting according to the UNGC is a "side product" to the GRI reporting, disregarding the UNGC Advanced Level as generally not being the key focus of reporting for organizations. Even more, SAP_3 suggested that the UNGC Advanced Level suits only a few publicly listed large companies, implying that most UNGC adopters rather report on the principle-based Active Level.

SAP_2 elaborated in more detail on the complexity and intangibility of the UNGC as reasons for the lower adoption rate among his clients. In general, the SAPs consider the GRI as more tangible but at the same time as very comprehensive, resource demanding, and not unproblematic in terms of comparability and interpretation. This holds in particular for the GRI A+ application level, which is the highest reporting level in the GRI guidelines and the main target level of many major clients in SAP_1's audit firm. On the contrary, a smaller group of clients consider it less important to cover all indicators required to receive the A+ application level. Rather, they focus on indicators which are usually more material for stakeholders, according to SAP_1. This trend towards materiality is also embedded in GRI 4 which incorporates a definition of materiality central to the framework for integrated reporting developed by the International Integrated Reporting Council (IIRC).

SAP_1 recognized GRI 4 as a move away from the traditional ABC levels and suggested that an increased materiality focus might lead to more accurate and differentiated reporting based on actual stakeholder interests. SAP_1 further predicted that supply chain sustainability would thus increasingly become a highly important topic that organizations will need to report on (Table 1).

Table 1 Quotes from interviews with audit firm sustainability professionals

Salient topic	Link to practice
Adoption of IAS	"I would say that most of the clients, maybe more than 90 %, follow GRI and a huge number of clients also follow the UN Global Compact" (SAP_3)
	"The UN Global Compact is not so much an issue of our clients" (SAP_2)

(continued)

Table 1 (continued)

Salient topic	Link to practice
Global reporting initiative	"So the GRI A+ is an ambition level for many clients, while I have few clients where this A level ambition is not the focus and it is more about materiality" (SAP_1)
	"If we look at GRI as a multi-stakeholder standard, the most challenging part is that of reporting boundaries; how to define what is the scope of reporting, what should be included" (SAP_3)
UN global compact	"The UN Global Compact reporting… is a side product of the overall GRI reporting process. It takes quite a lot of resources to collect all the information for GRI" (SAP_1)
	"The UN Global Compact is not as tangible as the GRI. It's more like a global framework… it's too complex for them and… doesn't offer them the benefit that they need either for their internal sustainability management or for the external position" (SAP_2)
Supply chain focus	"Supply chain sustainability is increasingly a topic for sustainability reports with a more materiality focus, in GRI 4 for example, that will shift even more the focus to supply chains" (SAP_1)
	"This is the key challenge, I would say, to get the information from the complete supply chain and with this it is difficult to say the data and the information is accurate within the sustainability report" (SAP_3)
Performance measures and indicators	"The energy consumption is lower compared to the prior year, but it depends on different factors so it is not always that clear whether the reporting standards help to identity the right things or to measure the progress in the right way" (SAP_3)
	"Companies put a lot of effort into collecting data and publishing these to get the A+ label. I think this is a misallocation of resources…you could talk about a counterproductive or misleading role of reporting" (SAP_1)

During the interview with SAP_1 the interviewee informed us that the GRI 4 aims to better address upstream supply chain tiers and that they currently contain some preparatory steps towards higher tiers. Although a supply chain perspective is not completely absent in previous GRI guidelines and despite the fact that supply chain related indicators can also be found in the UNGC Advanced Level, the progress towards greater specificity for more upstream tiers in GRI 4 can be considered a significant advancement which can provide stakeholders with new informative insights. GRI 4 thereby addresses an information gap as mentioned by SAP_1. According to this interviewee, supply chain reporting can be framed as reporting of company or product performance emerging between several business partners. The interviewee further indicates that this kind of reporting will rarely be conducted by using multi-stakeholder standards such as the GRI or the UNGC. It is rather a matter of assessing suppliers' sustainability performance based on a set of criteria defined by customers, according to SAP_1.

However, reporting to company-specific criteria is highly challenging for suppliers that usually supply multiple focal companies. For this reason, SAP_1's audit firm works towards more industry-wide applicability of reporting practices. SAP_2's audit firm represents mainly clients in the mid-market that are often suppliers to focal companies. She stated that her clients generally adapt the industry's specific standards because stakeholders often require it. Consequently, identifying standards relevant for clients and determining their importance for stakeholders remains a big challenge for SAP_2's firm. As a remedy, SAP_2 said this problem could be addressed with a standard that is specific for a particular industry and at the same time compatible with the most relevant of other industry-specific standards. However, such an initiative will still be challenged by the fact that most of the clients only acknowledge a need to address tier 1 of their supply chain and do not recognize their responsibility for actions including more distinct suppliers. With regard to potential benefits of integrating supply chain perspectives into reporting, SAP_3 commented that although guidelines, codes of conduct, and other monitoring and governance mechanisms might reduce reputation and quality risks, these issues remain difficult to measure and value for organizations. More directly, SAP_1 questioned whether measuring supply chain indicators is relevant and feasible at all, adding that if this were the case, the measurement would need to encompass a longer time horizon in order to capture long-term effects.

3.2 Towards an Improved Framework for Sustainability Reporting

The interviews indicate that although both IAS, GRI and UNGC, show high adoption rates among firms, GRI is clearly the IAS priority in corporate reporting practice. Yet, the actual degree of corporate implementation of IAS seems to vary significantly. Despite the positive aspect of their diffusion, there is still a need for an integration of an extended supply chain perspective across IAS such as the GRI and UNGC. Consequently, scholars have stressed the need "to look outside an organization's boundaries" when it comes to sustainability performance (Meehan and Bryde 2011, p. 95). Moreover, stakeholders have developed a distinctive awareness for these issues, hence nowadays firms, such as Nike, Apple and BP are often held liable for their supply chain partners when it comes to environmental and social incidents (Hartmann and Moeller 2014). Partly because of these developments, companies like Microsoft have started to actively embrace external partners in their supply chains by requiring annual sustainability reports. Thus, we suggest:

P_1: A dynamic standard for corporate sustainability reporting must require accounts for sustainability issues both within the boundaries of the corporation and in the corporate supply chain.

Moreover, an increasing trend towards defining materiality within sustainability reporting practices was identified in conformance with the debates in the academic literature. Although materiality is a fundamental concept stemming from modern accounting (Messier et al. 2005), due to the increased call from practice and scholars, materiality has diffused into the field of sustainability, where it continuously gains relevance. Nonetheless, many scholars point out a lack of materiality considerations within sustainability reporting and a high potential to develop the field further in this regard (e.g. Kanzer 2010). Although the materiality concept is well-known and called for within IAS and the reporting literature, we adopt this critical aspect and hence propose:

> P_2: A dynamic standard for corporate sustainability reporting must integrate the concept of materiality, accounting for whether the reported information concerns core business activities within the supply chain, that are material to the society, the corporation, or both, in an integrative manner.

Labeling the UNGC as complex and not tangible is also reflected in the IAS-related literature, where some of the criticism focuses on the lack of precision (Bigge 2004; Nolan 2005) in the description of UNGC principles and their generality (Deva 2006). These issues make the UNGC hard to understand and apply for organizations. The GRI is perceived as more tangible, but also comprehensive, resource demanding and—despite a large number of quantitative indicators—at least to some degree incomparable. Particularly this last point of critique is also present in scholarly literature (Dingwerth and Eichinger 2010; Levy et al. 2010). The overall lack of comparability weakens the UNGC and GRI as potential frameworks for measuring and reporting sustainability performance. This is especially problematic as the concept of comparability is central for standards in general and for IAS in particular. Part of the problem (and the solution) is touched upon in the interviews, which question the value of quantitative indicators if information about the factors influencing the annual decrease or increase is not reported in a comparable and uniform way. For instance, the increase or decrease of a company's energy consumption rate is influenced by external factors such as weather or demand and not only by its sustainability activities. Hence, certain aspects within sustainability reporting and measurement of sustainable performance require a differentiated view. Yet, these important issues are only scarcely discussed topics in the IAS literature, but have been known and debated in the accounting literature for decades under the concepts of "leading and lagging indicators".

Epstein and Roy (2001) describe the common understanding of leading indicators as input or process indicators that connect more closely to operations. Hence, in order to link corporate activities with corporate strategic objectives, sustainability performance measures "must include leading indicators that give insight into the organization's ability to improve its competitive position in the future and are predictors of future performance" (Epstein and Roy 2001, p. 600). This idea is in line with Peloza (2009, p. 1522) who identifies leading and lagging indicators as mediating metrics which are "those that capture the 'mediating variable' that generates business value" as opposed to intermediate or end state outcome metrics. Both leading and lagging indicators can be of a quantitative nature, but the important point is that considering a leading indicator as an *organizational ability*, means that it is most often a composite of a number of complementary corporate practices: best practices for sustainability performance.

Paradoxically, in 2010 the UNGC introduced the *Differentiation Programme* with an Advanced Level for reporting, based on corporate adherence to 100+ best practices, which allows for benchmarking and comparability on e.g. supply chain implementation of the UNGC principles. However, the adoption rate of this Advanced Level is relatively low, as it comprises a number of conceptual weaknesses. These weaknesses make the Advanced Level potentially subject to criticism, although its existence has been relatively undetected in the IAS literature which perceives the UNGC still as a solely principle-based IAS. Nevertheless, it is worth noticing that the best practices described in this framework do not differ much across the UNGC issues: Labor, Human Rights, Environment, and Anti-Corruption. This feature makes it interesting per se to have a closer look, particularly against the background of a need for a cross-industry standard as identified in the interviews. In other words, while some lagging indicators for sustainability performance will be more material in some industries than others, best practices as leading indicators will differ less across industries and be less subject to individual firms' or industries' materiality concerns. Taking buyer-supplier collaboration as a best practice in order to increase sustainability performance along the supply chain, the following example illustrates the advantage of capability approaches: The general characteristics of a focal company's collaboration with its suppliers on a certain environmental issue like for instance the reduction of CO_2-emissions will not differ much from the general characteristics of another company's collaboration with a focus on labor issues such as working hours. However, the mere numbers of CO_2-emissions and the amount of working hours of employees are in this case lagging indicators measured in very different ways. Traditional reporting practices just publish these numbers and leave stakeholders alone in interpreting it. This holds especially in situations where companies from different industries report on the same lagging indicators. It is self-evident that an IT provider will have lower numbers of CO_2-emissions than a coal power station. But where is the line between sustainable or unsustainable performance drawn then? This leads us to propose that a standard for supply chains should better be based on best practices as factors that describe organizations' capabilities for sustainability. Such a standard could overarch different industries and supplement industry standards, which use specific quantitative measures for lagging indicators. Thus, we postulate:

P_3: A dynamic standard for corporate sustainability reporting must allow for industry-specific indicators as well as cross-industry applicability.

P_4: A dynamic standard for corporate sustainability reporting must capture a firm's longitudinal learning and development of intra- and inter-organizational sustainability capabilities by integrating them as leading indicators.

In the following we focus on this last proposition as the capability approach, which is the most important aspect and the most promising compared to deficiencies of current IAS.

3.3 Best Practices for Improving Sustainability Performance

Building on a systematic account of scholarly literature in the area of sustainable supply chain management, we are able to identify and aggregate several practices driving sustainability performance along the supply chain. Following Beske et al. (2014) who find best practices to be a resemblance of a firm's capabilities, we consider the extracted practices to provide a first step towards an improved sustainability reporting framework which focuses on firms' intra- and inter-organizational capabilities in enabling sustainability performance (Table 2).

Table 2 Best practices and examples of sustainability measures

Best practices	Examples
Assessment and measurement of sustainability impact in the corporate supply chain	• Carbon management across supply chain (Gopalakrishnan et al. 2012) • Remediation projects (Gavronski et al. 2012) • Internal performance evaluation system (Zhu et al. 2013)
Stakeholder consultations on sustainability issues	• Communicating proactively with stakeholders (Carter and Easton 2011) • Stakeholder communication (Beske et al. 2014) • Stakeholder management (Pagell and Wu 2009)
Analysis of sustainability risks and opportunities in the supply chain	• Risk management in sustainable supply chain management (Foerstl et al. 2010; Hofmann et al. 2014; Klassen and Vereecke 2012; Leppelt et al. 2013) • Monitoring (Koplin et al. 2007) • Pressure group management (Seuring and Müller 2008)
Internal coordination and communication concerning sustainability issues	• Department ensuring social, economic and environmental considerations (Gopalakrishnan et al. 2012) • Being "part of the mission" (Pagell and Wu 2009) • Senior-/top-management involvement (Seuring and Müller 2008) • Sustainability rooted in organizational culture (Gopalakrishnan et al. 2012; Carter and Easton 2011; Beske and Seuring 2014) • Management support (Zhu et al. 2008, 2013) • Generate environmental reports for internal evaluation (Zhu et al. 2008, 2013) • Cross-functional cooperation for environmental improvements (Zhu et al. 2008, 2013) • Special training for work environmental issues (Zhu et al. 2008, 2013)

(continued)

Table 2 (continued)

Best practices	Examples
External communication and capacity building concerning sustainability issues to suppliers and other business partners	• Enhanced communication (Beske and Seuring 2014) • Supplier development (Seuring and Müller 2008) • Transparency (Seuring and Müller 2008) • Long-term and close relationships (Seuring and Müller 2008; Mollenkopf et al. 2010)
Development of policies and targets for sustainability in the supply chain	• Key performance indicators (KPIs) of sustainability initiatives (Gopalakrishnan et al. 2012) • Policy statement (Large and Gimenez Thomsen 2011) • (Supplier) code of conduct (Jiang 2009; Schleper and Busse 2013)
Integration of sustainability issues in processes for the selection and evaluation of suppliers and other business partners	• Supplier management and integration of supply chain (Gopalakrishnan et al. 2012) • Sustainability in supply chain partner selection (Gold et al. 2010; Beske and Seuring 2014; Pagell and Wu 2009) • Green purchasing (Zhu et al. 2008, 2013) • Monitoring (Klassen and Vereecke 2012)
Engagement in sustainability-oriented collaboration with suppliers and other business partners	• Collaboration to enhance sustainability performance (Vachon and Klassen 2008; Sarkis et al. 2011) • Joint development (Beske and Seuring 2014) • Common IT interfaces and database structure (Srivastava 2007) • Information sharing with supply chain partners and external stakeholders (Srivastava 2007; Seuring and Müller 2008; Carter and Easton 2011) • Cooperation (Zhu et al. 2008, 2013)
Innovation of sustainable products, services and processes which are technologically new or significantly technologically improved	• Sustainability-related innovation (Beske and Seuring 2014) • Adaptation of products and processes (Gavronski et al. 2012) • Eco-Design (Zhu et al. 2007, 2008, 2013) • Innovation (Klassen and Vereecke 2012)
Compliance through the adoption of, and adherence to, sustainability standards and certifications	• Certification (Gold et al. 2010) • Standards and certifications (Mueller et al. 2009; Beske et al. 2014) • ISO 14000 certification (Zhu et al. 2008, 2013) • Eco-labeling of products (Zhu et al. 2008, 2013)

In order to improve their sustainability performance, organizations must be aware of their sustainability impact along the supply chain (Beske and Seuring 2014). This can for example be done by measures such as integrated carbon management (e.g. Gopalakrishnan et al. 2012) or a general internal performance measurement system (e.g. Zhu et al. 2013). Furthermore, stakeholder management and regular consultations with external actors have been found to drive sustainability performance (e.g. Pagell and Wu 2009; Wu et al. 2014). Related to this, Beske et al. (2014) and Carter and Rogers (2008) emphasize that a focus should lie on proactive communication about sustainability related issues. These communication practices, internal as well as external, are considered to be a key factor driving sustainability performance (Beske et al. 2014).

Several authors highlight the ambivalence of both risks and opportunities associated with social and environmental issues along the supply chain (e.g. Foerstl et al. 2010; Hofmann et al. 2014); Leppelt et al. 2013. Hence, companies should carefully monitor and manage these risks and look for opportunities that emerge for example together with new environmental developments (e.g. Klassen and Vereecke 2012).

As organizational sustainability initially requires resources and sometimes investments, and needs to be spread throughout the entire organization, successful practices should rely on top management involvement (e.g. Seuring and Müller 2008; Zhu et al. 2008; Pagell and Wu 2009), an adapted organizational culture (e.g. Carter and Rogers 2008; Gopalakrishnan et al. 2012) and extensive internal environmental reporting (e.g. Zhu et al. 2013). Internally, trainings enhance worker sustainability competences and raise awareness (e.g. Zhu et al. 2007, 2013). Externally, open communication, transparency and supplier development with sustained long term relationships are considered to be essential (e.g. Seuring and Müller 2008; Beske and Seuring 2014).

Ensuring sustainability not only intraorganizationally but along the supply chain rests on organizations' capabilities in buying goods and services that are already sustainable as in practices like green purchasing (e.g. Zhu et al. 2007, 2013). As purchasing departments account for the entire input part of a company, improving the overall sustainability performance starts with carefully selecting and consistently monitoring and evaluating suppliers (e.g. Klassen and Vereecke 2012). For these purposes sustainability criteria need to be clearly defined and included in supplier selection processes (Gold et al. 2010; Gopalakrishnan et al. 2012; Beske et al. 2014). Additionally, in order to manage these processes effectively, companies need to set clear goals and targets (e.g. Gopalakrishnan et al. 2012; Large and Gimenez Thomsen 2011) and communicate those across the supply chain, for example via policies and a (supplier) code of conduct (e.g. Jiang 2009; Schleper and Busse 2013). Moreover, besides these governance and control-based processes, close collaboration and information sharing with customers and suppliers can foster environmental and social initiatives as well as innovations (e.g. Seuring and Müller 2008; Zhu et al. 2008; Klassen and Vereecke 2012; Beske and Seuring 2014).

In general, scholars stress the importance and opportunities of innovation (Klassen and Vereecke 2012). Establishing a sustainable supply chain can lead to new product developments and other innovations (Beske and Seuring 2014) and increased adaption of products and processes (Gavronski et al. 2012), for instance in the context of eco-design paradigms (e.g. Zhu et al. 2008; Hoejmose et al. 2012).

In conclusion, there are many practices that foster the establishment of a sustainable supply chain. The degree to which an organization internalizes and implements these practices hints towards their true sustainability capabilities and hence their potential in performing sustainably. Thus, in addition to P_4 we add:

> P_{4*}: A dynamic standard for corporate sustainability reporting must capture a firm's intra- and inter-organizational sustainability capabilities by incorporating a saturated set of best practices which have been empirically found to have a positive impact on sustainability performance.

4 A Path for Future Improvement: Towards a New Framework

So far we presented five propositions as requirements for corporate sustainability reporting. In this section, we briefly describe how these propositions might be included in a new framework. The identification of a set of intra- and inter-organizational best practices that have empirically proven to be relevant in the enhancement of sustainability performance is an important first step, yet needs to be linked to theory.

An adequate candidate for one such theory could be the notion of absorptive capacity (ACAP) which concerns a firm's ability to acquire, assimilate, transform, and apply external knowledge (Cohen and Levinthal 1990; Zahra and George 2002).

In times of dynamic markets and highly uncertain environments, knowledge is the dominant source for competitive advantage (Jansen et al. 2005) as learning mechanisms enable organizations to "create, extend or modify its resource base" (Helfat et al. 2007, p. 4) thereby equipping them with new routines to prepare for future uncertainties. Hence, a high ACAP can strongly influence the overall performance of firms and their long term survival as it describes a firm's potential not only to analyze changes and turbulences in markets in which they operate but also to simultaneously process, internalize and use the newly acquired knowledge (Cohen and Levinthal 1990). Yet, particularly in the context of sustainability, there is still a lack of ACAP considerations within scholarly research, although it is a very complex topic and prone to a high uncertainty environment.

Applying the ACAP concept as a theoretical base offers a novel focus for exploring corporations' implementation of IAS. In particular, this framework provides several aspects, which are requested in the above stated propositions. In the following, two of these aspects are characterized:

1. Through its strong orientation towards the absorption of external sources of knowledge, ACAP emphasizes supply chain integration as a factor for potentially enabling sustainable competitive advantage. Raising this awareness within companies might also foster a closer collaboration between supply chain partners, thereby achieving certain specific goals like for instance boundary-spanning CO_2-emission reductions (Ramanathan et al. 2014).
2. Focusing on ACAP allows reporting firms to invest in building internal capabilities in the area of sustainability. Through these capabilities highly dynamic and uncertain environments, present when dealing with sustainability issues, become more manageable for companies. For instance, Lichtenthaler (2009) points out that learning processes linked to ACAP guide organizations' innovative potentials; particularly if they align their internal combinative capabilities—systematization, coordination, and socialization of knowledge—with the absorbed new information.

To further utilize the ACAP concept we suggest structuring the best practices vertically along the four ACAP dimensions to include a progressive organizational capabilities and learning perspective. Yet, as extracted from the key informant interviews, the actual implementation of best practices among companies differs. Hence, an improved reporting guideline needs to allow for differentiation in this regard as this also enables measurability of sustainable performance and comparability across industries. Implementation levels that account for different degrees of effort in best practices allow for weighing the actual performance of organizations.

5 Conclusion

The emergence of various IAS following different approaches reflects the importance of sustainability reporting. However, current IAS suffer from several shortcomings which prevent communicating a firm's real and holistic value and which often leave reporting disconnected from an organization's actual operations. In this research note, we uncovered these deficiencies and provided five propositions which pave the road for an innovative and dynamic framework for corporate sustainability reporting.

A new standard should incorporate a firm's entire supply chain and ensure reporting outside the firm-boundaries, thereby meeting environmental and social requirements. We follow an ongoing debate about lagging and leading sustainability performance indicators by proposing an approach that focuses on an organization's intra- and inter-organizational capabilities. These capabilities can be best built and expressed by best practices that improve an organization's sustainability performance along the entire supply chain. Since current standards further lack an objective scale to value the degree of capability implementation achieved to make reports more accurate and enable comparability, we outline the need to develop mechanisms that allow the new reporting standard to weigh and differentiate organizations' efforts across industries. One theoretical basis for these features could be provided by utilizing the ACAP concept to structure the best practices we extracted from scholarly literature. However, much remains to be done to reduce complexity in the field and to ensure effective and efficient reporting for all stakeholders along the supply chain.

References

Beske P, Land A, Seuring S (2014) Sustainable supply chain management practices and dynamic capabilities in the food industry: a critical analysis of the literature. Int J Prod Econ 152:131–143

Beske P, Seuring S (2014) Putting sustainability into supply chain management. Supply Chain Manage Int J 19(3):322–331

Bigge DM (2004) Bring on the bluewash—a social constructivist argument against using Nike v. Kasky to attack the UN Global Compact. Int Legal Perspect 14(6):6–21

Carter CR, Easton PL (2011) Sustainable supply chain management: evolution and future directions. Int J Phys Distrib Logistics Manage 41(1):46–62

Carter CR, Rogers DS (2008) A framework of sustainable supply chain management: moving toward new theory. Int J Phys Distrib Logistics Manage 38(5):360–387

Cohen WM, Levinthal DA (1990) Absorptive capacity: a new perspective on learning and innovation. Adm Sci Q 35(1):128–152

Deva S (2006) Global compact: a critique of the UN's "public-private" partnership for promoting corporate citizenship. Syracuse J Int Law Commun 34(1):107–151

Dingwerth K, Eichinger M (2010) Tamed transparency: how information disclosure under the global reporting initiative fails to empower. Global Environ Polit 10(3):74–96

Epstein MJ, Roy M-J (2001) Sustainability in action: identifying and measuring the key performance drivers. Long Range Plan 34(5):585–604

Etzion D, Ferraro F (2010) The role of analogy in the institutionalization of sustainability reporting. Organ Sci 21(5):1092–1107

Foerstl K, Reuter C, Hartmann E, Blome C (2010) Managing supplier sustainability risks in a dynamically changing environment—sustainable supplier management in the chemical industry. J Purchasing Supply Manage 16(2):118–130

Gavronski I, Klassen RD, Vachon S, Machado do Nascimento LF (2012) A learning and knowledge approach to sustainable operations. Int J Prod Econ 140(1):183–192

Gold S, Seuring S, Beske P (2010) Sustainable supply chain management and inter-organizational resources: a literature review. Corp Soc Responsib Environ Manag 17(4):230–245

Gopalakrishnan K, Yusuf YY, Musa A, Abubakar T, Ambursa HM (2012) Sustainable supply chain management: a case study of British Aerospace (BAe) Systems. Int J Prod Econ 140(1):193–203

Hartmann J, Moeller S (2014) Chain liability in multitier supply chains? Responsibility attributions for unsustainable supplier behavior. J Oper Manage 32(5):281–294

Helfat C, Finkelstein S, Mitchell W, Peteraf M, Singh H, Teece D, Winter S (2007) Dynamic capabilities: understanding strategic change in organizations. Blackwell, Malden

Hofmann H, Busse C, Bode C, Henke M (2014) Sustainability-related supply chain risks: conceptualization and management. Bus Strategy Environ 23(3):160–172

Hoejmose S, Brammer S, Millington A (2012) "Green" supply chain management: the role of trust and top management in B2B and B2C markets. Ind Mark Manage 41(4):609–620

Jansen JJP, Van Den Bosch, FAJ, Volberda HW (2005) Managing potential and realized absorptive capacity: how do organizational antecedents matter? Acad Manage J 48(6):999–1015

Jiang B (2009) The effects of interorganizational governance on supplier's compliance with SCC: an empirical examination of compliant and non-compliant suppliers. J Oper Manage 27(4):267–280
Kanzer AM (2010) Toward a model for sustainable capital allocation. In: Eccles RG, Cheng B, Salzman D (eds) The landscape of integrated reporting. Reflections and next steps. Harvard Business School, Cambridge, MA
Kaptein M, Wempe J (1998) Twelve Gordian knots when developing an organizational code of ethics. J Bus Ethics 17(8):853–869
Klassen RD, Vereecke A (2012) Social issues in supply chains: capabilities link responsibility, risk (opportunity), and performance. Int J Prod Econ 140(1):103–115
Koplin J, Seuring S, Mesterharm M (2007) Incorporating sustainability into supply management in the automotive industry—the case of the Volkswagen AG. J Clean Prod 15(11):1053–1062
KPMG (2013) The KPMG survey of corporate responsibility reporting 2013. KPMG Special Services B.V, Netherlands
Large RO, Gimenez Thomsen C (2011) Drivers of green supply management performance: evidence from Germany. J Purchasing Supply Manage 17(3):176–184
Leppelt T, Foerstl K, Reuter C, Hartmann E (2013) Sustainability management beyond organizational boundaries—sustainable supplier relationship management in the chemical industry. J Clean Prod 56:94–102
Levy DL, Brown HS, Jong M (2010) The contested politics of corporate governance: the case of the Global reporting initiative. Business & Society 49(1):88–115
Lichtenthaler U (2009) Absorptive capacity, environmental turbulence, and the complementarity of organizational learning processes. Acad Manage J 52(4):822–846
Meehan J, Bryde D (2011) Sustainable procurement practice. Bus Strategy Environ 20(2):94–106
Messier WF Jr, Martinov-Bennie N, Eilifsen A (2005) A review and integration of empirical research on materiality: two decades later. Auditing J Practice Theor 24(2):153–187
Mollenkopf D, Stolze H, Tate L, Ueltschy M (2010) Green, lean, and global supply chains. Int J Phys Distrib Logistics Manage 40(1/2):14–41
Mueller MV, dos Santos VG, Seuring S (2009) The contribution of environmental and social standards towards ensuring legitimacy in supply chain governance. J Bus Ethics 89(4):509–523
Nolan J (2005) The United Nations' compact with business: hindering or helping the protection of human rights? Univ Queensland Law J 24(2):445–466
Pagell M, Wu Z (2009) Building a more complete theory of sustainable supply chain management using case studies of 10 exemplars. J Supply Chain Manage 45(2):37–56
Peloza J (2009) The challenge of measuring financial impacts from investments in corporate social performance. J Manage 35(6):1518–1541
Ramanathan U, Bentley Y, Pang G (2014) The role of collaboration in the UK green supply chains: an exploratory study of the perspectives of suppliers, logistics and retailers. J Cleaner Prod 70:231–241
Sarkis J, Zhu Q, Lai K (2011) An organizational theoretic review of green supply chain management literature. Int J Prod Econ 130(1):1–15
Schleper MC, Busse C (2013) Toward a standardized supplier code of ethics: development of a design concept based on diffusion of innovation theory. Logistics Res 6(4):187–216
Seuring S, Müller M (2008) From a literature review to a conceptual framework for sustainable supply chain management. J Cleaner Prod 16(15):1699–1710
Srivastava SK (2007) Green supply-chain management: a state-of-the-art literature review. Int J Manag Rev 9(1):53–80
Vachon S, Klassen RD (2008) Environmental management and manufacturing performance: the role of collaboration in the supply chain. Int J Prod Econ 111(2):299–315
WBCSD (2013) Reporting matters. Improving the effectiveness of reporting. WBCSD Baseline Report 2013. Retrieved from: http://en.vbcsd.vn/upload/attach/WBCSD_Reporting_matters_2013_INT_PDF.pdf. Accessed 1 March 2015

Wu I-L, Chuang C-H, Hsu C-H (2014) Information sharing and collaborative behaviors in enabling supply chain performance: a social exchange perspective. Int J Prod Econ 148:122–132

Zahra SA, George G (2002) Absorptive capacity: a review, reconceptualization, and extension. Acad Manage Rev 17(2):185–203

Zhu Q, Sarkis J, Cordeiro JJ, Lai K-H (2008) Firm-level correlates of emergent green supply chain management practices in the Chinese context. Omega 36(4):577–591

Zhu Q, Sarkis J, Lai K-H (2007) Green supply chain management: pressures, practices and performance within the Chinese automobile industry. J Cleaner Prod 15(11–12):1041–1052

Zhu Q, Sarkis J, Lai K-H (2013) Institutional-based antecedents and performance outcomes of internal and external green supply chain management practices. J Purchasing Supply Manage 19(2):106–117

Sustainable Fuels for the Transport and Maritime Sector: A Blueprint of the LNG Distribution Network

Simon K. Thunnissen, Luke G. van de Bunt and Iris F.A. Vis

Abstract Liquefied Natural Gas (LNG) is one of the alternative sustainable fuels available to the transportation sector. A new distribution network needs to be in place to enable better accessibility and more efficient usage of this fuel type for all modes of transport. Currently, a chicken-and-egg problem can be noted where users wait for the infrastructure to be in place and suppliers do not initiate new investments due to uncertainty in future usage. In this paper we aim to fill this void by visualizing the supply chain of LNG and bio-LNG and derive essential issues to be addressed in developing LNG networks. These issues form the blueprint of the LNG distribution network, supported by the empirical input of stakeholders through interviews and expert sessions. We classify the challenges found in terms of supply chain network design literature and indicate future research issues.

Keywords LNG · Sustainable logistics · Distribution network · Blueprint

1 Introduction

Industry and governments have gradually increased their resources for the development of LNG (liquefied natural gas) or LBG (liquefied bio gas produced from biomass) as an alternative fuel for freight transport. Although a fossil fuel, LNG has significant benefits for emissions of NO_x, SO_x, and particulate matter. LNG trucks and barges currently in operation also realized a reduction of 15 % in CO_2 emissions compared to diesel-driven ones. Apart from these societal benefits, natural gas reserves are large and operational cost reductions are possible with LNG (Fuelswitch 2015). Additionally, LBG can be used for the trucking industry. However, a new transport fuel is not introduced overnight as we know from electric

S.K. Thunnissen · L.G. van de Bunt · I.F.A. Vis (✉)
Faculty of Economics and Business, Department of Operations, University of Groningen, Groningen, The Netherlands
e-mail: i.f.a.vis@rug.nl

H. Zijm et al. (eds.), *Logistics and Supply Chain Innovation*,
Lecture Notes in Logistics, DOI 10.1007/978-3-319-22288-2_6

cars and compressed natural gas (CNG). To make LNG a realistic option for more logistics service providers, more gas stations for trucks and bunkering locations for shipping are needed. With current low, but growing demand, the question for suppliers of LNG is which locations offer the best possibility to have a profitable business. In the Dinalog R&D project "Design of LNG Networks" (Dinalog 2015) partners from industry and academic researchers aim to identify the requirements of a distribution network for (bio-) LNG fuel and derive tools to enable location decisions for gas and bunker stations with efficient distribution for replenishment.

A key issue in fulfilling growing demand for new transport fuels is deriving a robust distribution network. So far in literature, studies have been published into the design of distribution networks for new transport fuels such as hydrogen and biofuels. Several authors present an overview of problems to be addressed in designing a hydrogen infrastructure (e.g., Agnolucci and McDowell 2013; Dagdougui 2012). The solution approaches available in this stream of literature can be classified as mathematical modeling, geographic information system analyses, and assessment plans for supply chain planning. Note that these studies have looked primarily at passenger transport and locations for gas stations, and include production. An LNG distribution network on the other end, should serve all modes of freight transport, including seagoing vessels. An et al. (2011) present an overview of topics and methods in the design of networks for biofuels. The authors conclude that biofuel studies mainly focus on upstream processes. In LNG distribution networks, upstream production is only relevant for bio-LNG, not for fossil LNG. Additionally, papers that focus on down- and midstream processes include only one mode of transportation, mostly cars (Awadu and Zhang 2012; An et al. 2011).

While acknowledging many similarities in the problems and processes surrounding biofuel, hydrogen, and LNG distribution networks, the first two streams of literature do not address the complexities that LNG distribution networks experience. LNG can be used as transport fuel for road transport (e.g., trucks, buses), barges, seagoing vessels, and even trains. Designing (shared) networks for all modes is an additional complexity. Literature on biofuel and hydrogen infrastructure, however, mainly focuses on cars. Secondly, the distribution of LNG lacks an option that biofuel and hydrogen have, namely distribution via pipelines. Lastly, the main problems for the introduction of LNG as transport fuel do not include upstream production issues, as we will show in this paper. The goal of this paper is to define a blueprint of the LNG distribution network, including relevant research questions, constraints, and data. The blueprint pictures all current issues in the development of LNG as a transport fuel. We expect that this blueprint can serve as input for network simulation studies and can support business cases for investments in LNG infrastructure as well.

To derive a blueprint of the LNG distribution network, we perform an explorative study. In Sect. 2, we describe our methodology in more detail. Section 3 presents our blueprint of the LNG distribution network based on a literature study, explorative interviews, and expert sessions with industry. The main outcome of the interviews is a list of bottlenecks experienced for the development of an LNG distribution network. Short live surveys during expert sessions among logistics service providers verify the outcomes of the interviews. The outcomes are then visualized and translated into logistics decision problems, and organizational and governance issues in the network. In Sect. 4, we define research challenges for designing an LNG distribution network by comparing the results of Sect. 3 with data from literature. Section 5 presents conclusions.

2 Methodology

In working towards our goal to derive a blueprint for a LNG distribution network we combine literature studies, interviews, and expert sessions. The following sequential steps form our research methodology:

1. literature review of grey literature;
2. market consultation through semi-structured interviews;
3. ranking of factors through live surveys and expert sessions;
4. definition of blueprint of LNG distribution network;
5. comparison with literature to derive new research challenges.

In steps 1 and 2, we aim to identify requirements for designing a distribution network for LNG. We started with a review of grey literature on LNG (i.e., non- or semi-scientific studies, performed by practice or governmental institutes) and LNG in transportation. We performed a market consultation by conducting semi-structured interviews with all partners in the Dinalog R&D project "Design of LNG Networks". Those partners together represent the major stakeholders involved in the supply chain, varying from LNG suppliers to logistics service providers. Through these interviews, we identified which factors reflect potential bottlenecks for the development of an LNG distribution network as experienced by the market. In step 3, we asked all partners to rank the factors in order of importance. This was done scale-wise. The most important factor was ranked at the lowest number (i.e. 1) and the least important factor was ranked at the highest number present on the scale. The highest number present on the scale was dependent on the amount of factors to be ranked at the particular node. Additionally, we held short live surveys during expert sessions among logistics service providers, LNG suppliers, and a general public interested in logistics to verify the ranking outcomes. The outcomes of those first three steps are combined to design a visualization of the LNG distribution network. Based on these data, we define in step 4 a blueprint of logistics decision problems, organizational and governance issues in a LNG distribution network.

In steps 1–4, we deliberately take this 'consult market first' approach, because there is little academic literature available on LNG distribution networks and supply chains. Most publications related to this topic concerns industry reports. By consulting the market and grey literature first, we are able to narrow down industry relevant problems that provide potential bottlenecks in the development of an LNG distribution network. With these 'empirical problems' summarized in the blueprint, we conclude our study in step 5 with a short literature survey of supply chain network design (SCND) and location decision literature to identify potentially available solution methods and indicate research challenges for LNG distribution network problems (LNGDNP).

3 Blueprint

3.1 Identification of Factors

In this section, we describe the main outcomes of our grey literature study on the LNG market as well as the outcomes of the interviews of consortium partners and expert sessions. We consider the following supply chain stakeholders: short sea, inland shipping, trucks, gas stations, bunker facilities, and bio-LNG producers. Consequently, we study whether the interests differ per stakeholder.

First we present the critical factors identified from grey literature on a stakeholder level (for a full overview and our sources refer to Appendix A). Several reports indicate that (potential) bunker facilities and gas stations encounter a number of problems in planning for LNG infrastructures. First of all, *bunkering guidelines, regulations, and equipment* are not yet fully in place in European ports. There is a strong need for harmonization and *standardization* such that network alignment can be achieved in further developing (bio-) LNG networks. Secondly, several reports indicate that the public still has a *negative perception* on LNG, which might influence bunker location decision-making. *Boil-off effects* at bunker facilities, gas stations, and trucks need to be taken into account, i.e. guidelines should be formulated how to deal with boil-off in a sustainable and cost efficient way. Lastly, a stable and high *quality* LNG is required for efficiency in engines as well as for more substantial reductions in emissions.

Availability is a key concern for the demand side of the supply chain. Ship and truck owners show their interest in LNG, but question the availability of supply along their routes. In Table 1, we summarize all critical factors identified per node of the supply chain as identified in the literature.

Table 1 Critical factors identified in the literature

Critical factor	Supply chain node(s)
Regulations/safety guidelines	Bio LNG, bunker facilities
Standardization	Bunker facilities, gas stations
Boil-off	Bunker facilities, trucks
Quality	Bunker facilities, gas stations
Negative public perception	Bunker facilities
(Re-)Training	Short sea
Availability	Short sea, inland shipping, trucks
Loss of space/time	Short sea, trucks
Reliability	Trucks

Table 2 Critical factors identified in the interviews

Critical factor	Supply chain node(s)
Interchangeability with LNG	Bio-LNG
Footprint	Bio-LNG
Regulations/safety guidelines	Bio LNG, bunker facilities, short sea, gas stations, inland shipping, trucks
Capacity	Bunker facilities, gas stations
Standardization	Bunker facilities, gas stations, Bio-LNG, inland shipping, short sea
Quality	Bunker facilities, gas stations, trucks, short sea, inland shipping
Negative public perception	Bunker facilities
Training	Short sea, bunker facilities, inland shipping, trucks
Availability	Short sea, inland shipping, trucks, bio-LNG, gas stations
Reliability	Trucks
Boil-off	Trucks, bunker facilities, inland shipping, gas stations
Power of engine	Trucks

Secondly, semi-structured interviews were organized with all partners in the Dinalog project "Design of LNG Networks". Each interviewee was asked to accept or reject each factor as mentioned in Table 1 based on the relevance of the factor on their (potential) LNG activities. Additionally, each interviewee could add new factors to the blueprint. Rejected factors were not removed but used in step 3 in a ranking procedure to assess their importance. Additional to the interviews, the relevance of the factors in Table 1 was also tested during a few expert sessions, including fuel owners/traders, truck owners/users, barge owners/users, barge (parts) suppliers, and regulatory authorities. The participants were asked to prioritize the factors based on the relevance of their (potential) LNG activities and add new factors to the blueprint. We then ranked the factors relative to each other (see Appendix B).

In Table 2, we show the critical factors as identified by the interviews and the expert sessions. Refer to Appendix B for a more detailed description of each of the factors, and their rankings as a result of the interviews and the expert sessions. If we compare the results of Table 2 with Table 1, we conclude that multiple factors are important for a variety of stakeholders.

3.2 *Visualization of LNG Supply Chain Network*

Based on the outcomes of steps 1–3, Fig. 1 represents a visualization of the LNG distribution network and specific stakeholder factors. Bio-LNG and LNG distribution networks can be designed separately. Typically, we observe mainly local initiatives in bio-LNG networks so far. At this stage, the potential options and/or needs for an

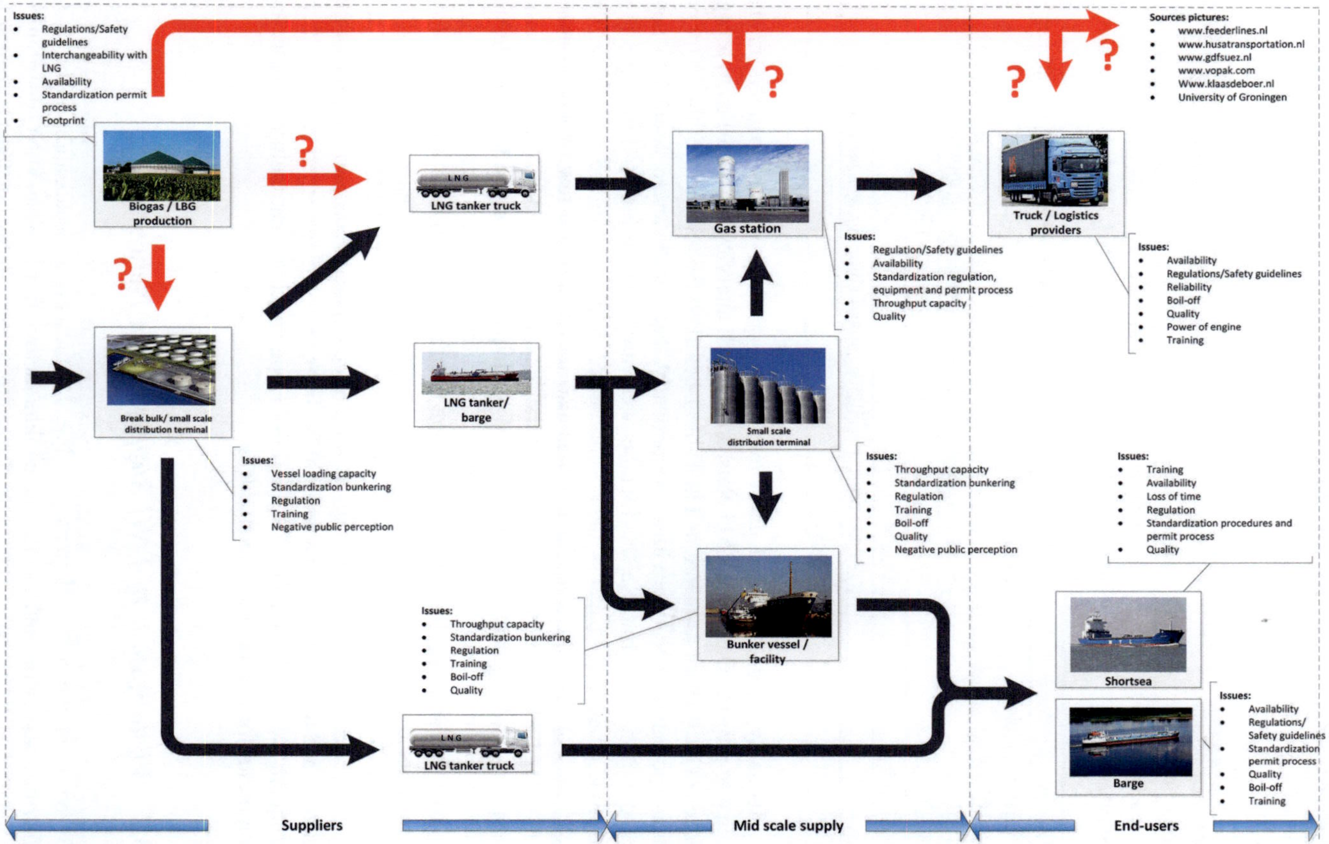

Fig. 1 Visualization of (bio-) LNG supply chain network and stakeholder factors

integration of the bio-LNG supply chain and the LNG supply chain remains unclear. Further research in benefits and production problems need to be performed to enable an answer to this question. This study is not included in this paper.

3.3 Blueprint Logistics Decision Problems

In this section, we move from the listing of issues for each separate node as presented in Fig. 1, to overarching decision problems that induce cost and time in the whole distribution network. We distinguish logistics decision-making problems, and organizational and governance issues. In the latter category, the following overarching issues have been found:

- Quality of LNG. What will be the specification of LNG as a transport fuel? What will be the role of bio-LNG in the specification?
- Standardization of regulations for bunkering and transport of LNG
- Standardization of equipment for LNG
- End-user factors (e.g., investment, range, engine power)
- Permit processes for gas and bunker stations
- Training
- Negative public perception

The outcomes of discussions in the field on the above issues need to be monitored closely and serve as important input in scenario definition for testing of the solution approaches that need to be designed for logistics decision-making. Nonetheless, the aim of the remainder of this paper is to focus on logistics decision-making issues that can increase the overall efficiency of the network.

The results of the previous sections show many similar issues between upstream and downstream stakeholders. Upstream stakeholders are concerned with the supply of LNG and downstream stakeholders represent the users of LNG. For example, among upstream stakeholders, the replenishment of LNG is an issue. Suppliers want to make LNG available at filling stations at the lowest possible cost. As demand grows, barges and ships will become more economical and sustainable modes for distribution (economies of scale). This way of distributing also requires a number of smaller terminals from which trucks are used for last mile logistics. Considering demand rates, questions raised by upstream stakeholders are: how and when to replenish filling stations with which mode of transport? Where to build a gas or bunker station? What kind of bunkering options to use given a certain demand?

At the downstream part of the network, the demand of LNG can be grouped in trucks, barges, and ships (short sea). Trains are a possible fourth group, but not included in this paper. End-users have indicated that the most important bottlenecks in the development of an LNG distribution network are availability, range, engine power (trucks), methane waste, and the specification of LNG as transport fuel. To these issues can be added, the high investment costs for an LNG truck or vessel. Together, logistics issues of LNG and the business case make a hard comparison to the alternative (diesel) in terms of total cost of ownership.

Summarizing, the logistics network design problems for LNG distribution networks that we identified and that constitute essential parts of our blueprint are:

- Demand forecasting
- Location and overall capacity of bunker facilities and gas stations
- Network utilization and bunkering options
- Choice of mode of transportation for replenishment individual facilities
- Storage capacity at individual LNG storage facilities dealing with boil off effects

4 Research Challenges

In this section, we show how to relate the logistics network design problems derived in Sect. 3 to literature on supply chain network design. Supply chain network design decisions typically face a planning horizon of a few years and primarily include facility location decisions concerning the locations, number and size of facilities and assignment of sales points to stock locations (e.g., Melo et al. 2009; Klose and Drexl 2005). Facility location models available in the literature often assume that decisions at the strategic level impact decisions at the operational level. However, a design of a new network for a new transport fuel is highly impacted by operational decisions of potential customers (e.g., truck routing). A crucial aspect of many practical location problems regards the existence of different types of facilities, of which each one plays a specific role (Melo et al. 2009; Sahin and Süral 2007). Each set of facilities of the same type and with the same role is usually denoted by a layer or an echelon, thus defining a level in the hierarchy of facilities. Facilities can serve an infinite amount of demand (i.e., uncapacitated problems) or a restricted amount of demand (i.e., capacitated problems). Melo et al. (2009) provides a classification of facility location models by means of number of layers, number of commodities, planning horizon and type of data (deterministic/stochastic). Most solution approaches available tackle single horizon networks with one or two layers with deterministic demand.

In LNG network design, decisions need to be made on the types of facilities used, their number, size, and location. A newly introduced network with increasing demand and high investments asks for different stages in storage and infrastructural development. In an early stage network, intensity might be scarce with a limited number of facilities. Mobile bunkering options (e.g., vessel to vessel, or truck to vessel) might be added to static facilities to serve a larger group of customers. If demand grows, more static facilities might be added to the network. Important in this context is that demand density and need for storage capacity will be uncertain and dynamic.

A selection needs to be made between different types of facilities including LNG terminals, gas stations, and bunker stations for different types of customers at different layers of the supply chain. Each facility in the network can have a different size and resulting storage volume. An additional complexity factor is that over time the volume of LNG decreases with 0.1–0.5 % of the total volume of LNG per day. The boil-off rate will increase as the level of LNG in the storage tank decreases.

Consequently, a multi-period setting needs to be considered with a continuous modeling approach. In literature, mostly discrete location models assume that parameter values are constant over time (Melo et al. 2009; Owen and Daskin 1998). However, in an immature market, parameters will show a very dynamic behavior. Scenario planning models can consider a generated set of possible future parameter values. Uncertain parameters, which are relevant for the design of the LNG-network, are demand, demand intensity, and boil-off rates. Demand for LNG will grow, but at which rate and at what locations is difficult to forecast. Secondly, the demand characteristics for trucks, barges, and seagoing vessels are different, which adds to the uncertainty.

The LNG facility location problem can be classified as a continuous, capacitated, multi-period, multi layer, single commodity problem with stochastic demand. Different types of modes of transportation can be used for replenishing those facilities. The modal choice for replenishment is dependent on demand at LNG terminals, affects cost of transportation, and impacts location choices due to accessibility of a location for a certain mode of transportation. The latter is also relevant for the demand side as the network will serve different modes of transportation. As a result, studying the integrative problem of mode selection and LNG facility location decisions is to be recommended. In the literature, hardly any papers deal with the integrative problem of mode selection and facility location decisions (Melo et al. 2009). We suggest to extend this line of research by also studying multi-objective models for this problem considering both time and costs. Each mode of transportation involves different replenishment times and costs. Costs are also associated with the type of terminal selected.

5 Conclusions

In this paper, we have derived a blueprint of the LNG distribution network and defined associated logistics decisions problems. Our blueprint illustrates all stakeholders and their current issues in the development of LNG as a sustainable transport fuel alternative. In defining the blueprint the following methodology was followed: reviews of grey literature, market consultations, live surveys and expert sessions and definition of research challenges. Consequently, we gathered input from major stakeholders in the LNG supply chain. We have visualized the LNG distribution network and indicated relevant critical factors for broad adoption at each supply chain node. We note that the issues mentioned vary per stakeholder. However, a large group of organizational and governmental issues is encountered by each of the stakeholders. Those issues are quality, regulations and safety guidelines, standardization, availability, training and boil-off effects. The role of bio LNG in the LNG distribution network, integrated or separate networks, mainly remains unclear at this point, due to the limited availability. We have grouped those factors into overarching decision problems to be tackled in the design of LNG distribution networks. The main logistics decision problems defined vary from forecasting demand to selecting the right locations for bunker facilities and gas stations, integrated with mode selection for replenishment operations. We show that these problems can be classified in the facility location literature and indicate future research challenges to solve these problems. The blueprint definition can be used as input for network simulation studies, the design of new planning and control tools for the decision problems mentioned as well as to support further business case development for each of the stakeholders.

Appendix A: Identification of Critical Factors in the LNG Network from Grey Literature

Supply chain	Critical factors	Grey literature	Description
Bio-LNG	Regulations/safety guidelines	Groengasnl (2013)	Ambiguity about production and, tank and buffer locations cause delay in the development of bio-LNG initiatives
Bunker facility	(Standardization) Bunkering guidelines and equipment	DNV (2013), DMA (2012), Argos (2014), BPO (2014), European Commission (2013), MIIP (2013), GLE (2013)	Bunkering guidelines and equipment are not yet fully in place in European ports. Harmonization of these guidelines and equipment is, according to most of these reports, just as important
	Regulations/safety guidelines	Creatieve Energie (2009), DMA (2012), Argos (2014), European Commission (2013), SGMF (2014), GLE (2013)	"Much of the techniques and procedures proposed for application in the LNG bunkering infrastructure development refer to downscaling of established techniques and procedures applied in the large-scale LNG import industry. In order to be able to assess the risks associated with small-scale LNG handling and bunkering operations and to outline adequate regulations, it is important to be size specific." (DMA 2012, p. 195)
	Boil-off	DNV (2013), DMA (2012)	Boil-off management represents the guidelines and regulations to minimize vapor gas

(continued)

(continued)

Supply chain	Critical factors	Grey literature	Description
			release during bunkering, and therefore minimize the cost and environmental damage
	Quality (methane number)	SGMF (2014), TNO et al. (2013)	The quality of LNG is important regarding two aspects: the quality should be standardized and the quality should be high • Standardization: Engine manufactures are able to develop more efficient engines • High quality: Low quality may delay the development of higher efficiency engines and lead to additional greenhouse gas emission
	Negative public perception	European Commission (2013), DMA (2012)	LNG is a dangerous product in the public perception. Few residents are therefore happy to find a LNG storage location in their neighborhoods
Short sea	Training	DNV (2013), DMA (2012), SGMF (2014), IBIA (2014), European Commission (2013)	"A potential problem for LNG to take a larger share of the market for maritime fuels is the need for well-trained crew. Today, the crew of an LNG bunkering/feeder vessel (ocean-going) would need the same competence as the crew of a large IGC tanker, a competence which few people have and which takes a long time to achieve." (DMA 2012, p. 205)
	Availability	Creatieve Energie (2009), Ministry of I&E (2014), SGC (2011), European Commission (2013)	"Many ship owners and ship operators have stated their interest in switching to LNG fuel, but withhold their investment and conversion plans due to missing LNG supply at

(continued)

(continued)

Supply chain	Critical factors	Grey literature	Description
			their preferred ports of call." (European Commission 2013, p. 2)
	Loss of time	European Commission (2013), GLE (2013)	Ships sailing on LNG will spend more non-productive bunker time in ports compared to ships sailing on oil. Mainly because it is in Europe, with the exception of the Port of Stockholm, not allowed to bunker LNG during cargo loading or while passengers are on board. This causes problems for ports, which maintain tight port schedules and shippers, who lose valuable time compared to oil-bunkering
Inland shipping	Availability	Creatieve Energie (2009), Ministery of I&E (2014)	Refer to 'Availability' at 'Short sea'
Gas stations	Quality	TNO et al. (2013), Simon Loos (2012), GLE (2013)	Refer to 'Quality' at 'Bunker facility'
	Standardization regulations	Vos Logistics (2012), GLE (2013)	Lack of harmonized regulations for gas stations constraints the development of LNG as a fuel for road transport
	Standardization of equipment	GLE (2013)	Lack of harmonized equipment (e.g. refueling nozzles) results in a sub optimization of the LNG network. Trucks are not able to refill at every gas station
Trucks	Availability	DAF (2011)	Refer to 'Availability' at 'Short sea'
	Reliability	DAF (2011)	The maturity of and experience with LNG-technology in heavy transport is low, therefore DAF considers LNG not yet reliable enough
	Boil-off	Simon Loos (2012)	Boil-off due to evaporation of LNG in the tank should be minimized

Appendix B: Identification of Critical Factors in the LNG Network from Interviewing Consortium Partners

Supply chain	Critical factors	Description	Interviews	Expert session 1: inland shipping	Expert session 2: fuel owners/traders	Expert session 3: short-sea, inland shipping and trucks
Bio-LNG	Interchangeability with LNG	To accelerate the development of the bio-LNG market, bio-LNG should be fully interchangeable with LNG. Because, at the moment, the supply of bio-LNG is too low to satisfy all demand. The difference between bio-LNG and traditional LNG is the methane number (respectively 99 and 89 %)	3	–[a]	–	–
	Availability	At the moment, bio-LNG is produced in low volumes and therefore most of the times a local initiative. Thus, bio-LNG is only available locally. Subsequently, local availability is dependent on waste flows in the surrounding area to produce bio-LNG	4	–	–	–
	Standardization permit process	Because bio-LNG production sites are new, governmental institutions do not use standards methods to grant a permit. Standardization will decrease the waiting time of a permit	1	–	–	–
	Footprint LBM	The footprint of LBM is a very important advantage compared to traditional LNG. Requirements should be formulated to remain this advantage	2	–	–	–

(continued)

(continued)

Supply chain	Critical factors	Description	Interviews	Expert session 1: inland shipping	Expert session 2: fuel owners/traders	Expert session 3: short-sea, inland shipping and trucks
Bunker facility	Regulation/safety guidelines	Regulations should be defined for any type of bunkering (truck-to-ship, ship-to-ship and pipeline-to-ship), otherwise investments in bunker facilities will lag behind	2	–	1	–
	Capacity	Since demand is hard to forecast, also the capacity of a bunker facility is difficult to determine. In particular because LNG warms up, pressure increases in the tank and the owner of the bunker facility has to cool the gas down or to boil-off	3	–	–	–
	Standardization bunker guidelines and bunker equipment	Refer to '(Standardization) Bunkering guidelines and equipment' at 'Bunker facility' in Appendix A	1	–	–	–
	Training	Personnel should be qualified to work with LNG	N/A[b]	–	2	–
	Quality	Refer to 'Quality' of 'Bunker facility' in Appendix A	4	–	2	–
	Boil-off	Refer to 'Boil-off' at 'Bunker facility' in Appendix A	N/A[b]	–	3	–

(continued)

(continued)

Supply chain	Critical factors	Description	Interviews	Expert session 1: inland shipping	Expert session 2: fuel owners/traders	Expert session 3: short-sea, inland shipping and trucks
Short sea	Regulation/safety guidelines	It is crucial that important ports in Europa create regulations for ships on LNG and that these regulations are harmonized among these ports	2	–	–	2
	Standardization procedures and permit process	Standardization refers in this case to the standardization of port and ship procedures, and the standardization of the permit process of designing a LNG-ship	3	–	–	–
	Availability	LNG should at least be available at all important European ports. From there, other ports will follow	1	–	–	1
	Quality	Refer to ‘Quality’ of ‘Bunker facility’ in Appendix A	4	–	–	4
	Training	Personnel should be qualified to work with LNG	–	–	–	3
Inland shipping	Regulation/safety guidelines	Regulation and safety guidelines should be clear and harmonized in ports and along the routes	–	1	–	1
	Standardization permit process	At the moment the development of an LNG inland vessel is very long due to lack of a standard permit process. There should be introduced a standard permit process for LNG inland vessels to create a shorter lead time for end users	1	2	–	–

(continued)

(continued)

Supply chain	Critical factors	Description	Interviews	Expert session 1: inland shipping	Expert session 2: fuel owners/traders	Expert session 3: short-sea, inland shipping and trucks
	Availability	Refer to 'Short sea' at 'Availability' in Appendix A	2	1	–	1
	Quality	Refer to 'Quality' of 'Bunker facility' in Appendix A	3	4	–	2
	Boil-off	Not only during the bunkering of inland vessels boil-off should be minimized to safe costs and prevent environmental damage, but also during operating an inland vessel	4	5	–	–
	Training	Personnel should be qualified to work with LNG	–	3		1
Road transport (gas stations)	Regulation/safety guidelines	Regulation about gas stations and tanker trucks should be clear and fully based on the specifications of LNG. For example, at the moment many regulations are based on LPG, which unnecessary restrict owners of gas stations and tanker trucks	1	–	2	–
	Standardization gas filling-guidelines and -equipment	Just like bunker guidelines and equipment for the shipping industry, also the gas filling-guidelines and equipment for trucks should be fully standardized (throughout Europe)	1	–	–	–
	Standardization permit process	The permit process for gas stations is very long. Standardizing the permit process reduces the payback period, and therefore makes the investment more interesting	2	–	3	–
	Capacity	Refer to 'Bunker facility' at 'Capacity'	3	–	–	–

(continued)

(continued)

Supply chain	Critical factors	Description	Interviews	Expert session 1: inland shipping	Expert session 2: fuel owners/traders	Expert session 3: short-sea, inland shipping and trucks
	Quality	Refer to 'Quality' of 'Bunker facility' in Appendix A	–	–	4	–
	Availability	Refer to 'Short sea' at 'Availability' in Appendix A	–	–	1	–
Road transport (end users)	Regulation/safety guidelines	Regulation and safety guidelines should be clear and harmonized around truck design, at gas stations and along the routes	–	–	–	2
	Quality	Refer to 'Quality' of 'Bunker facility' in Appendix A	4	–	–	5
	Availability	LNG should at least be available at major highways throughout Europe and the Netherlands	1	–	–	1
	Power of engine	The power of the LNG engine is lower than a diesel engine. Therefore a LNG truck needs more time to accelerate and has more trouble in, for example, mountain terrain	2	–	–	4
	Boil-off	LNG warms up in the tank of the truck. Therefore a truck is not able to stand still for a longer time period	3	–	–	–
	Training	Truck drivers should be qualified to work with LNG	–	–	–	3

[a]Symbol '–' is used when the factor was not mentioned during the interviews or expert sessions
[b]Identified as a critical factor by a consortium partner, but due to an error in the ranking process, not possible to result in a ranking position

References

Agnolucci P, McDowall W (2013) Designing future hydrogen infrastructure: Insights from analysis at different spatial scales. Int J Hydrogen Energy 38(13):5181–5191

An H, Wilhelm WE, Searcy SW (2011) Biofuel and petroleum-based fuel supply chain research: a literature review. Biomass Bioenergy 35(9):3763–3774

Argos (2014) LNG Bunkering development: technical considerations. Presentation at the 2nd annual LNG Bunkering summit, Amsterdam, 28–29 January 2014

Awudu I, Zhang J (2012) Uncertainties and sustainability concepts in biofuel supply chain management: a review. Renew Sustain Energy Rev 16(2):1359–1368

Baltic Ports Organization (BPO) (2014) Case: LNG in Baltic Sea Ports. Presentation at the 2nd Annual LNG Bunkering Summit, Amsterdam, 28–29 January

Creatieve Energie (2009) LNG als scheepsbrandstof: Ervaringen en perspectieven uit Noorwegen. Creatieve Energie. http://wikimobi.nl/wiki/images/4/4b/LNG_als_scheepsbrandstof._Ervaringen_en_perspectieven_uit_Noorwegen,_juli_2009.pdf. Accessed 22 Nov 2013

DAF (2011) Alternatieve aandrijfvormen en brandstoffen in het wegtransport: Van schoon naar duurzaam. DAF. http://www.daf.eu/NL/News-Media/Documents/DAF-trucks-van-schoon-naar-duurzaam-diesel-VS-gas.pdf. Accessed 8 Jan 2014

Dagdougui H (2012) Models, methods and approaches for the planning and design of the future hydrogen supply chain. Int J Hydrogen Energy 37(6):5318–5327

Dinalog (2015) http://www.dinalog.nl/en/projects/r_d_projects/design_of_lng_networks/

DMA (2012) North European LNG infrastructure project: a feasibility study for an LNG filling station infrastructure and test of recommendations. The Danish Maritime Authority (DMA). http://www.dma.dk/themes/LNGinfrastructureproject/Documents/Final%20Report/LNG_Full_report_Mgg_2012_04_02_1.pdf. Accessed 8 Oct 2013

DNV (2013) LNG fuel bunkering in Australia: infrastructure and regulations. Det Norske Veritas (DNV). http://www.dnv.com/binaries/public%20version%20-%20jip%20on%20lng%20bunkering%20in%20australia%20dnv%20andpartners%20january%202013_tcm4-539390.pdf. Accessed 9 Dec 2013

European Commission (2013) Actions towards a comprehensive EU framework on LNG for shipping. European Commission. http://ec.europa.eu/transport/themes/urban/cpt/index_en.htm. Accessed 8 June 2014

Fuelswitch (2015) Vloeibaar groengas en aardgas. http://www.fuelswitch.nl/index.php?mod=pages&item=53. Accessed 20 April 2015

GLE (2013) GLE Position Paper: Overcoming barriers in the Small Scale LNG development. Gas LNG Europe (GLE). http://www.gie.eu/index.php/news/gle-news/16-news/gle/240-gle-publishes-position-paper-overcoming-barriers-in-the-small-scale-lng-development. Accessed 9 Dec 2013

Groengasnl (2013) Een nieuwe horizon: Productie van bio-LNG, kansen in een opkomende LNG-markt. Groengasnl. http://groengas.nl/report/position-paper-bio-lng/. Accessed 8 Jan 2014

International Bunker Industry Association (IBIA) (2014) LNG—It's the economy, stupid!. Presentation at the 2nd Annual LNG Bunkering Summit, Amsterdam, 28–29 January

Klose A, Drexl A (2005) Facility location models for distribution system design. Eur J Oper Res 162(1):4–29

MIIP 2013 (2013) LNG distributiemodel via het water: Eindrapportage. Maritiem Innovatie Impuls Project 2013 (MIIP 2013). Accessed 17 Feb 2014

Melo MT, Nickel BC, Saldanha-da-Gama F (2009) Facility location and supply chain management—A review. Eur J Oper Res 196(2):401–412

Ministry of I&E (2014) Een duurzame brandstofvisie met LEF: De belangrijkste uitkomsten uit het SER-visietraject naar een duurzame brandstoffenmix in Nederland. http://www.energieakkoordser.nl/~/media/01303b49eab0429b88c79c3353683e9e.ashx. Accessed 11 June 2014

Owen SH, Daskin MS (1998) Strategic facility location: a review. Eur J Oper Res 111(3):423–447
Şahin G, Süral H (2007) A review of hierarchical facility location models. Comput Oper Res 34(8):2310–2331
Simon Loos (2012) Simon Loos: municipal logistics (Netherlands). Presentation at Natural Gas Vehicle Summit: FC Business Intelligence, Brussels, 24–25 October
Society for Gas as a Marine Fuel (SGMF) (2014) Update on regulations and best practice. Presentation at the 2nd Annual LNG Bunkering Summit, Amsterdam, 28–29 January
TNO, ECN, CE Delft (2013) Natural gas in transport: an assessment of different routes. TNO, ECN & CE Delft. http://www.fuelswitch.nl/files/files_news/natural_gas_in_transport_tno_ce_delft_ecn_4818.pdf. Accessed 9 Dec 2013
Vos Logistics (2012) Vos Logistics: opening up the market for LNG transportation in the Netherlands. Presentation at natural gas vehicle summit: FC Business Intelligence, Brussels, 24–25 October

Efficiency Optimization for Cold Store Warehouses Through an Electronic Cooperation Platform

Raphael Heereman von Zuydtwyck and Holger Beckmann

Abstract The cold store market is characterized by strong competitive thinking which impedes efforts to foster transparency. Theoretically, higher efficiency (viz. a reduction of annual kWh consumption per stored pallet) could be gained through smarter allocation of available capacity. A future electronic logistics marketplace (ELM) for cold store capacity (CSC) can realize this task and therefore contribute to an improved cost performance. After building a theoretical model for the positive effects of an ELM, this study aims at identifying the market's attitude towards such a virtual warehouse. Taking the results of two Dutch-German market researches as a basis, we identify current market behaviour, constraints of market participants as well as opportunities for different stakeholders of an ELM for CSC. An electronic logistics marketplace (ELM) for CSC would find positive response among a large group of CSC seekers and suppliers. The e-business concept of e-connection is most suitable for an ELM for CSC. However, distrust in new media, a strong competitive attitude and the sensation that higher transportation costs are an inevitable consequence are issues that apparently hinder stronger national and international collaboration within the cold storage warehouse sector. Cost savings are pursued as a short-term goal and are the most important argument for participants in favour of an ELM for CSC.

Keywords Cold chain · Green logistics · Energy efficiency · CO_2 reduction

R.H. von Zuydtwyck (✉) · H. Beckmann
Institute GEMIT, University of Applied Sciences Niederrhein, Moenchengladbach, Germany
e-mail: raphael.heereman@hs-niederrhein.de

H. Zijm et al. (eds.), *Logistics and Supply Chain Innovation*,
Lecture Notes in Logistics, DOI 10.1007/978-3-319-22288-2_7

1 Introduction

1.1 The Project Green²—Green Logistics in Agrobusiness

Within the European-funded project *Green²—green logistics in agrobusiness*,[1] led by the research institute GEMIT of the University of Applied Sciences Niederrhein (Germany), 21 partners from the German-Dutch cross-border region Rhine-Maas-North have been working on greening the supply chain of agrilogistics and on improving the environmental performance of companies from the food and food logistics sector between summer 2013 and spring 2015. In four different work packages the project dealt with issues such as modal shift actions (Waaden 2015), warehouse optimization (Gruner et al. 2015), energy efficiency (Alsmeyer et al. 2015) and electronic marketplaces (Heereman von Zuydtwyck and Krehl 2015).

In work package IV on *electronic marketplaces* scientists and practitioners focused on the utilization rates of cold store warehouses and attempted to improve them by allocating demand and supply for cold store capacity in a more efficient way. The realization of new cold store capacity consumes high amounts of financial and ecological resources. Identifying idle capacity and prioritizing its usage through a future electronic marketplace before building new warehouses might lead to a higher efficiency and therefore to an improved cost performance. Whereas electronic logistics marketplaces (ELMs) are increasingly implemented since the 1990s (Wang et al. 2007; Zellerhoff 2008), allocation of idle CSC is still not addressed satisfactorily.

In this chapter, we identify the impediments of warehouse operators that prevent the realization of an ELM for CSC. The overall objective is to reduce the energy consumption per pallet in cold store warehouses in Germany and the Netherlands. The goal was to identify what an ELM for CSC is able to offer in real life and what participants might expect. Finally the project participants attempted to define a sound business case for an ELM for CSC in Germany and the Netherlands.

1.2 Environmental Effects of the Cold Store Sector

The cold chain is responsible for 2.5 % of global greenhouse gas emissions (Evans et al. 2014, p. 697) due to its consumption of about 15 % of globally available energy (IIFIIR 2003). The volume of cold store capacity in Germany and the Netherlands amounts to 22 and 13 million m^3, respectively, while the processes of refrigeration and cooling amount to 1.1 billion kWh in Germany and 0.65 billion kWh in the Netherlands. This finally translates into 6,765 Mt of CO_2e in Germany and 39,975 Mt of CO_2e in the Netherlands per year.[2] Being able to slow down the

[1] Cp. www.green2logistics.eu.

[2] Based upon 615 g CO_2 per produced kWh by conventional fleet of power stations in Germany in 2012 calculated with GEMIS (IINAS 2013, p. 3).

acceleration of emitting CO_2e clearly requires to keep the number of cold stores as low as possible.

1.3 Characteristics and Trends of the Cold Storage Sector

Global cold store capacities show different growth rates, depending on a market's level of development. India and China account for more than half of global capacity growth with more than 100 % annual growth rates in the years 2008 and 2009 (Salin 2010). Nevertheless, according to Salin (2010), also highly industrialized countries such as Germany and the U.S. register almost double-digit growth rates (7 and 8 %, respectively) in the period from 2008 to 2010. Another benchmark can be cold storage market penetration, which is measured in cubic metres of cold capacity per person. In an international comparison the Netherlands rank highest with 1.15 m^3 per capita and Germany shows a ratio of about 0.36 m^3 per capita, thereby belonging to the middle range group with India (almost 0.30 m^3 per capita). With demand of frozen and chilled food increasing globally from year to year, this sector steadily gains importance within the food industry (Salin 2010).

The demand for chilled and frozen food rises steadily in most parts of the world (Salin 2010). In terms of revenue, it can be said that the consumption of frozen food will continue to increase. Forecasts speak of 3.4 % annual growth in the field of ready meals to 2019 (the largest segment within the frozen food market), thereby reaching a global turnover of 84.60 billion euros (Transparency Market Research 2013).

New technology has led to slightly improved utilization rates of cold store capacity in mature markets. Surveys for the German market, for example, registered improved average utilization from 74.4 to 77.6 % from 2011 to 2012 (VDKL 2013).[3] Companies, however, are also expanding their capacity at an average rate of 8.37 % (Germany) and 6.05 % (the Netherlands), thus still leaving more than 20 % idle capacity[4] (viz. space for more than 600,000 euro-pallets). In Germany, for example the utilization rate grew by 3.2 % from 2011 to 2012 and reached 77.6 % on average (VDKL 2013).

Assuming that in mature markets as Germany and the Netherlands technological optimization is realized where possible, the missing 20 % can be attributed to a lack in cooperation since companies show highly different utilization levels throughout

---

[3]For the 2013 annual report of the German cold store Association (Vdkl) 788 cold stores were asked for their monthly utilization rates.

[4]The utilization rate always applies to the available and reasonable capacity. On average, a utilization of 80 % of overall floor space is regarded to be reasonable. An average utilization rate of 77 % thus implies that 77 of 80 % overall floor space is used. Consequently this leads to a 61.6 % utilization of overall floor space.

the year. Evans et al. (2013) indicate that the cold store market has a diversified structure. Sizes of cold stores vary. They state that Europe has 1.7 million cold stores totalling 60–70 million m^3 of storage volume. Thereof, 67 % are small stores with a volume of less than 400 m^3 (Evans et al. 2013, p. 39).

1.4 *Information and Knowledge Management in the Cold Store Sector*

Cold store warehouses can basically be grouped in two categories. On the one hand there are public cold store warehouses that offer the storage and handling of cold and frozen goods as a core business. On the other hand there are private cold store warehouses operated by companies to aid their core business, e.g. food processing companies. Both groups together form the cold store sector.

For small and medium-sized companies within the cold store sector matching demand and supply is difficult (Howells 2006) due to limited access to vital knowledge of available capacity in regions of interest. The information gap between demand and supply is one challenge that especially small and medium sized companies face (Bougrain and Haudeville 2002). Consequently these companies need specialized Third-Party Logistic providers (3PLs) to cover this gap. Therefore, public cold stores do not only help to balance cold store demand and supply but also offer additional services such as order picking, flash freezing, professional thawing, and cool transport solutions. Current research shows that shippers put growing emphasis on extra-regional trade and 3PLs are increasingly approached (Capgemini Consulting 2014).

Companies currently seeking cold store capacity that exceeds their own CSC follow the process shown in Fig. 1. However, this process may be very time consuming. The standard process of seeking additional CSC starts by contacting a public cold store warehouse. The majority of companies would activate an existing network of 3PLs or own branches as cooperation partners. In case the partners do not have the necessary capacity, they in turn would contact another partner network. Depending on the available capacity on this second-level market, the initially approached company will be able to offer capacity to its client sooner or later. A company experiencing peak utilization with high regularity or not disposing of a reliable network of cooperation partners might think of building a new cold store warehouse of its own in order to keep business.

When it comes to offering CSC, public rather than private cold store warehouses have the standard possibility to contact either by phone or mail their existing clients or to perform marketing campaigns to increase their client base.

The idea of cooperating and sharing information to generate synergies for the participants has merged into very diverse e-business concepts. A short insight is given below.

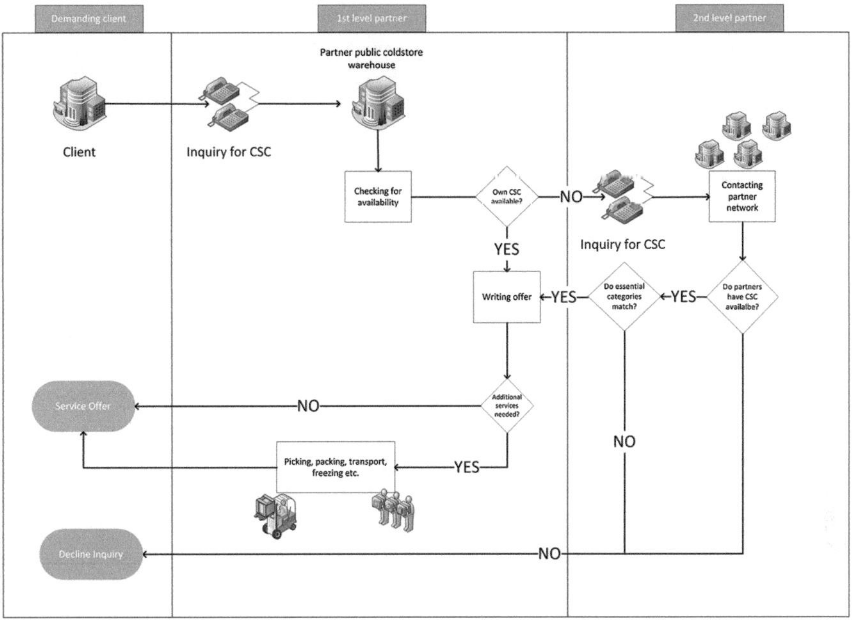

Fig. 1 Conventional process of acquiring CSC

1.5 E-Business Models that Have Proven to be Successful

Since the Internet became widely accessible almost 20 years ago, several e-business models have proved to be suitable (Bauer and Hammerschmidt 2004, p. 91). Depending on the pursued aim, four concepts can be distinguished according to Heinemann (2009, p. 29), namely e-content, e-commerce, e-context and e-connection (see Table 1). Business transactions are mostly realized through the platforms e-procurement, e-shop and e-marketplace. Of these, an e-marketplace is widely defined as an "interorganizational information system that allows the participating seekers and suppliers to exchange information about prices and product offerings" (Bakos 1991, cited in Rask and Kragh 2004, p. 270).

Whereas every model of e-business concepts provides the added value of getting an overview of the concerning good or service, the most appropriate model for an ELM for CSC must be found. E-marketplaces prevail in the concepts of e-commerce, e-context and e-connection. To elect the most appropriate concept, a quantitative market research as well as a qualitative investigation involving practitioners in both the Dutch and the German cold store market has been undertaken. The results of both the quantitative and qualitative study were taken into account to identify an appropriate concept for a future successful ELM for CSC and to define a sound business case.

Table 1 E-business concepts (according to Heinemann 2009)

	E-content	E-commerce	E-context	E-connection
Definition	Collecting, selecting, compiling and offering content via Internet	Preparing, realising business transactions via Internet	Classifying, systematisation and matching information that is available through Internet	Creating the possibility of information interchange via Internet
Aim	Offering consumer-oriented and customized information via Internet	Completing or substituting traditional transaction phases via Internet	Reducing complexity, aiding navigation and offering matching functions via Internet	Creating technological, commercial and communicative relations via Internet
Platforms	E-shop, E-community, E-company	E-shop, E-procurement, E-marketplace	E-community, E-marketplace	E-marketplace, E-company, E-community
Added value	Overview, selection, cooperation, processing, transaction realisation	Overview, selection, transaction realisation	Overview, selection, negotiating	Overview, selection, negotiating, exchange, transaction realisation
Example	www.nytimes.com, www.guenstiger.de	www.amazon.com	www.google.com, www.alibaba.com	www.t-online.de

1.6 Electronic Marketplaces in the Transport and Storage Sector

In contrast to the transport sector, the storage and warehouse sector does not have an abundance of electronic marketplaces. This is due to the fact that transport services are fast-moving services whereas storage and warehouse services are generally characterized by a slower rotation. In the core business of this market, this comprises the action of storing and retrieval as well as the physical storage of goods. Currently, available storage capacity is usually not offered via internet, thereby causing a short-term gap in available storage. Peak demand and capacity deficits of companies are thus not compensated via a spot marketplace.

A transparent system would likely lead to the statement that nowadays there is an excess of capacities that varies from region to region. At the same time, a more transparent information situation might lead to savings in kilometres driven. Transport and storage go hand in hand and are closely intertwined. Additionally, one has to consider that the price is the most important factor and determines decisions on the market since participants are primarily motivated by price and not by route optimization (Bretzke and Barkawi 2012, p. 328). In the future, however, this factor might become more important than low storage rates due to the ongoing rise of transportation costs by truck. If the success of e-marketplaces could be transmitted from the transport sector to the cooling and freezing storage sector, in theory storage capacity might be used more efficiently, energy savings might be realised and greenhouse gas emissions were to be reduced. If and to what extent this is possible in real life is the subject of the project Green².

2 Theoretical Model

Motivational reasons for seekers and suppliers to act in e-marketplaces were first investigated. It is generally accepted among scientists that time savings (less time needed to look for alternatives) and the advantages of being able to contact, select, communicate and evaluate suppliers are the primary reasons to participate in an e-marketplace for seekers. Furthermore, the so-called 'electronic brokerage effect' that describes the significantly lowering transaction costs is widely recognized. On the supplier side, it is well known that price transparency and increased competition via an e-marketplace leads to lower prices and attracts ever more seekers (Rask and Kragh 2004; Smart and Harrison 2003).

It is assumed that average capacity utilization among cold store shows a huge gap to the maximum available capacity due to huge differences in utilization throughout the year and among companies.

In theory, the annual maximum of existing CSC is increased continuously due to the already identified market growth. The dotted line in Fig. 2 indicates the average utilization rate on the market. Although capacity utilization improves throughout the years augmenting the annual maximum impedes an overall improvement.

Taking individual utilization rates into account, the reason for the current practice becomes clear: cold store capacity is generally adjusted to the calculated maximum need of the individual company. This generally urges market participants to build new capacity whenever their maximum is exceeded (Fig. 3).

Virtual allocation of the peaks and lows in individual utilization rates reduces the overall variance to mean ratio. As a result, these rates show less intense oscillations (Fig. 4).

Finally, virtual allocation via an ELM leads to a smoothened annual average rate. This observation in turn might change the general attitude of investing in new capacities. The CSC market is as a virtual warehouse that gives clear possibilities of flexibly increasing efficiency (viz. available and effectively used chilled or frozen warehouse space) (Zellerhoff 2008). Investments will consequently be postponed (Fig. 5).

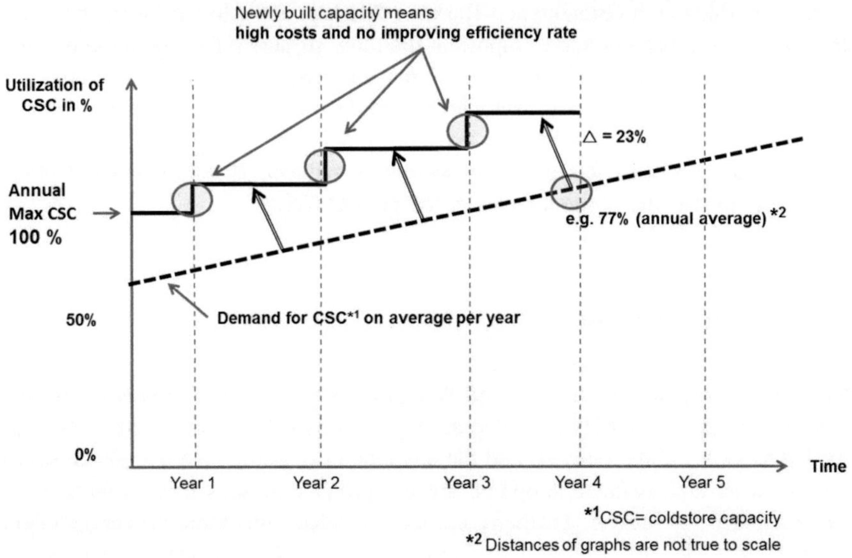

Fig. 2 Current ratio of annual average and available CSC

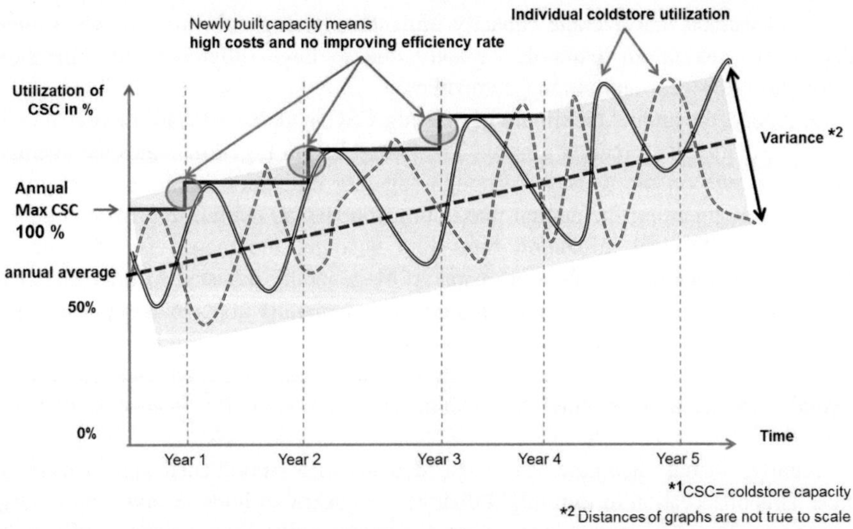

Fig. 3 Individual utilization rates in the course of the years

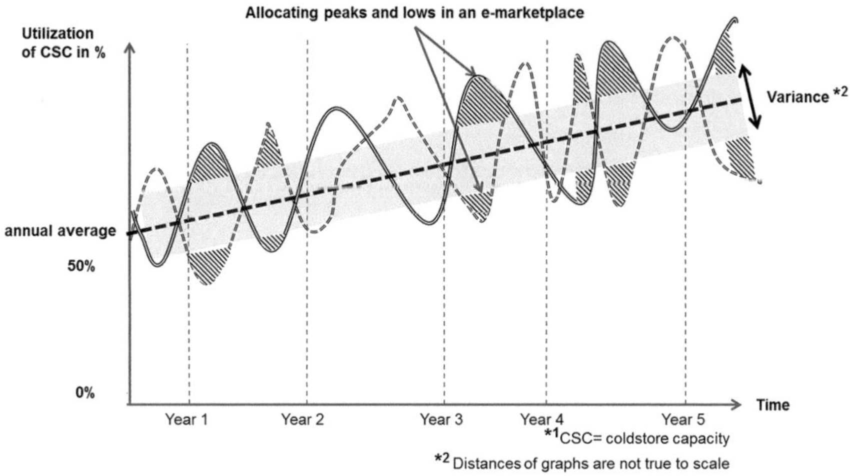

Fig. 4 Matching peaks and lows of market participants

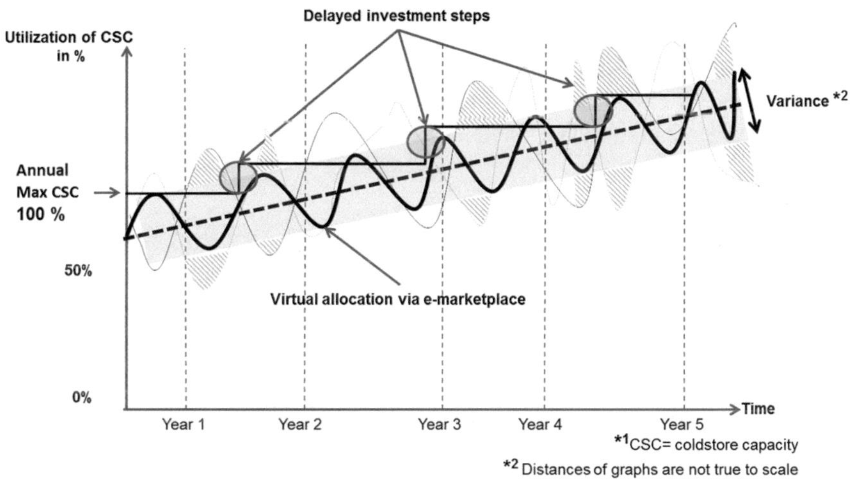

Fig. 5 Allocation effect on annual investments

3 Testing Theory to Real Market Behaviour

To check whether the established theory is applicable to the current situation at the Dutch and German cold store market in reality, opinions and attitudes of cold store operators towards the participation in a future ELM for CSC must be investigated. Thus, three research questions were formulated and translated into an online survey. The first research question addresses the current opinion of cold store operators. For a study that is closely linked to market relevance, this question has to be the starting point.

RQ1: How do market participants currently see the establishment of an e-marketplace for cold store capacities?

The second research question concentrates on the characteristics of utilization rates of cold store warehouses.

RQ2: How are demand and supply of cold store capacities distributed throughout the year?

This research question aims at detecting the balance between supply and demand of CSCs within the researched area. Moreover, being able to characterise the cold store capacity market makes it possible to draw conclusions on future energy needs or saving potentials through an e-marketplace.

The third and final research question focuses on existing constraints with respect to e-marketplaces in the cold store warehouse sector.

RQ3: Which restraints for a future e-marketplace do cold store providers see?

RQ3 aims at identifying main current restraints against an e-marketplace for cold store capacities. To name the principal reservations about such an e-marketplace would be essential for being able to start designing a sustainable and customer-oriented e-marketplace. Identifying basic constraints regarding an e-marketplace of CSCs increases the acceptance and thereby the chance of successfully implementing such a solution. To find answers to the established research questions, a mixed method approach was pursued, applying both quantitative and qualitative approaches.

3.1 A Quantitative Research Approach

Quantitative data was collected via two surveys among Dutch and German cold store operators based upon the formulated research questions.

The first survey was carried out in the winter of 2013 and the second survey lasted from May to July of 2014. Both surveys were conceptualized on the basis of expert talks within the project Green^2. Whereas the first survey directly addressed Dutch and German cold store operators, the second survey was realized through the Green^2 corporate project partner TimoCom, a company that operates as an Europe-wide IT service provider managing a freight exchange platform running in over 30 countries with more than 35,000 clients. In general, the target population consists of Dutch and German public and private cold store warehouses with chilled and frozen food products in the range from −18° to +7 °C. Within the project, partners from each of these groups contributed and therefore provided a sound basis for the following qualitative research. For the first survey an intensive Internet research (based on the information provided by the German association of cold stores, VdKL) identified the basic population and detected 244 companies as public

Table 2 Survey details

Parameters	Survey	
	1st survey	2nd survey
Approach	Direct	Indirect
Response	16 out of 156	49 out of 501
Response rate	10.2 %	11.7 %
Participants	Dutch and German cold stores	Dutch and German cold stores
Time frame	Winter 2013	Spring 2014

cold stores in Germany and the Netherlands. The sample was chosen by excluding businesses with less than 250 chilled and frozen loading positions (calculated in pallets), because smaller companies are considered to show insignificant storage movements. As a result, the survey was based on information from a (non-probabilistic) selection of 156 companies in Germany and the Netherlands. The second survey aimed at TimoCom's clients in Germany and the Netherlands that showed 'storage' on their profile, resulting in a sample size of 501. Table 2 gives a summary of the two surveys.

The participants were offered a maximum of nine questions, asking for their attitude towards an ELM for CSC, their behaviour in case of peak and low utilization, utilization throughout the months of the year and their company's size.

3.2 A Qualitative Research Approach

The results of the quantitative research were presented to the experts in the project and individually discussed. This yielded valuable insights not only into the cold store sector but also into the sector of trading and producing chilled and frozen food. The qualitative research followed the Delphi approach, thus discussing experts' opinions anonymously with other experts and building up on the new findings. In a final meeting all experts from the project consortium were brought together to design unanimously the requirements of a future ELM for CSC.

3.3 Results from the Quantitative Surveys

The surveys indicates that market participants have no general bias against an ELM for CSC with 89 % (1st survey) and 77.6 % (2nd survey) saying 'yes' or 'maybe' to the usage of an e-marketplace for cold store capacity in the future.

Figure 6 shows the behaviour of survey participants in case of low utilization of their cold stores (multiple answers allowed).

Fig. 6 Behaviour at low utilization

In times of low utilization rates, 44.62 % of the respondents actively engage in acquiring new customers, which can be interpreted as being the conventional business practice. This top answer is followed by *offering space to cooperation partners*, which indicates that networks already exist and are used. However, it is not known whether these cooperation partners are solely 3PLs or also private cold store warehouse operators. It is surprising, however, that *not reacting at all* also ranks among the top three answers and implies that they thereby are missing the chance of increasing their customer basis or business activities.

In case of peak utilization, survey participants chose from a range of possible answers. Figure 7 shows the combined result from the two surveys (multiple answers possible).

In such an event, more respondents obviously engage in business relations with cooperation partners (36.9 %) then when experiencing low utilization rates. The

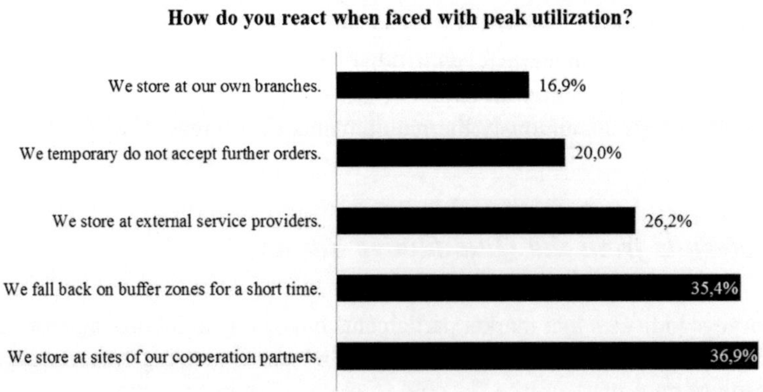

Fig. 7 Behaviour at peak utilization

surveys furthermore confirm real life observations that goods are unconventionally stored in buffer zones for a short time due to pressure in daily business. It is remarkable that external service providers are more frequently consulted during peak time 26.2 % than at low utilization rates (15.4 %). Note that in our surveys the answer *storing at own branches* does not rank as second best answer, which might be traced back to the mere fact that participants to not have numerable branches to fall back on.

The annual average compared to the monthly utilization rates indicates a wide dispersion in utilization between cold stores and confirms the theoretical deductions made in advance. This characteristic is more obvious when taking into account the upper and lower quartile showing an interquartile range of 20–25 % (march as an exception with a range of 15 %).

Having observed this, one might ask why market participants do not engage in virtual interaction of CSC more actively. When asked for their restraints, 1st survey-participants showed the following answers.

Figure 8 shows that the sense of competitiveness and rivalry seems to be quite firmly set among respondents. However, transparency is considered to be a must in order to be able to show healthy corporate growth in the future.

Furthermore, participants remark that additional transport costs to warehouses with free capacity are unwanted. This argument has to be taken into account when designing a future ELM for CSC since cost efficiency will certainly be one of the most important motives to participate in an electronic marketplace. Nevertheless expert interviews also show that through ELMs transport routes and thus cost can be reduced depending on the situation.

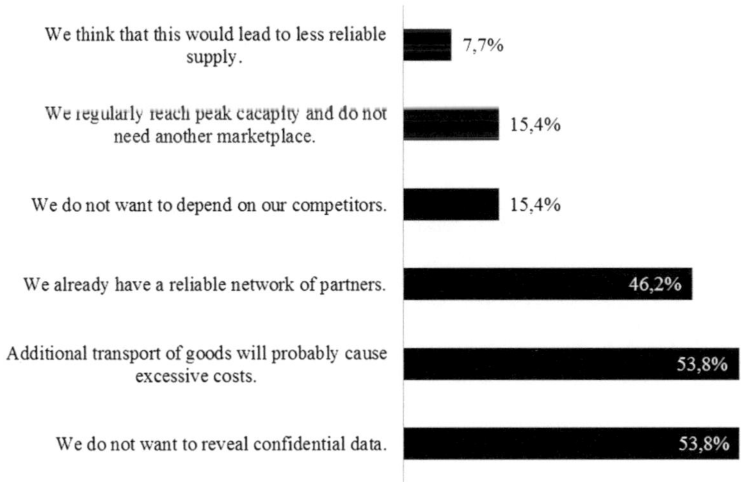

Fig. 8 Restraints against participating in an ELM for CSC

3.4 Results from the Qualitative Research

Firstly, through role-playing the common indicators were identified that the supplier of CSC is willing to reveal and the seeker of CSC needs to have access to in order to realize business relations. These indicators are shown in Table 3.

Table 3 stands for the minimum information that a future ELM for CSC must offer in order to attract seekers and suppliers. Only if this information is provided, both parties can see added value in such a platform. From an additional point of view, this basic information also guarantees an equal starting point for both parties, thereby lowering entry barriers. Clement and Schreiber also refer to this as the service of coordination (Clement and Schreiber 2013, p. 6) of an electronic marketplace and regard this service to be crucial for a successful ELM.

To build customer loyalty, the participants expressed that they would like to see further information in order to consult the ELM more regularly. This information exchange, however, was not agreed upon by both parties to the same degree since it was perceived as being of a highly sensitive nature.

Especially 'benchmarking' and 'customer evaluation' would signify an attractive added value to build seeker loyalty to the ELM. The qualitative research showed that cold store operators see themselves able to rise to the challenge of being publicly evaluated by customers.

Table 3 Common indicators for suppliers and seekers

Data that seekers want to see from suppliers and suppliers are willing to show	• Available storage space • Available temperature • Available storage period • Company details (name, location; contact) • Method of payment • Certificates • Probable hazardous material • Provided/needed temperature • Type of product
Data that suppliers want to see from seekers and seekers are willing to show	• Needed storage space • Needed temperature • Needed storage period • Company details (name, location; contact) • Location, accessibility • Product insurance • References • Required storage space • Security • Warehouse conditions (added value services)

Table 4 Sensible data in an electronic marketplace

Data that seekers want to see from suppliers and suppliers are not willing to show	• Benchmarking • Customer evaluation • Pictures of premises • Price
Data that suppliers want to see from seekers and seekers are not willing to show	• Payment morale • Price the seeker is prepared to pay

To capacity suppliers, seeker's solvency and payment morale is of utmost importance. A seeker's credit assessment would be of high value to suppliers. Also experiences from other CSC suppliers with the client in concern could be very valuable. On the other hand offering pictures and prices in an ELM are viewed with scepticism. Pictures cannot be accepted as legally binding and prices are too sensitive to publish them directly. These data will rather be optionally offered, for example within a premium account for participants.

The data shown in Table 4 is of high value for both sides and in most cases also a unique selling proposition (USP). Thus, revelation of that data in an anonymous marketplace to unknown prospective clients and competitors is against due business interests. Consequently, participants are only willing to interchange these data in a more private setting, e.g. via direct phone-calls, which was also confirmed in the final meeting with project partners. The experts put forward that in order to generate trust further in-app services (chat application within the ELM or other communication services) might be helpful to bind the participants to the marketplace and resolve this issue to some extent.

4 Bridging the Gap: Applying the Findings to Design a Feasible ELM for CSC

The surveys illustrate that the majority of participants will take part or are at least interested in a future ELM for CSC. Furthermore, survey participants expressed concerns regarding data security in an ELM. These concerns referred mainly to data transparency issues and issues of competitor relations. Security concerns in terms of insurance cover, however, were not mentioned by cold store operators. A future ELM for CSC will not bear the risk of insurance cover since effective business transactions are not realised on the e-marketplace but offside. This is due to the fact that the ELM provider will not have insight and influence in the deals to be concluded.

4.1 A Future E-Marketplace for Cold Store Capacity—A Reliable Information Tool

It has to be assured that the information provided by CSC seekers and suppliers will be kept within the marketplace. Fraud can be prevented by establishing registering processes, e.g. with the upload of personal IDs or company credentials (Zellerhoff 2008).

Qualitative research among both cold store and e-marketplace experts show that an ELM for CSC manages to bring different parties together and supervise the allocation of cold store capacity. The exchange of still some highly sensitive data is necessary before a transaction for a fixed period of time of renting or lending cold store capacity is concluded.

Figure 9 illustrates the ELM effect of shortening the ways of information and bringing the participants directly together.

The most obvious advantage compared the conventional processed displayed in Fig. 1 lies in the shortened processes in terms of time and levels. The ELM therefore streamlines the relevant processes. Whether CSC is requested or offered, the relevant information for both parties can be retrieved at the ELM. In case no match is found at the first intent, the ELM client can either change his search or opt out. In the case demand and supply match both parties are able to immediately start business negotiations.

4.2 Changing the Current Mind-Set as a Challenge for the Future

Protectionism is still firmly set among at least some cold store operators. Nevertheless, project partners stated that the market is willing to enlarge business opportunities via an e-marketplace and do not ask an ELM for covering insurance

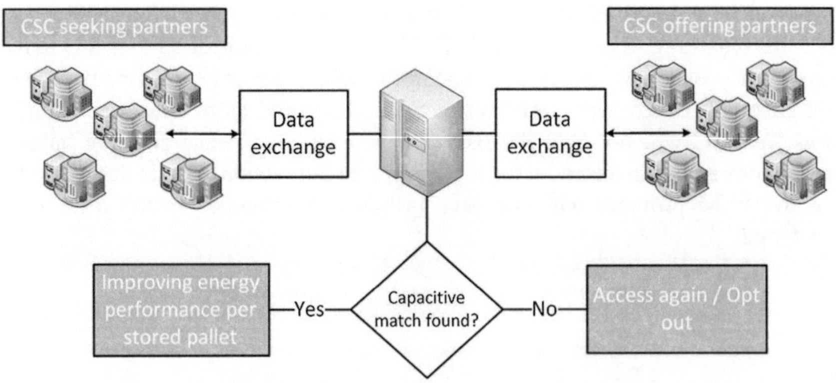

Fig. 9 ELM effect on CSC allocation

for transactions. An ELM for CSC does not need to procure orders, which is clearly neither desired by participants nor marketable for the ELM itself. The former have insurance contracts for their business activities anyway and do not need further cover and the latter would bear too much responsibility for actions it cannot control.

It is not a question of how to ensure access to sufficient storage capacity through an ELM for participants but rather a question of offering the right information at the right time. The advantages provided by an ELM in the sense of e-context (Table 1) are information security within the marketplace and direct exchange between relevant parties.

4.3 Next Step: Identify the Market and Attract Clients

The next area to identify is the expansion of the prospective market. An ELM for CSC could be an asset especially for small and medium-sized companies to bring their services of cooling and chilling agri-products to a regionally and product-wise broader clientele. Relevant sectors are engaged in chilled and frozen products processing and wholesaling. Also pharmaceutical companies and basically all parties forming part of the cold chain might be interested in such an e-marketplace. In a next step, this potential has to be evaluated for European ELM for CSC.

Furthermore the research reveals that especially owner of small and medium-sized cold stores may benefit from an ELM for CSC. Due to the fact that this target group often has no experience with exchanging capacity it should be evaluated how entry to this market can be eased for this group. From the experts' point of view the development of a guideline and a solid information tool including e.g. standardized contracts could be handy for this type of companies.

4.4 Risks: Diversity Versus Relevance

Bringing in a *diverse* customer range, however, bears risks too. The more heterogeneous the clientele is the less matching opportunities are likely to be detected. However, an ELM that is only accessible to cold store operators might not lead to the desired *relevance* in terms of customer numbers. In such a specialized segment the regional players know each other and are either already working together or have proven to be strong competitors. In that case, an ELM would only make sense when operating internationally. Thus, the ELM for CSC will have to be open to all types of companies dealing with chilled and frozen goods that are in need for or are able to offer CSC.

Clement and Schreiber (2013) add *instability* as another risk factor which is strongly linked to *relevance*. They mention that a balanced ratio between capacity sellers and seekers is as important as a high number of marketplace members.

4.5 *Sustainability of an ELM for CSC—Independence Versus Insignificance*

Qualitative research discussions showed that an ELM for CSC would most probably not be able to operate self-sufficient and self-sustaining unless a membership with fee is introduced, to ensure a steady revenue. A solution might be that an ELM for CSC is linked to one of the many existing freight exchange platforms thereby guaranteeing visibility to a sufficient number of prospective clients. It was concluded that as a start, an ELM for CSC would be most viable as an additional service rather than a stand-alone solution. After having gained sufficient clientele and significance on the market, independence should be reconsidered.

Discussion of different economic models led to the conclusion that the most common model of '*suppliers pay and seekers don't*' (cf. marketplaces such as www.immobilienscout24.de or www.rightmove.co.uk) should be applied to generate a revenue for healthy growth.

Summarizing, an ELM for CSC makes sense when it is established according to the following definition:

The ELM for CSC is both a vertically and a matching oriented open marketplace, on which seekers and suppliers of CSC directly interchange relevant information on desired services via a dynamic pricing process (spot-pricing).

The market operator should play a neutral role and can therefore add value for all parties involved. The operating company will have several revenue models at its disposal. Market research shows, however, that a direct revenue model that is not linked to the actual usage behaviour of the client (viz. subscription) is most adequate and also fairly common for exchange markets of the logistics sector.

References

Alsmeyer F, Newiadomsky C, Bruchmann J, Schäfer C (2015) ARBEITSPAKET III—Energieeffiziente Lager- und Kühltechnik. [Online] Available at: http://www.green2logistics.eu/fileadmin/userfolders/downloads/pdf/News_Dokumente/Abschlussbericht_Green__AP3_Druck.pdf. Accessed 14 Mar 2015

Bauer H, Hammerschmidt M (2004) Beziehungsmarketing für elektronische Business-to-Business-Marktplätze. Der Markt 3+4:91–105

Waaden T (2015) Arbeitspaket 1—Nachhaltiger Güterverkehr im Agrarsektor. [Online] Available at: http://www.green2logistics.eu/fileadmin/userfolders/downloads/pdf/News_Dokumente/Abschlussbericht_Green__AP1_Druck.pdf. Accessed 20 July 2015

Bougrain F, Haudeville B (2002) Innovation, collaboration and SMEs internal research capacities. Res Policy 31(5):735–747

Bretzke W-R, Barkawi K (2012) Nachhaltige Logistik: Antworten auf eine globale Herausforderung, 2nd edn. Springer, Berlin

Capgemini Consulting (2014) 2014 Third-party logistics study. [Online] Available at: www.capgemini.com/resources/2014-third-party-logistics-study-the-state-of-logistics-outsourcing. Accessed 17 Mar 2015

Clement R, Schreiber D (2013) Internet-Ökonomie. Springer, Berlin
Evans J et al (2013) Specific energy consumption values for various refrigerated food cold stores. Energy Build 74:141–151
Evans J et al (2014) Assessment of methods to reduce the energy consumption of food cold stores. Appl Therm Eng 62:697–705
Gruner S, Köppen-Spielmann N, Kreetz J, Tißen K (2015) *ARBEITSPAKET II—Lageroptimierung*. [Online] Available at: http://www.green2logistics.eu/fileadmin/userfolders/downloads/pdf/News_Dokumente/Abschlussbericht_Green__AP2_Druck.pdf. Accessed 15 Mar 2015
Heereman von Zuydtwyck R, Krehl F (2015) ARBEITSPAKET IV—Elektronischer Marktplatz für Kühlkapazitäten. [Online] Available at: http://www.green2logistics.eu/fileadmin/userfolders/downloads/pdf/News_Dokumente/Abschlussbericht_Green__AP4_Druck_v02.pdf. Accessed 14 Mar 2015
Heinemann G (2009) Der neue Online-Handel—Erfolgsfaktoren und Best Practices. Gabler, Wiesbaden
Howells J (2006) Intermediation and the role of intermediaries in innovation. Res Policy 35:715–728
IIFIIR (2003) International Institute of Refrigeration—How to improve energy efficiency in refrigerating equipment. [Online] Available at: http://www.khlim-inet.be/media/ice-e/publications/How_to_improve_energy_efficiency_in_refrigerating_equipment.pdf. Accessed 14 Aug 2014
IINAS (2013) Der Nichterneuerbare kumulierte Energieverbrauch des deutschen Strommix im Jahr 2012. [Online] Available at: http://www.iinas.org/tl_files/iinas/downloads/GEMIS/2014_KEV-Strom-2013_HEA.pdf. Accessed 17 July 2015
Rask M, Kragh H (2004) Motives for e-marketplace participation. Electronic Markets 14(4): 270–283
Salin V (2010) Global cold storage capacity report, Alexandria, Virginia: International Association for Refrigerated Warehouses
Transparency Market Research (2013) Frozen Food Market (Vegetables & Fruits, Potatoes, Ready-to-eat Meals, Meat, Fish/Seafood and Soups)—Global Industry Analysis, Size, Share, Growth, Trends and Forecast, 2013–2019, Transparency Market Research, Albany, NY
VDKL (2013) Jahresbericht. Bergisch Gladbach: Heider Druck GmbH
Wang Y, Potter A, Naim M (2007) An exploratory study of electronic logistics marketplaces and its impact on customised logistics. Cardiff University Innovative Manufacturing Research Centre, Dallas, TX
Zellerhoff J (2008) Modellierung eines Informationssystems für ein "Virtuelles Lager". Universität Dortmund, Dortmund

Part II
Urban Logistics

Urban Freight Transportation: Challenges, Failures and Successes

Goos Kant, Hans Quak, René Peeters and Tom van Woensel

Abstract In this paper, we present the challenges, failures and successes on urban freight transportation. We first identify the various involved stakeholders with their interests. Then we evaluate a large number of urban freight transport initiatives and identify lessons learned, which are distinguished in policy, logistics and technology based views. Further, we present a vision for urban freight transportation, which is not only based on the lessons learned, but also on actual market research reports and recent findings.

Keywords Urban freight transportation · Lessons learned · Success · Vision

1 Introduction

The OECD (Organisation for Economic Co-operation and Development) Working Group on Urban Freight Logistics (OECD 2003) defines urban goods transport as the delivery of goods in urban areas, including the reverse flow of waste. Many large cities face significant challenges related to the congestion and pollution generated by the number of vehicles within urban areas.

Urban transportation includes also a significant proportion of passenger transport, including not only residents and shoppers, but also service and other vehicle trips for commercial purposes, which are essential to the urban functioning. Within

G. Kant (✉) · R. Peeters
Faculty of Economics and Business Administration, Tilburg University,
PO Box 90153, 5000 LE Tilburg, The Netherlands
e-mail: g.kant@tilburguniversity.edu

H. Quak
TNO Mobility and Logistics, PO Box 49, 2600 AA Delft, The Netherlands

T. van Woensel
Industrial Engineering and Innovation Sciences, Eindhoven University of Technology,
PO Box 513, 5600 MB Eindhoven, The Netherlands

H. Zijm et al. (eds.), *Logistics and Supply Chain Innovation*,
Lecture Notes in Logistics, DOI 10.1007/978-3-319-22288-2_8

the European Community, many cities have managed the transportation of people by developing public transportation networks that are generally integrated. Freight transportation is completely different and is very immature. Urban transportation of goods is almost completely managed by private Logistics Service Providers (LSPs) and/or shippers who manage their own transport without any coordination, leading to many unnecessary movements of underutilized vehicles in congested areas.

There is also a strong necessity for more efficient urban transportation. The amount of people living in cities is increasing rapidly (OECD 2003), the sizes of freight packets are decreasing and their shipment frequencies increasing, due to internet and the focus on more convenience type of stores. Additionally, there is a continuous pressure on emission reduction and logistic efficiency in the city. An important contemporary challenge for large cities is to improve the air quality: to satisfy the European norms for NO_x, cities have to improve their air quality considerable. High concentrations of NO_x and PM-$_{10}$ have negative consequences for the residents' health.

Many stakeholders are involved. Specifically, authorities, carriers, receivers, residents, shippers and traffic participants all make use of the same scarce resources available in the urban areas. This makes it difficult to develop sustainable urban freight transport solutions, as a wide variety of (often conflicting) problem perceptions, stakes and solutions exist (Browne and Allen 1999). This, of course, has a significant impact and sound logistics solutions are required. The majority of goods in urban areas are delivered via road haulage, with significant air quality impact as indicated above. This demonstrates the need for better control of urban freight transportation in order to reduce its impact on adverse living conditions in cities.

This contribution outlines the different challenges, successes and failures as observed in this urban environment. The remainder of the paper is structured as follows. In Sect. 2 we identify the different stakeholders with their interests. In Sect. 3 we review the known urban freight transport projects, distinguished in policy, logistics and technology driven initiatives. In Sect. 4 we describe the lessons learned, again according to policy, logistics and technology based views. We present in Sect. 5 our vision for urban freight transportation including various innovative aspects, based on the finding and more recent developments. We identify various opportunities for successful urban freight transportation, but also see a strong need for it, given the current trends in urbanization, higher customer demands, more restrictions at receiver level and increasing cost.

2 The Different Stakeholders in Urban Freight Transportation

Stakeholders differ in their urban freight transport interests, their resources and possibilities, as well as their decision power. Since the focus is on freight, we do not explicitly consider some other, important, stakeholders in the city, such as inhabitants. In all cases, (local) government is concerned with the stakes of this type of stakeholders.

2.1 Government

Governmental stakeholders are roughly divided into higher and local authorities. Higher governments tend to deal with urban freight transport problems as local issues, and are therefore usually not directly involved. Interestingly, higher governments take measures with a significant impact on local issues, like air quality. Examples are the EURO-norms for truck engines, which have resulted in a considerable decrease in local pollutants. Higher government (national and, more and more, the European Union) determines the playing field of all private stakeholders: regulations on truck sizes, working hours, infrastructure, etc.

Local authorities have the autonomy to determine the context for urban freight transport for their specific city. This implies that the context in which carriers are active with their urban freight transport operations differs not only per country, but also even per city (or, in the extreme but realistic case, even per street, see e.g. Dablanc 2007).

In general, (local) government looks after the interests of the impactees, since these are their 'voters' in elections. Impactees are these involved parties that notice the impacts of urban freight transport, both positive and negative impacts (Ogden 1992). We discern three groups among the impactees: the residents (who live in the city centers), the shopping public (a collective term including the stakeholders that make use of the facilities in city centers) and traffic participants (those actors that are confronted with urban freight transport, for example passenger transport for commuter traffic).

In Dablanc (2007) it is argued that most authorities plan and regulate urban freight transport similar as they did over two decades ago. These regulations aim at restricting the urban freight traffic in order to reduce negative impacts (Allen et al. 2000). This means that the increased possibilities due to the recent ICT developments and the changing urban environments are ignored.

2.2 Shippers, Carriers and Receivers

The relationships between carriers, shippers and receivers seem straightforward: the receiver (e.g. a store) orders goods at a supplier, the supplier hires a carrier to transport the goods to the store and then the carrier delivers the goods. These groups however consist of very heterogeneous stakeholders.

2.3 Shippers and Shops

There is a strong interaction between the receiver (e.g. a store) and the shipper. From a shipper's perspective there is usually no direct involvement in urban freight transportation if they do not own stores themselves. The receiver's interest is especially in minimizing the perceived inconvenience caused by trucks and in creating a nice shopping environment. In the case a receiver is part of a retail chain, deliveries are usually coordinated by the retailer's headquarters, which then acts as shipper. The result is that the retail chain is responsible for the transport and not the receiver. Their supply chain strategy of Just-In-Time store deliveries leads to an increase of frequency for delivering the goods and, hence, to a less sustainable solution.

2.4 Carriers

Quak (2008) distinguishes between regional carriers, functional carriers and generalists. The regional carriers often cooperate with other regional carriers in a network (for example, TransMission in the Netherlands). Regional carriers usually have a depot in the region in which they are active. As a consequence, it takes only a limited time to reach cities in this region. The regional carriers see local time-windows (and other restrictions) as an opportunity, since their vehicle route planning and their vehicle fleet is adapted to the local situation. Whereas, these local restrictions are a threat for other carriers that have to deal with longer distances to cities and several local regulations, and cannot easily adapt to various local situations.

3 Review of Urban Freight Projects

In this section, we review the large number of city logistics initiatives and the more limited number of research projects executed over the past years. A number of private and public sector initiatives have been proposed. Next to solutions initiated solely by (local) authorities and companies, there is a huge number of initiatives involving several stakeholders and combining two or all solution directions (i.e. technology, logistics and policy based). These initiatives are system initiatives, since they require changes in more than one part of the urban freight transport system.

In his dissertation, Quak (2008) presented 106 unique Urban Freight Transport initiatives, undertaken between 1998 and 2006. This review from 2008 is updated in this paper with all relevant recent initiatives and projects.

Three different solution directions are distinguished, based on policy, logistics and technology considerations. Usually (local) authorities use policy initiatives in order to regulate urban freight transport operations. Most technological initiatives aim at reducing nuisance without changing the actual operations. Technology focuses more on technological advances, e.g. electrical vehicles. Logistical solutions include the organization of transport and logistics within one company and also supply chain collaboration or other forms of collaboration. We use this classification for reviewing the variety of initiatives aiming to improve urban freight transport.

3.1 Policy

Authorities have several means available in realizing its urban freight transport policies, i.e. regulating, coordinating, facilitating, and stimulating measures. Most policy initiatives are in the area of regulation. Using regulation, local authorities aim at changing carriers' operations in order to become more sustainable (or at least to cause less nuisance) by obliging legislation (vehicle restrictions, vehicle utilization controls, low emission zones, and time-access window). Usually these are not applied in isolation, but to make other solution directions work (for example, a low emission zone forces carriers to use cleaner vehicles). Licensing and regulation initiatives are quite common in European countries (Ogden 1992).

Local authorities use *vehicle restrictions* to improve traffic safety, reduce traffic problems, and protect buildings and infrastructure from being damaged. Besides, these restrictions are implemented to reduce nuisance caused by large trucks and hence to improve quality of life in cities. As a result, carriers have to use more small trucks to deliver the same overall volume as with larger trucks. Therefore, many vehicle restrictions result in a negative impact on the accessibility of cities, on the environment, and on the logistical costs. Local authorities use *low emission zones* (or environmental zones) to improve local air quality in cities by excluding pollutant trucks from entering city (centers), i.e. local pollutants, such as PM and NOx. Only trucks that fulfill engine requirements are allowed in the low emission zone. *Time-windows* restrict access to areas, usually the city center, during periods of the day, in which residents are not bothered and shopping public is not hindered.

The key action to coordinate urban freight transport is road pricing. Well-known road pricing examples are found in London (congestion charge) and Stockholm. Road pricing aims at all traffic participants and not only at urban freight transport. Whether road pricing changes the actual urban freight transport operations depends on the type of carrier, the price charged, and the actor that determines the delivery time.

Authorities can also play a role in facilitating urban freight transport. Initiatives focusing on the locations for (un)loading in the (dense) city centers are often taken. Most parking and unloading initiatives focus on the creation (e.g. using bus bays) or reservation of dedicated unloading areas. This is relatively easy and not expensive. Many of these initiatives are implemented in practice.

We see different ways to stimulate more sustainable urban freight transport. The EU provides research and demonstration subsidies. Another example, is the Urban Freight Transport Award, which is offered annually in the Netherlands by the Ambassador Urban Freight Transport.

3.2 Logistics

Shippers, carriers and receivers also initiate actions towards improving urban freight transport. The first type of action contains initiatives in which carriers cooperate (Cruijssen et al. 2007). The main aim of carrier cooperation initiatives, that require competitors to cooperate, is to improve urban freight deliveries' efficiency. This means that unit costs decrease and more products or services can be offered. As a side effect, city congestion and pollutant emissions are reduced. Other drivers for carrier cooperation in urban freight transport are improved customer service, overcoming legal and regulatory barriers and accessing new technology. Carriers can cooperate in different ways, e.g. by consolidating goods at one's premises or by using a neutral carrier (for example by shared participation) in order to prevent two half-filled vehicles to visit the same street within a limited time span.

Another way is to use environmentally friendly transportation modes. Some applications of intermodal transport exist, such as waterborne or rail concepts in city distribution. Intermodal urban freight transport is only feasible in specific circumstances and for a marginal part of the total urban freight transport volume. Hence, this is not a generic sustainable solution for urban freight transport issues.

System initiatives are often considered in urban freight transport: the development and use of urban consolidation centers (see for example Taniguchi and Thompson 2002). The rationale for city distribution centers (CDC) is to divide the freight transport in two parts: the part inside the city and the part outside the city. City Logistics models have been proposed in the academic literature. A number of demand models have been proposed for evaluating the demand for freight movements within urban areas (see Gentile and Vigo 2007 for a review). The decisions with respect to planning of CDCs in terms of business models and other characteristics have not been extensively studied (Quak and Tavasszy 2011).

3.3 Technology

Technological initiatives can be pushed or stimulated by (vehicle) manufacturers or authorities, but also by shippers or carriers. We make a distinction between vehicle-technology innovations and initiatives related to ICT and ITS applications in urban freight transport. Most technological vehicle innovations reduce some of the nuisance caused by vehicles, such as noise, emissions and even safety. Electric trucks, hybrid propulsion are CNG trucks are experimented with.

There exist several initiatives in which (usually at a limited scale at this moment) electric vehicles are used for urban freight transport. The use of these vehicles often requires also changes in logistics, more in particular in positioning a decoupling point close to the city. This is required because of the limited radius of action of the electric vehicles (see also Quak and van Duin 2010). Although electric transport is often mentioned as a solution for urban freight transport (silent vehicles and no local emissions), there are hardly (large scale) experiments with this type of urban freight transport using trucks, since reliable vehicles only recently have become available. One of the few pilot projects that were fully evaluated is published in (Leonardi et al. 2011).

Many (city logistics) academics belong to the VRP-field. Vehicle routing improvements are an example of technological solutions based on IT. Typical city problems, i.e. congestion, render classical vehicle planning routines infeasible. Routing improvement initiatives better incorporate real-life problems and thereby reduce the number of vehicle kilometers travelled and the penalty costs. Therefore, these initiatives show positive results on carriers costs and on the environment.

4 Lessons Learned

As the number of related industry-based projects and initiatives is increasing, there is a necessity to carry out sound scientific research in this area which addresses the actual base problem faced by those involved in these activities. This necessity stems from the fact and the need that scientific research can easily be disseminated between interested parties. In this section, we draw some overall conclusions, based on the various elements presented.

4.1 Policy

Governmental policies often result in a deterioration of the carriers' logistical performance. Many policy actions strongly focus on regulations and sanctions to force carriers to cooperate. Regulations, like vehicle restrictions and time-window access, result in a negative impact on the accessibility of cities, on the environment, and on the logistical costs. For road pricing, parking and unloading, and some dedicated infrastructure initiatives, carriers and local authorities share the same problem perception, i.e. congestion versus reduced city accessibility. If this is the case the results of the initiatives appear to be more positive, indicating that enforcement is not the only way to success.

Policy initiatives show that the knowledge levels of government of logistical operations is usually limited. On the other hand, carriers also lack knowledge about the sustainability objective of cities. The lack of interaction between local authorities and carriers prevents an increase in understanding of each other's issues. At this moment there is little cooperation -or harmonization between the various local authorities, which makes urban freight transport regulations quite disordered. A complicating factor within a local authority is that urban freight transport responsibility is often spread across several departments. In some initiatives the local authorities also play a more private role, e.g. they offer or finance transport services from the city distribution center (CDC) or manage the CDC.

Although city distribution centers seem very appealing (from a city-perspective), there are almost no examples running a successful business without financial support from governments. For example, of the approximately 200 planned or realized city consolidation center schemes in the nineties in Germany at most five are actually operating in 2005 (Browne et al. 2005). In many studies, the carrier participation is estimated higher than it turned out to be in practice. This implies less bundling of goods and fewer scale advantages than planned for the participating carriers. From the different consolidation center initiatives we learn that many carriers consider supporting policy measures as a way for local authorities to keep the non-viable center alive, instead of the consolidation center as a way to deal with the (usually restricting) policy measures, such as time-windows. Currently, subsidies are necessary to operate urban consolidation centers. So there should be sufficient societal gains (e.g. less pollution) to justify for the subsidies. These gains are not always clear. Overall, consolidation centers seem to be most feasible, if feasible at all, for historical cities that suffer from restrictive and inhibitory conditions for urban freight transportation anyhow, next to potential governmental restrictions (Visser et al. 1999).

4.2 Logistics

Carrier cooperation initiatives mainly focus on improvements in the logistical operations and have only an economic incentive rather than a sustainability incentive. Based on the limited number of implemented initiatives in practice, a real incentive for carriers appears to be lacking. This type of initiative asks for alignment between the logistical operations of more than one carrier. In Kawamura (2006) it is argued that this type of cooperation, due to the costs for reaching an agreement, can be quite costly from a business perspective, which is the opposite of an incentive. A survey (Cruijssen et al. 2007a) reveals that one of the main impediments to cooperation of LSPs is the size of their company. The market of LSPs is very fragmented which makes cooperation only practical for the larger firms. To make it more likely for carrier cooperation initiatives to be successfully implemented there should be a clear problem and a sound business case. Further potential success factors are: making sure a company does not lose its identity, include social costs in the initiative, support from the public sector, make all gains clearly visible, distribute them fairly and be as transparent as possible, and appeal to an organization's social and environmental reputation (Cruijssen 2007b).

To actually improve urban freight transport, carrier co-operation is expected to increase in the near future (Capgemini 2016). These urban (or regional) bundling activities could, but not necessarily should, be supported by some (innovative) type of city distribution centers (CDCs). The cooperation and exact use of a CDC depends on the efficiency of the logistic processes, the characteristics of the city, the regulations, and the goods and involved stakeholders.

The number of retailer initiatives documented in the reviewed literature is limited. Retailers and local shops attempt to attract consumers to buy goods in the most convenient way. For consumers, sustainability is a growing concern in their selection process, though they are less willing to pay extra. This challenges the retailers to continuously improve their supply chain, without increasing the overall costs. For the involved environmental pressure on the retailer, we have to consider the replenishment transports. In general the execution of those transports is outsourced to logistic service providers (LSPs). LSPs have the lead in combining different retailers as much as possible in one truck, in as far as this is acceptable for the retailer. In general, big beverage suppliers accept this less. Here, the qualitative aspects of competition and not losing the identity seem to be more important than the quantitative aspects of sharing benefits. This also holds for food retailers, but here the volumes are in general quite large, which also makes it less necessary to combine.

Very interesting from a retailer perspective is whether replenishment and delivery strategies of their products can take sustainability into account. Research in this area is focused on combining replenishment with transportation planning and is called inventory routing. Retailers nowadays in general have fixed delivery patterns and routing schemes for a certain period, rather than applying dynamic delivery. Also for replenishment days there is hardly any alignment currently between stores with the same specific brands or products to make a joined delivery possible at all. Hence the challenge for the retailers and urban shops is to optimize their logistics, given the context of local government rules and policies, and the capabilities of logistic service providers. This means, optimizing their drop size, delivery sequence and delivery time, while improving sustainability by collaborative planning.

4.3 *Technology*

There are different initiatives that are based on ICT and ITS developments, also outside the urban freight transport area. The main idea of vehicle routing improvement initiatives is to better incorporate real-life problems and, by using the newest optimization technology, to reduce the number of kilometers travelled and the related emission. The initiatives in this direction are implemented by commercial VRP-software programs (like route planning and advanced planning systems) and show positive results on carriers' costs and on the environment. Some of these initiatives do not attempt to improve the planning, but use real time information to find the best possible solution after an event or congestion actually occurs. This is a crucial element to improve the service (meeting service time agreements). ICT also plays an important role to have a transparent and open overview of the transportation, when multiple partners are involved.

The technological vehicle innovation initiatives that were found (in academic literature) show positive environmental results. The advantage is that this type of initiatives usually does not require serious changes in urban freight operations; however the investment costs are generally high for carriers. Therefore, there could be an active role for governments to stimulate (or at least facilitate) investments by subsidies or other advantages for clean trucks by means of licensing and regulation initiatives. The use of these vehicles often requires also changes in the logistics, such as a decoupling point close to the city because of the limited radius of action of the electric vehicles.

Finally, most technical solutions require investments, usually from carriers. Carriers are more likely to invest if in return they are granted some advantages themselves, such as longer time-windows, the allowance to enter a low emission zone, etc. In other words, policy solutions could stimulate or facilitate the actual implementation of technical solutions in practice.

5 A Vision for Urban Freight Transportation

The previous sections show the need for a systemic view on efficient urban freight transportation systems, resulting in a sustainable distribution network. Significant gains can be achieved through better coordination and consolidation of the urban freight distribution resulting in fewer vehicles in cities, better utilization of these vehicles and less emissions. Coordination reflects the fact that shippers, LSPs and retailers consider each other's activities when planning their own. Consolidation leads to combining different loads and carriers in the same vehicles. In order to achieve these goals, new organizational urban freight models based upon policy, logistics and technology considerations are necessary.

Clearly, the development of a global solution to urban freight transport needs to incorporate promising new concepts and solutions (covering all three aspects), while considering the characteristics of cities, also leading to a powerful definition of the role of governmental regulation. In the public sector, just as in the private sector, accurate modeling of operations and logistics functions is a necessary precondition to effective operational planning and control for society as a whole. Since policy measures have a fundamental impact on the costs of logistics activities of firms, the private sector will be affected as noted above. While the public sector needs to take important decisions to improve the air quality, like with 'green deals' or zero-emission agreements, we argue that the same tools that have brought so many gains to the private sector are equally critical in the public sector to analyze the impact. This is perhaps nowhere more evident than in the area of transportation policies where many local authorities fail to propose extensive and well-motivated freight transport policies (Allen et al. 2000). Besides, most urban freight transport regulations or initiatives are not evaluated on their effects in practice.

Most solutions for LSPs in urban freight transportation are in cooperation or collaboration. Important success criteria are hard constraints like revenue or profit improvement and a fair sharing model of the costs. However, as reflected in the lessons learned, also soft criteria like not losing the identity and finding an eligible partner are crucial success factors. But also business models for setting up possible cooperations by introducing City Distribution Centers or efficient ways of cross-docking or exchanging resources are crucial for success. Moreover, LSPs will share transportation requests if and only if this leads to improved solutions from their point of view, without losing their identity. The operator of the City Distribution Center and/or the last mile delivery should not have any competition aspects with the (national) carriers they like to attract. If this operator is a white-label or local specialist, or when there is a neutral operator in between, the success rate will improve considerably. It is generally accepted that City Distribution Centers will play a major role in the future to support efficient urban freight transportation (Quak and Tavasszy 2011). They are crucial for bundling the last mile, in general supported by smaller electric vehicles for which in turn larger time windows are appropriate. But they can also create local storage for managing the inventory.

The last mile of the retail supply chain up to the shelf represents both the highest supply chain cost and the biggest citizen service risk. Retailers are often located relatively close to each other, e.g. in the same shopping area. In almost all cases, retailers act independently of each other. Specifically, it would be interesting from a transport consolidation point of view, to coordinate the inventory ordering process among different retailers. Delivery schedules need to be adapted among different retailers employing different inventory policies to increase the delivery efficiency

Our discussed vision is in line with the Future Supply Chain Capgemini and Future Supply Chain 2016 report (Capgemini 2016) which argues that the future supply chain architecture should anticipate new collaborative models for city distribution that need to be applied in urban infrastructures. The key element in their concept lies in merging different streams into one infrastructure via the implementation of city hubs, with a collaborative cross-dock operation. The final solution will differ per shipment category, implying consolidation of different delivery streams (e.g. all for the same shopper) via city hubs. For this, retailers have to consider their replenishment strategies, jointly maximizing the resource utilization, leading to efficient deliveries into the urban stores.

From a technology perspective, the highest focus is on electric and hybrid vehicles. For example the ambition level of Rotterdam is to have zero emission freight transportation in the inner city in 2020. This plan is strongly based on the use of electric vehicles. Various successful initiatives have started recently, where the electric technology has become more mature and the cost level has become commercially attractive.

Finally, we view a good cooperation between academia and industry bing essential. For the qualitative and quantitative aspects of collaboration between LSPs more fundamental research is required, e.g. by investigating both market and literature. Also a solid business case model to show cost and benefits for all stakeholders (public and private) is important for creating a sustainable solution. For a successful implementation, and related validation and valorization, a strong link with the appropriate stakeholders (local governments, carriers and retailers) is necessary.

References

Allen J, Anderson S, Browne M, Jones P (2000) A framework for considering policies to encourage sustainable urban freight traffic and good/service flows. Transport Studies Group, University of Westminster, London

Browne M, Allen J (1999) The impact of sustainability policies on urban freight transport and logistics systems. In: Meermans H, Van de Voorde E, Winkelmans W (eds) 8th World conference on transport research (WCTR), vol 1., Transport Modes and SystemsElsevier, Antwerp, pp 505–518

Browne M, Sweet M, Woodburn A, Allen J (2005) Urban freight consolidation centres—final report. Transport Studies Group, University of Westminster, London

Capgemini, Future Supply Chain 2016 (2011). Global Commerce Initiative, Report

Cruijssen F, Cools M, Dullaert W (2007a) Horizontal cooperation in logistics: opportunities and impediments. Transp Res Part E 43:129–142

Cruijssen F, Dullaert W, Fleuren H (2007b) Horizontal cooperation in transport and logistics: a literature review. Transp J 46:22–39

Dablanc L (2007) Goods transport in large European cities: difficult to organize, difficult to modernize. Transp Res Part A 41(3):280–285

Gentile G, Vigo D (2007) Movement generation and trip distribution for freight demand modeling applied to city logistics. Technical Report, Unversita di Bologna

Kawamura K, Lu Y (2006) Effectiveness and feasibility of innovative freight strategies for the U.S. urban areas. In: Taniguchi E, Thompson RG (eds) Recent advances in city logistics: proceedings of the 4th international conference on city logistics, Elsevier, Amsterdam, pp 269–282

Leonardi J, Browne M, Allen J (2011) Before-after assessment of a logistics trial with clean urban freight vehicles: a case study in London. In: Taniguchi E, Thompson RG (eds) City logistics VII, City Logistics proceedings, Kyoto, pp 158–170

OECD (2003) Delivering the goods: 21st century challenges to urban goods transport. Organisation for Economic Co-Operation and Development, OECD Publishing

Ogden KW (1992) Urban goods movement: a guide to policy and planning. Ashgate, Aldershot

Quak HJ (2008) PhD dissertation, Sustainability of Urban Freight Transport: Retail Diribution and Local Regulations in Cities. Rotterdam Erasmus University

Quak HJ, Tavasszy LA (2011) Customized solutions for sustainable city logistics; The viability of urban freight consolidation centres. In: van Nunen J, Rietveld P, Huijbregts P (eds) Transitions towards sustainable mobility, Springer, Berlin

Quak HJ, van Duin R (2010) The influence of road pricing on physical distribution in urban areas. Procedia Soc Behav Sci 2(3):6141–6153

Taniguchi E, Thompson RG (2002) Modeling city logistics. Transportation Res Record 1790: 45–51

Visser J, van Binsbergen A, Nemoto T (1999) Urban freight transport policy and planning. In: Taniguchi E, Thompson RG (eds) City Logistic. I, 1st International Conference City Logistics. Institute of Systems Science Research, Kyoto, pp 39–69

The Role of Fairness in Governing Supply Chain Collaborations—A Case-Study in the Dutch Floriculture Industry

G. Robbert Janssen, Ard-Pieter de Man and Hans J. Quak

Abstract Supply chain collaboration, in which two or more autonomous firms work together to plan and execute supply chain operations, is becoming ever more important to remain competitive in business. Yet, through collaboration concerns arise about whether the benefits and risks of collaboration are split in an acceptable and fair manner. This research illustrates the role of fairness (organizational justice theory) in creating and appropriating value from supply chain collaborations. We therefore analyze an extensive case study in the Dutch floricultural industry, in which six companies enter a supply chain collaboration. We conclude that fairness considerations are very important for explaining the outcomes of supply chain collaborations. Asymmetries in perceived value appropriation can be offset if the collaboration is deemed fair on distributive, procedural, interpersonal and informational justice dimensions. Firms may improve the success rate of supply chain collaborations if the fairness perceived is considered to be adequate.

Keywords Supply chain collaboration · Organizational justice theory · Value appropriation · Case study

1 Introduction

Firms are increasingly looking beyond their firm boundaries attempting to leverage their customers', partners' and suppliers' resources, capabilities and knowledge to improve supply chain effectiveness and efficiency through supply chain collaboration (Cao and Zhang 2011). This is no less the case in the Netherlands where we are witnessing new supply chain collaboration initiatives emerging constantly. For

G.R. Janssen (✉) · H.J. Quak
TNO Mobility and Logistics, Van Mourik Broekmanweg 6, 2628XE Delft, The Netherlands
e-mail: robbert.janssen@tno.nl

G.R. Janssen · A.-P. de Man
VU University Amsterdam, Boelelaan 1105, 1081HV Amsterdam, The Netherlands

H. Zijm et al. (eds.), *Logistics and Supply Chain Innovation*,
Lecture Notes in Logistics, DOI 10.1007/978-3-319-22288-2_9

instance, road transportation companies are joining their transportation resources in order to combine their routes and reduce empty-running (Cruijssen et al. 2007). Also, the Rotterdam-based deep-sea terminals are confronted with the need to compete for more cargo volumes to be transshipped and at the same time collaborate in order to keep rail services to the hinterland viable (Weissink 2013). There, too, is broad government support for promoting supply chain collaboration, for instance in the inland navigation industry (Rijkswaterstaat.nl 2014), and through improved coordination and control in global supply chains by means of control towers (Dinalog.nl 2014).

Yet, despite the popularity of supply chain collaboration, previous research has indicated that the failure rate among alliances is quite high and failure rates over 50 % have been reported (De Man 2005; Kale and Singh 2009). For this study, we are primarily concerned with a key aspect of collaboration that is often considered difficult: sharing the value created in the collaboration. In collaboration, partners seek to jointly create value but each of them may also have individual concerns about how to appropriate (a part of) that jointly created value (Lavie 2007). The need to balance value creation and value appropriation puts emphasis on the management of the collaborative relationship and the perception of fairness in the relationship. This paper investigates such fairness considerations and the ability of firms to create and capture value from supply chain collaborations. Our main research question, therefore, is formulated as: *How do fairness considerations govern the ability of firms to create and appropriate value from supply chain collaborations?*

We do so by using a case-study methodology in which we analyze the formation of supply chain collaboration among six firms in the floriculture industry, aimed at offering a joint delivery service to their customers. Our contribution lies in the fact that we provide evidence on the role of fairness and the ability of firms to create and capture benefits that accrue from supply chain collaboration. Also, the supply chain collaboration is shaped according to a structured process that may be easily replicated in other settings and business contexts.

2 Literature Review

In this section we examine the literature on how fairness and justice theory influence the extent to which firms can create and capture the value from supply chain collaboration.

2.1 Supply Chain Collaboration

Supply chain collaboration means that two or more autonomous firms work jointly to plan and execute supply chain operations (Simatupang and Sridharan 2002; Cao and Zhang 2011). The rationale behind these supply chain collaborations exists in the expectation that collaboration can bring a broad range of benefits, that can't be gained by a single firm on its own (Simatupang and Sridharan 2002) such as gaining access to complementary resources, creating economies of scale, higher flexibility, lower costs, lower emissions, higher service levels, improved information integration, lower inventories, and improved end-customer satisfaction (Min et al. 2005; Tjemkes et al. 2012).

The concept of supply chain collaboration is closely related to that of (strategic) alliances. Alliances are generally horizontally oriented; however we use the term supply chain collaboration here to refer to both vertical as well as horizontal collaboration. There exist many famous examples in the logistics and supply chain industry which could be termed supply chain collaboration or a strategic alliance. For example, the strategic alliance of Royal Dutch Airlines KLM and Northwest Airlines has served as a role model and blueprint for other airline alliances. It has endured multiple organizational restructurings, and currently consists of KLM, Delta Airlines (former Northwest), Air France, and Alitalia (KLM.com 2010).

Accordingly, supply chain collaboration is becoming more and more widespread in the logistics and supply chain industry (Nyaga et al. 2010). However, many questions remain, for instance to secure the benefits of supply chain collaboration (Dyer et al. 2008; Cao and Zhang 2011), that is, how to create and how to appropriate value.

2.2 Value Creation and Value Appropriation

There is wide evidence supporting the idea that collaboration is an important source of value creation (Nyaga et al. 2010). Yet, it is necessary to consider how collaborating firms split the value that is generated in the collaboration, as we do not know why some firms are able to extract more value than others (Dyer et al. 2008).

Then what is that value that can be created and appropriated? Tangible and monetary forms of value are such examples as pooled profits, revenues, and costs. Luo (2009) adds that collaboration partners often limit the contractual value sharing specifications to monetary value. Also, the operations research (OR) discipline has investigated various tangible value allocation mechanisms. For example, cooperative and non-cooperative game theory concepts are used for determining the tangible pay-off for collaborating partners (Frisk et al. 2010) such as the Shapley Value (Cruijssen et al. 2010). Non-monetary types of value can also be of importance to the collaborating partners. Examples of non-monetary value types are gaining access to new markets and the enhancement of their reputation (Luo 2009). Other examples of non-monetary value include the ability to exchange valuable resources, improve service levels, reach information integration, reduce inventories, and to bring in additional capacity.

Essentially, value creation and value appropriation are part of the complex interaction that exists between collaboration (joint value creation) and competition (value appropriation) (Tjemkes 2008). Value creation is the total value generated by a focal firm in collaboration with its partners (Lavie 2007). Value appropriation is defined as the relative amount of benefits (Lavie 2007), or the outcomes, benefits and gains (Jap 2001) that each partner is able to extract from collaboration. Simply put, it's that part of the total 'value pie' that a firm claims successfully.

The disparity between value creation and value appropriation is akin to the distinction between common and private benefits (Lavie 2007; Dyer et al. 2008). Joint value creation is a collaborative process leading to common benefits shared by all parties in the alliance. Value appropriation then determines the distribution of the common benefits to individual partners. Also, value appropriation is concerned with appropriation of private benefits from the collaboration that are not available to other partners. The extent to which partners can appropriate common and private benefits from the collaboration and maintain creating new value is dependent on perceptions of fairness of the supply chain collaboration.

2.3 Fairness and Justice Theory

Fairness, as part of organizational justice theory (Greenberg 1987), is a foundation for all types of economic and social exchanges and relationships (Liu et al. 2012). Research on justice theory follows from the observation that not only purely rational arguments impact the success of relationships between organizations, but also more sociological non-rational concepts. For instance, increased fairness has been shown to improve operational performance and (through higher inter-firm trust) can increase financial performance (Luo 2008). With roots in equity theory (Adams 1965) and social exchange theory (Homans 1958) fairness has traditionally been treated as a three-dimensional construct consisting of distributive, procedural, and interactional justice. More recently, Colquitt (2001) empirically established four distinct dimensions of justice that have been successfully applied in a supply chain context (Liu et al. 2012):

1. *Distributive justice*—concerns whether partners are satisfied that the expected value produced by the alliance is proportional to their contribution. E.g., are outcome allocations perceived as being fair?
2. *Procedural justice*—entails the fairness of the alliance's strategic decision making process and that the allocation of decision making rights establishes fair procedures to make future decisions that influence value creation and value appropriation. E.g., are the procedures and processes used to make allocations perceived as being fair?
3. *Interpersonal justice*—involves whether people are treated with respect and sensitivity during implementation of procedures. E.g., is interpersonal treatment bestowed on individuals during the implementation of procedures perceived as being fair?
4. *Informational justice*—comprises the kinds of information and the way parties share information. E.g., is the foundation for decisions explained adequately?

Many studies examine the role of fairness on collaboration. For instance a study on social exchange theory in supply chain relationships indicates that displays of procedural and distributive justice by a supplier enhance the long-term orientation and relational behavior of its distributor (Griffith et al. 2006). Also, Jap and Anderson (2007) show that informational exchange norms (i.e., informational justice) are important in the formation stage of a supply chain relationship. We postulate that fairness is important for all stages in the lifecycle of a supply chain collaboration initiative. Already in the pre-formation stage, ex-ante fairness perceptions are important as firms are judging future value appropriation potential, and how the created value will be split—a matter of distributive (in)justice. Also, in the post-formation stage, firms will experience interpersonal (in)justice and procedural (in)justice, which effect the long-term orientation of the collaboration.

Still, to the best of our knowledge there has been no evidence on the role of fairness on the ability of firms to appropriate value that accrue from supply chain collaboration, which is therefore the subject of this work.

2.4 Theoretical Model

Figure 1 shows the conceptual model for this study, integrating the insights obtained from the literature review. We use a causal loop diagram in order to

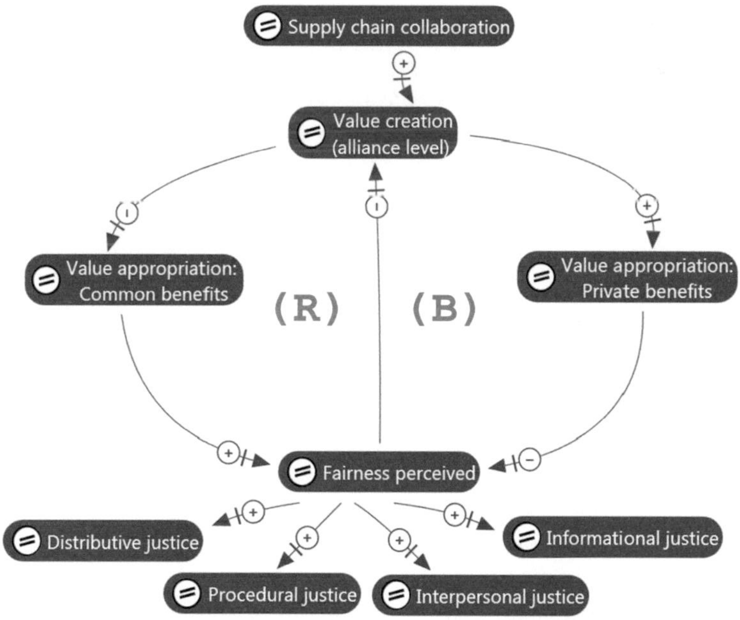

Fig. 1 Theoretical model

recognize the assumed non-recursiveness of the relationships among the variables (MARVEL 2014). For brevity of argument, we consider fairness as a second-order construct, aggregating perceptions of distributive justice, procedural justice, interpersonal justice, and informational justice.

The model reads as follows: supply chain collaboration creates value at the alliance level. Subsequently, a focal firm will try to capture common or private benefits from the collaborative relationship. In the case of common benefits that accrue to all collaborating partners, this leads to a higher level of fairness perceived among the partners. A higher level of fairness perceived finally increases the willingness to continue the cooperative relationship and allow further value creation thus establishing a *reinforcing* causal loop (indicated by the 'R' sign). Alternatively, if a firm focuses on appropriation of private benefits, this may leads to fairness to be negatively perceived by the partners. This can have a negative impact on further value creation, establishing a *balancing* loop (indicated by the 'B' sign). In the next section, we illustrate the relationships in this model by means of a case-study.

3 Case Study Methodology

Case research lends itself to study supply chain collaboration in actual practice (Voss et al. 2002; Quak and De Koster 2007). For our case study, we use the formation of a joint delivery service of six companies in the floriculture industry in the Netherlands as an illustration to our research problem. Such practices of setting up horizontal supply chain collaboration are rather common in the Dutch floriculture industry. Our case setting therefore constitutes a representative case (Yin 2009); a typical situation that we observe relatively often in the Netherlands in recent years. In order to allow others to replicate our results, and fully exploit the transfer potential of these results, we have included our research protocol in Appendix A: Research protocol (Yin 2009).

3.1 Case Study Design

Case study research is well suited to study supply chain collaboration in detail but may generate considerable problems in ensuring sufficient rigor and reliability. Table 1 briefly summarizes the case study design, and how validity and reliability concerns are addressed.

3.2 Case Description

The Dutch floriculture industry is world-famous and sometimes even referred to as the Wall Street of Flowers. This case study reports on a research project in which

Table 1 Case study design, addressing validity and reliability concerns

Case study design	Implementation in this study
Construct validity	
Use correct operational measures for the concepts studied	Construct operationalizations for fairness and value appropriation are taken from established literature
Establish chain of evidence	We employ a hands-on action research design (Akkermans et al. 2004) as one of the authors was consulting for the companies in the case study
Use multiple sources of evidence	During the case study, there were many plenary sessions in which the company's executives participated. Also six one-to-one interviews have been conducted in order to elicit the respondents' perceptions on sensitive subjects such as fairness, and value appropriation. Interviews lasted between 30 and 60 min. The interviews were digitally recorded and transcribed afterwards, leading to a total of over 50 pages of transcribed text
Use multiple sources of evidence	Throughout the project additional data (research reports, minutes, slide presentations, emails, etc.) have been collected and analyzed that were drafted during the research project—in total exceeding 100 pages
Allow respondents to review draft case-study reports	All participants were asked to review the interview reports and they generally appraised the quality of the transcribed interview reports
Internal validity	
Establish logical causal relationships	We used justice theory to guide our analysis of the case materials (Ariño and Ring 2010)
Using analytical techniques	Qualitative interview data was coded in several iterations (open, axial, selective coding). Also, tabular displays were used to assist in pattern matching
External validity	
Establishing the domain to which a study's findings can be generalized	We used a representative case (Yin 2009), easing the need to conduct a multiple case study, meaning replication logic does not apply. Still, findings are limited to industry-specific and culture-specific settings
Reliability	
Replicability and transfer potential is ensured by using consistent processes	The collaboration formation process used in this research is based on the step-based approach of Tjemkes et al. (2012). Also business model analysis was carried out using the business model canvas of Osterwalder et al. (2010)

Table 2 Case company characteristics

Company	Company type	SME	Type of product/service	No. of employees (FTE)
Alflora	Wholesaler	Yes	Accessories (pots, vases), packaging material	450
Basics&Trends	Wholesaler	Yes	Accessories (pots, vases), packaging material	50
FleuraMetz	Wholesaler	No	Cut-flowers	1100
GMBloemen	Wholesaler	Yes	Cut-flowers	14
Plantion	Flower and plants auction, cooperative	Yes	Cut-flower, tree nursery, plants	120
VianenvanVliet	Wholesaler	Yes	Cut-flowers	10

five wholesale firms and one flower and plants auction participated, that sought to explore the potential of a joint delivery service to their customers, i.e., setting up a supply chain collaboration (see Table 2). The wholesale firms are all but one small and medium-sized enterprises (SMEs), that operate by the cash-and-carry business model[1], and therefore did not provide any formalized delivery services beforehand.

Over time, some wholesalers had initiated ad hoc delivery services, but these were considered very inefficient and thought to yield high costs that could not be offset by the delivery fees charged. All companies realized that the cash-and-carry business model is not sustainable over the longer term, as customers are now demanding e-commerce ordering and included deliveries. Therefore, it was decided that the firms needed to innovate their business models by starting to move away from the basic cash-and-carry model towards an ordinary wholesale model including delivery services.

3.3 Stages in the Case-Study Project

Figure 2 shows four stages that can be distinguished in the formation of the supply chain collaboration. The main data collection moments are logged in italics.

1. *Pre-formation.* At the start of the study, the companies were very eager in setting up a joint delivery service. They were already clustered at the same

[1]Wholesalers' cash-and-carry business model: downstream customers (e.g., small retailers, florists, kiosks, etc.) visit either the physical store or online store, buy the products they need and collect the products at the physical store. The wholesale firm essentially provides break-bulk services, as it procures large volumes upstream, and repackages these into very small bundles, customized to the buyer.

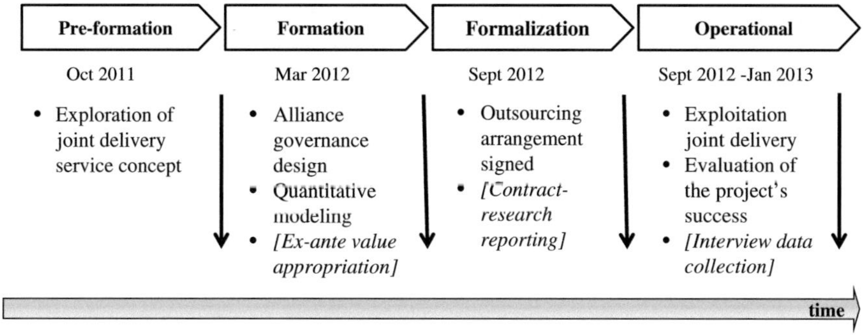

Fig. 2 Stages in formation of the supply chain collaboration

physical location, the Plantation marketplace in the Eastern part of the Netherlands and also shared the same parking lots, electricity, loading docks et cetera, however previous attempts in setting up a joint delivery service were unsuccessful so far.

2. *Formation.* Various plenary workshops were held in order to develop a thorough understanding of the problems that could emerge during design and implementation of the joint delivery service. Individual goals were clarified and the shared project goal was established. Quantitative modelling was employed to estimate ex-ante how the total value created in the collaboration could be shared among the partners.
3. *Formalization.* The partners needed to jointly decide upon the preferred method of organization. They could either (1) plan and control the delivery service themselves or (2) outsource the bundled delivery volumes to a logistics service provider, effectively leveraging the collaboration to obtain better rates. After an intensive process of evaluating the alternatives, they decided to outsource the combined volumes to a logistics service provider, on a pay-per-use basis.
4. *Operational.* In this stage, the supply chain collaboration became operational. All parties now offered formalized delivery services to their customers. No actual contracts were signed between the project participants, except an overarching contract with the logistics service provider to formalize the procedure.

3.4 Ex-ante Value Creation and Appropriation

In stage two, we used vehicle routing software to quantitatively estimate the costs reductions from moving from individual distribution processes to the joint delivery service. We estimated multiple scenarios with variations in transport capacity, spatial demand change patterns and low/high autonomous volume growth per partner. For brevity reasons, we only present outcomes for the high volume scenario. Within the high volume scenario, 539 carts of flowers and plants were

Scenario	Kms driven (weekly)	Hours (weekly)	Total Costs (weekly)
No collaboration (539 carts)	18,252	312	€ 18,471
With collaboration (539 carts)	12,278	213	€ 12,529
Savings from collaboration	**5,974**	**99**	**€ 5,942**

Fig. 3 Main scenarios: key metrics with and without supply chain collaboration

transported in the representative week chosen. Figure 3 shows the total costs, number of kilometers and hours driven for the collaboration and non-collaboration models. Compared to the baseline scenario of no collaboration, the modeling work yielded total annual cost savings of 297k €, or approximately 32 % per cart delivered. Put differently, the monetary value created through supply chain collaboration was estimated to be about €5942 per week.

Subsequently, we employed a simple allocation mechanism for determining the division of cost savings among the partners; the value they could appropriate based on the ex-ante calculations. Figure 4 shows the amount of monetary value appropriated per partner. Essentially, the operational surplus in the collaboration scenario was allocated to the partners based on their contribution as an equally weighted function of the number of carts transported and number of kilometers and hours driven. Basically, this allocation of value ensures that partners who contribute more also benefit more from the supply chain collaboration. It shows that in the

		No Collaboration	Collaboration		
Partner	Carts	Total Cost (no collaboration)	Weighted contribution (kms, hours, carts)	Savings appropriated	Total Cost (with collaboration)
Alflora	53	€ 2,720	13%	€ 778	€ 1,942
Basics&Trends	29	€ 2,231	10%	€ 585	€ 1,646
FleuraMetz	23	€ 1,587	7%	€ 425	€ 1,162
GM Bloemen	99	€ 2,655	16%	€ 933	€ 1,721
VianenvanVliet	105	€ 5,328	26%	€ 1,528	€ 3,800
Plantion	230	€ 3,951	28%	€ 1,693	€ 2,258
Total	**539**	**€ 18,471**	**100%**	**€ 5,942**	**€ 12,529**

Fig. 4 Ex-ante value appropriation based on weighted contribution of partners

collaboration model, FleuraMetz, for instance, would gain only 7 % of the total operational surplus. Moreover, it indicates that especially Plantion (28 %) and VianenvanVliet (26 %) would gain significantly more from collaboration than the other partners, as they contribute more towards the collaboration. During the project, the partners also received these numbers—albeit with less detail for reasons of confidentiality—in order to assess whether they would want to renegotiate alternative splits. Little discussion on this topic ensued as the partners were primarily interested in creating the value first rather than addressing appropriation concerns at that early point in time of the formation process.

3.5 *Ex-post Value Creation and Appropriation*

Note, in stage four, we posed interview questions relating to the actual quantitative ex-post results of the supply chain collaboration. Unfortunately, due to the timing of the interviews and lack of monitoring systems in place, there was no such information available about ex-post savings and allocation of those benefits over the partners. We were also unable to obtain this information months later, as one of the partners filed for bankruptcy. The remaining supply chain collaboration was structurally so different that any comparisons between ex-ante modelling and ex-post results were not meaningful.

4 Discussion of the Results

From the previous section, it appears that ex-ante estimations of total monetary value creation were significant. This section, therefore, investigates how fairness materialized in the case-study and what other monetary and non-monetary values were created and appropriated in the collaboration.

4.1 *Distributive Justice*

The partners experienced strong differences in input/output ratio from the collaboration benefits, meaning their contributions were not always in line with the value appropriated. VianenvanVliet noted that benefits for their own firm were insufficient and they were unsatisfied so far, but this was partly due to the fact that performance monitoring and measurement just started up: “No, these [cost savings] are not clear for me yet. I have the feeling we are saving costs, but we have to measure for a longer period of time”. This indicated that firms judge the level of distributive fairness as a longitudinal concept, allowing the desired level of justice to develop over time.

At the same time, partners perceived their own benefits to compare unfavorably against the others' parties benefits. This implies that there is a real distributive justice issue here: even though the firms acknowledge that real measurement of cost savings is limited, they already implied that they are being deprived from benefits by their partners. Consequently, they easily assigned blame and mentioned free-riding as a common concern: "I sense we are saving costs, but often I got the idea that other partners didn't do anything to deserve those benefits. [...] especially two partners did not do anything while they were enjoying the new service as well." Still, the inertia displayed by some partners did not influence the perceptions on an interpersonal level—interpersonal respect remained high and as such *interpersonal justice* was not a concern from a *distributive perspective* (Griffith et al. 2006).

Also, the ex-ante value appropriation calculations did not significantly influence distributive justice levels. GMBloemen, the partner that gained disproportionally little from collaboration, noted that an asymmetrical allocation of value was to be expected in any case. "We still want to pursue setting up the collaboration to profit from lower transport rates and reduce our carbon emissions". Here, the argument of jointly expanding the 'value pie' dominates the 'pie splitting' argument (Jap 2001), effectively limiting distributive justice concerns.

4.2 Procedural Justice

All partners explained they were well able to voice their concerns and had their feedback incorporated in the decision-making process. This was very satisfactory for all involved and entails virtuous consideration of *procedural justice* issues. The partners agreed that their counterparts were well aware of the stakes and challenges everybody faced, which helped the process of collaboration formation significantly: "[...] indeed in the beginning of the project, we took care to address everybody's issues and now we know quite good what these are." Further, in terms of procedural fairness, the chairman of the project team was praised for setting up clear decision-making procedures. These were developed in good harmony between all partners, for the operational stage, an indication of *procedural and informational fairness*. We did not find very explicit asymmetries in perceptions of procedural justice, as suggested by Luo (2005) in this collaboration, but the perceptions were not fully symmetrical either. As one partner noted that all issues surrounding ethicality and morality were adequately treated, however "nobody was delaying the progress, but some were not helping it either." Yet in fact, these issues regarding differentiated procedural justice perceptions may have been mitigated by the presence of a neutral third party, see Klein Woolthuis et al. (2014) who note that fairness may also be created by mediation and interference by a third party. Plantion noted: "[the research firm] imposed professionalism on us. They forced us to prepare each meeting, and guided us through the decision-making process".

4.3 Interpersonal Justice

The partners had already very good working relationships with each other at the start. All partners unanimously agreed that their working relationships were very positive and they respected each other highly; a display of *interpersonal justice*. Even competitive tendencies between the cut-flower wholesalers were suppressed in order for the joint delivery service to become a reality.

Another indication of high *interpersonal fairness*: partners spend a lot of time together during working hours, in an informal but polite and professional manner. The executives trusted each other and followed up upon mutual request. They jointly developed mutual interpersonal justice, leading to continuous relationship development (Liu et al. 2012).

4.4 Informational Justice

In the formation stage, two partners were concurrently setting up a bi-lateral collaboration. Partner Alflora noted that this did not help the larger collaboration to take shape: "I didn't think it was very beneficial for the project that Basics&Trends and VianenvanVliet were setting up a joint initiative in the background, without us knowing it". That information was largely kept confidential, until the other parties found out at the end—something that could be considered *informational unfairness*. On the other hand, Plantion noted that "[…] the fact that that bi-lateral collaboration was initiated, made sure everyone else got a wake-up call", suggesting that perceptions of informational fairness can be incongruent, even with respect to the same incident. Still, this argument is consistent with previous research by Luo (2007), as performance improvements are expected if informational fairness expands.

Likewise, informational fairness was in danger as one partner observed that information sharing was not always timely and complete: "I wouldn't say they sabotaged the progress, but we had to work really hard to convince them of the benefits this could bring. And even then, they were slacking in delivering the necessary data". On the other hand, the partners contended that all partners shared the required sensitive customer data quite easily. The partners surprised themselves in how open they were towards one another, even while collaborating with direct competitors, suggesting their excellent relations (i.e., interpersonal justice) to account for that. All in all, informational fairness seemed to have been addressed adequately.

4.5 Common and Private Benefit Appropriation

The partners explicated how supply chain collaboration could create value for them at the start of the project such as achieving lower total costs, higher flexibility, opportunities for learning from each other, sustainability improvements, and enhanced customer service. Moreover, our ex-ante modelling showed that the supply chain collaboration had the potential for significant common benefits, and all parties could appropriate a favorable if not equal share of the value. However, as mentioned, actual measureable results were quite limited at the time of data collection and the executives could not clearly indicate the realized ex-post savings so far. The partners indicated to have "a feeling of cost reduction", "hopefully a fixed asset cost reduction" and noted that the "awareness for potential of joint delivery" has increased; examples of common benefits appropriated.

Moving beyond common benefits into private benefit appropriation terrain was hardly even considered at this stage by the executives. One cut-flower wholesaler explained that they were now able to always offer delivery services because of the "virtually unlimited available capacity" available at the logistics service provider.

This could either be a timing issue (e.g., data collection took place too early in the collaboration's maturation process), or it indicates that private benefit appropriation matters are not important if fairness concerns are adequately addressed. That is, private benefit appropriation concerns are mitigated if distributive, procedural, interpersonal, or informational fairness increases. Indeed, the partners did not consider unilaterally extracting private benefits but rather still focused on gaining common benefits (Dyer et al. 2008). Since the future value creation potential of the collaboration was considered very high, considerations of private benefit appropriation were attenuated once more. That is, expanding the 'value pie' on its own already improves perceptions of fairness, which leads to further additional value creation.

5 Conclusions and Future Research

Building on recent literature and an empirical case in the Dutch floriculture industry, we have illustrated a conceptual model that explains how fairness governs whether firms in a supply chain collaboration appropriate common and private benefits. We conclude that fairness considerations are very important for explaining the outcomes of supply chain collaborations. Firms may improve the success rate of supply chain collaboration, if the fairness perceived on the four dimensions is considered adequate. Also, at the same time, expectations about future value creation mitigate concerns about private benefits appropriation as these issues subside by virtue of increases in the total value pie. This leads us to the most important direction for future research indicated by this study: does growth of the total value pie directly improves the fairness perception, or is the causal relation displayed by the inverse link?

Future research should seek to validate our conceptual model, by explicitly collecting data on both ex-ante and ex-post value created and division of benefits over the partners in the context of organizational justice. The role of fairness can then be unpacked, examining whether fairness perceptions may be different if the total value pie increases or whether fairness perceptions change dynamically over time under the influence of common and private benefits appropriation. In particular, future research should investigate whether explicitly improving fairness perceptions can mitigate concerns about common and private benefits appropriation, especially if significant asymmetries exist in the amount of value appropriated. Furthermore, to circumvent limitations from this single case-study design, future research should try to replicate the findings here in other conditions and industries, or by means of large-scale quantitative or experimental research. Indeed, we feel the concept of value appropriation is a very interesting subject that warrants further investigation in the context of supply chain collaboration.

Appendix A: Research Protocol

The following protocol guided the research data collection phases. It consists of four phases. It may be used for replication purposes (Yin 2009).

1. Initial discussions and blue-print development

   We collected the following data:

- Governance structure
- Delivery strategy, current structure
- Requirements to transportation (loading units, temperature)
- Products, markets, segments
- Information processing capabilities (IT solutions, connectivity, web-based access)
- Future strategy and goals, objective, mission, wishes

2. Transaction data collection

This step consisted of various site visits and discussions with logistics managers and IT-staff to gather quantitative transaction (sales) data for the quantitative estimation of the synergetic potential of a joint distribution service. The focus was on:

- Current distribution strategy
- Current cost structure and delivery tariffs used
- Customer addresses, volumes, delivery dates, times
- Transportation infrastructure (assets, vehicles)
- Financial metrics (cost of capital, depreciation period, etc.)

3. Semi-structured interviews about fairness and value appropriation

After the collaboration become operational, we investigated how firms experienced fairness and value considerations. We used semi-structured interviews to pose questions regarding:

- Justice dimensions
- Alliance formation
- Relationship satisfaction

4. Data completion

In this final step, we ensured data to be available in full. We resolved any case of indistinctness, or if additional information was needed, we contacted the parties by telephone or email.

References

Adams J (1965) Inequity in social exchange. In: Berkowitz L (ed) Advances in experimental social psychology. Academic Press, New York, pp 267–299

Akkermans H, Bogerd P, van Doremalen J (2004) Travail, transparency and trust: A case study of computer-supported collaborative supply chain planning in high-tech electronics. Eur J Oper Res 153(2):445–456. doi:10.1016/S0377-2217(03)00164-4

Ariño A, Ring PS (2010) The role of fairness in alliance formation. Strateg Manag J 31:1054–1087

Cao M, Zhang Q (2011) Supply chain collaboration: Impact on collaborative advantage and firm performance. J Oper Manag 29:163–180

Colquitt JA (2001) On the dimensionality of organizational justice: a construct validation of a measure. J Appl Psychol 86:386–400

Cruijssen F, Bräysy O, Dullaert W (2007) Joint route planning under varying market conditions. Int J Phys Distrib Logist Manag 37:287–304

Cruijssen F, Borm P, Fleuren H, Hamers H (2010) Supplier-initiated outsourcing: a methodology to exploit synergy in transportation. Eur J Oper Res 207:763–774

De Man A-P (2005) Alliance capability: a comparison of the alliance strength of European and american companies. Eur Manag J 23:315–323

Dinalog.nl (2014) 4C: cross chain control centers. http://www.dinalog.nl/en/themes/4c__cross_chain_control_center/

Dyer JH, Singh H, Kale P (2008) Splitting the pie: rent distribution in alliances and networks. Manag Decis Econ 29:137–148

Frisk M, Göthe-Lundgren M, Jörnsten K, Rönnqvist M (2010) Cost allocation in collaborative forest transportation. Eur J Oper Res 205:448–458

Greenberg J (1987) A taxonomy of organizational justice theories. Acad Manag Rev 12:9–22

Griffith D, Harvey M, Lusch RF (2006) Social exchange in supply chain relationships: the resulting benefits of procedural and distributive justice. J Oper Manag 24:85–98

Homans GC (1958) Social behavior as exchange. Am J Sociol 63:597–606

Jap SD (2001) "Pie sharing" in complex collaboration contexts. J Mark Res 38:86–99

Jap SD, Anderson E (2007) Testing a life-cycle theory of cooperative interorganizational relationships: Movement across stages and performance. Manage Sci 53:260–275

Kale P, Singh H (2009) Managing strategic alliances: what do we know now, and where do we go from here? Acad Manag Perspect 23:45–62

Klein Woolthuis R, Nooteboom B, de Jong G (2014) Roles of third parties in trust repair: an empirical test in high tech industries. In: Wick A, Harris J, Moriarty B (eds) Public trust in business. Cambridge, Cambridge University Press, pp 290–325

KLM.com (2010) http://nieuws.klm.com/alitalia-joins-air-france-klm-group-delta-air-lines-in-industry-a-s-leading-trans-atlantic-joint-venture/

Lavie D (2007) Alliance portfolios and firm performance: a study of value creation and appropriation in the U.S. software industry. Strateg Manag J 28:1187–1212

Liu Y, Huang Y, Luo Y, Zhao Y (2012) How does justice matter in achieving buyer–supplier relationship performance? J Oper Manag 30:355–367
Luo Y (2005) How important are shared perceptions of procedural justice in cooperative alliances? Acad Manag J 48:695–709
Luo Y (2007) The independent and interactive roles of procedural, distributive, and interactional justice in strategic alliances. Acad Manag J 50:644–664
Luo Y (2008) Procedural fairness and interfirm cooperation in strategic alliances. Strateg Manag J 29:27–46
Luo Y (2009) From gain-sharing to gain-generation: the quest for distributive justice in international joint ventures. J Int Manag 15:343–356
MARVEL (2014) Principles of a method for semi-qualitative system behaviour and policy analysis. In: Proceedings of the international conference on system dynamics. System Dynamics Society, pp 1–21
Min S, Roath AS, Daugherty PJ (2005) Supply chain collaboration: what's happening? Int J Logist Manag 16:237–256
Nyaga GN, Whipple JM, Lynch DF (2010) Examining supply chain relationships: do buyer and supplier perspectives on collaborative relationships differ? J Oper Manag 28:101–114
Osterwalder A, Pigneur Y (2010) Business model generation, pp 1–288
Quak H, De Koster R (2007) Exploring retailers' sensitivity to local sustainability policies. J Oper Manag 25:1103–1122. doi:10.1016/j.jom.2007.01.020
Rijkswaterstaat.nl (2014) Impuls Dynamisch Verkeersmanagement Vaarwegen. http://www.rijkswaterstaat.nl/zakelijk/verkeersmanagement/idvv/
Simatupang TM, Sridharan R (2002) The collaborative supply chain. Int J Logist Manag 13:15–30
Tjemkes B (2008) Growing and sharing the pie a study of performance in strategic alliances, pp 1–295
Tjemkes B, Vos P, Burgers K (2012) Strategic alliance management, p 368
Voss C, Tsikriktsis N, Frohlich MT (2002) Case research in operations management. Int J Oper Prod Manag 22:195–219
Weissink A (2013) Terminals Rotterdam: vechten om vracht, en toch samenwerking. Financ Dagbl 2
Yin RK (2009) Case study research: design and methods, 4th edn. pp 1–104

Reducing the Environmental Impact of Urban Parcel Distribution

Theodoros Athanassopoulos, Kerstin Dobers and Uwe Clausen

Abstract In the framework of rising requirements concerning the environmental impact of urban parcel delivery, UPS has tested various approaches to reduce emissions with steady or even enhanced operational efficiency. Three different alternatives to classic diesel drive technology were tested in the city of Hamburg, i.e. parcel vehicles with electric drive, and e-assisted cargo tricycles either solely or in combination with a parcel storage container. The corresponding costs, greenhouse gas, SO_2, NO_x, CO and particulate matter emissions have been assessed and discussed. The analyses show that none of the isolated pilot tests result in overall reduced costs and emissions; however they need to be evaluated thoroughly. Clear advantages of the alternatives are outlined and their embedding in the overall UPS urban network in combination with an increased purchase of renewable energy promise a future reduction of costs and the environmental impact of UPS' urban parcel distribution.

Keywords Carbon footprint · Distribution · Emission · Green · Logistics

T. Athanassopoulos (✉)
United Parcel Service Deutschland Inc. & Co OHG, Robert-Bosch-Str. 4,
60386 Frankfurt, Germany
e-mail: theodoros.athanassopoulos@europe.ups.com

K. Dobers · U. Clausen
Fraunhofer Institute for Material Flow and Logistics, Joseph-von-Fraunhofer-Str. 2-4,
44227 Dortmund, Germany
e-mail: kerstin.dobers@iml.fraunhofer.de

U. Clausen
e-mail: uwe.clausen@iml.fraunhofer.de

H. Zijm et al. (eds.), *Logistics and Supply Chain Innovation*,
Lecture Notes in Logistics, DOI 10.1007/978-3-319-22288-2_10

1 Introduction

Urban areas are burdened with rising individual and freight transport volumes today, accompanied by increasing environmental effects such as local air pollutants and noise, as well as competition for the use of space (e.g. congested roads, parking shortage). At the same time parcel service providers face a dynamic growth in their market. The courier, express and parcel (CEP) market grew by 57 % from 2000 to 2013 and this trend is expected to continue over the next years (BIEK 2014). Therefore, parcel service providers are challenged to design and implement "green" logistics services that are efficient in terms of resources as well as energy, and at the same time fulfill the high quality standard of parcel delivery (e.g. express, hour definite).

As one of the world's largest parcel service provider, UPS accepts its social, economic, and ecological responsibility: the company is testing delivery vehicles with gas and electric drive for reducing the dependency on fossil fuels as well as emissions in inner cities (UPS 2014). Besides, alternative distribution concepts have been developed and analyzed in real life conditions.

2 Initial Situation of Urban Parcel Delivery

CEP services are generally organized by means of comprehensive transport networks, often as hub-and-spoke-networks which are subdivided into pre-, main and on-carriage (see Fig. 1) (Arnold et al. 2008). Within pre-carriage the parcels are collected at the shippers' site and transported to the respective hub, usually by means of short distance traffic and vehicles. At the starting hub the parcels are transshipped onto the main carriage mode. At the destination hub an additional transshipment is necessary for the add-on distribution, again by means of short

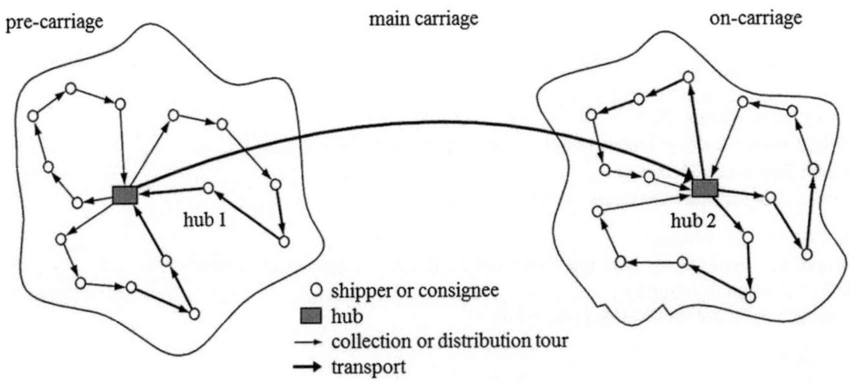

Fig. 1 Parcel networks (hub-and-spoke) (Arnold et al. 2008)

distance vehicles. Distribution and collection of parcels often are combined within one tour.

UPS was founded in 1907 and started its development across the US in the 1930s. While UPS took its first steps in going international and expanded to Canada in 1975, it is, today, providing services to more than 220 countries and territories worldwide, including Eastern and Western Europe, the Americas, the Middle East, Africa, and Asia Pacific. In Germany, UPS is covering its service with 73 package centers.

To keep the service on a high efficiency level, UPS developed a standard for the delivery operation. To keep distances economically short and to empower the drivers to manage their daily portion of work, countries are subdivided in center areas, which are separated in several loops, normally built as spokes around the center. Geographical borders, such as rivers, railways, etc., are often used as loop boundaries. Commonly each loop contains three to six delivery routes.

Package cars will be loaded during preload, early in the morning according to the area trace which was defined in the loop planning. The routes will be loaded with a specified amount of packages and stops which will define the planned day of the UPS service providers. The planned day will be revised daily to ensure efficiency on the road.

In urban areas with a high stop density UPS integrates delivery helpers, who assist the service providers using their own delivery information acquisition device in these areas. They meet the drivers in the cities for parcel transfer and deliver those by means of carry aid (e.g. tricycles, hand trucks). This is a common way to improve the delivery efficiency.

Due to restrictions like low-emission zones, expansion of pedestrian zones and traffic calming in the cities UPS is working on and testing new delivery concepts, which will help to meet these challenges and to work with continued efficiency.

3 Alternative Concepts for Urban Parcel Delivery

Today, diesel drives are the most common technologies of light and heavy duty vehicles in Germany (Rüdiger and Clausen 2014). The same situation exists within parcel networks (see UPS' fleet in Table 1); however, CEP service providers are interested to integrate other drives, as e.g. electric, gas and dual fuel (diesel/gas), or concepts into their existing systems to reduce the environmental impact of their urban parcel delivery.

Since 70 years, UPS has been testing innovative vehicle technologies, on the main and pre- or on-carriage. The company uses vehicles with various drives for gaining hands-on experiences and supporting further improvements in this area. Today, UPS owns more than 2688 vehicles worldwide with alternative drives covering electric drive, compressed and liquefied natural gas, and hybrid solutions (UPS 2013). One main driver is the reduction of the environmental impact of UPS's own activities, such as greenhouse gas, nitrous oxides emissions, particulate matter

Table 1 Share of drive technologies in UPS' fleet: Light and heavy duty vehicles (March 2014)

Max load (t)	Type	Diesel (%)	Gas (%)	Electric (%)	Area of use
2.8	P10	2.4			Collection/delivery
3.5	P45	14.2			
5.5	P45e			0.2	
7.5	P60/P80/P100	77.3	1.8	0.5	
7.5	Van	3.6			
	Total	*97.5*	*1.8*	*0.7*	
26	Feeder	100			Main carriage

or noise. Further drivers are the enhancement in process efficiency and the reduction of costs and other "soft" aspects such as avoidance of congestions and multiple delivery attempts.

The city of Dortmund was defined as the model area for pilot testing and improving the findings, concepts, and new technologies, as the e-assisted tricycle "cargo cruiser" (see Fig. 2, right), the e-driven package cars P80e and the package storage container concept for the city. After positive "Green Logistics" test results and experiences in Dortmund (Green Logistics 2015), UPS decided, due to the local political conditions and the high stop density, to implement the new concepts and technologies in the city of Hamburg and integrated these in the operational process of the center Hamburg-East. This center is responsible for the delivery areas in the city center of Hamburg and the quarter Hammerbrook, where three different pilot tests were implemented.

In the frame of these pilot tests alternative technologies and concepts have been implemented in urban parcel delivery and their environmental and economic effect quantified. The "cargo cruiser" was tested as an aid for the delivery helper, to improve his efficiency, the range of activity, and to meet several restrictions in the city center and pedestrian zones. The P80e replaced the diesel driven P80 (see Fig. 2, left) to test the possibility of direct emission reduction in the city.

Fig. 2 Parcel delivery by means of P80 (*left*) and cargo cruiser (*right*)

To reduce traffic, parking space problems and to meet requirements and restrictions, a package storage container concept was also tested.

In the following, selected pilot tests are described in more detail including relevant changes as well as obstacles met during the implementation.

3.1 *Parcel Delivery by Means of Diesel and Electric Drive*

One of the common package car sizes UPS is using in Germany is the P80, which represents 55.7 % of the fleet. The P80 has a gross vehicle weight of 7.49 t, a payload of 3.9 t, a gross load capacity of 23.4 m^3, with a diesel drive and an aluminum chassis. For this package car size, the company EFA-S GmbH (EFA-S 2015) has developed an electric drive. A 15 years old package car was converted and tested as a pilot in 2010. The results of the tests were used for further developments and lead to the new package car size P80e. The only change of the car specification is a payload reduction of 500 kg. The drive range is between 80–100 km, depending of the total vehicle weight. Essential advantages of the converted P80 to a total new car model construction are the following three: the remaining ergonomics for the service provider, savings of (primary) material resources and corresponding reduced environmental impacts because of the "re"-use of the existing old chassis (aluminum), and controllable acquisition costs.

In the frame of the research project "ELMO" (ELMO 2014), a field test with six P80e was performed in the UPS center Herne, which is responsible for the delivery area Dortmund. After the successful project process of "ELMO", UPS decided to implement a small fleet of electrically driven P80e's in the daily operational routine of UPS Hamburg-East.

Hamburg-East covers the service for the city center of Hamburg. The center area has a lot of small distance range routes. The parameters for choosing the operational area for e-driven package cars were: the same package car size, short distance with less than 75 km, high density and city center area, with the aim to reduce direct emissions in the city. Loop 12 was identified as the area which is fulfilling all the parameters and so the routes 12B, 12D and 12G were chosen to switch from diesel to electric drive. The only operational process change for this switch was the replacement of diesel fuelling by battery charging.

3.2 *Parcel Delivery by Means of E-Assisted Tricycle "Cargo Cruiser"*

UPS has developed a sophisticated and efficient methodology for the delivery process over the years. One proper method to improve the efficiency of a delivery route is to install an additional delivery helper on the route. This will result in a higher utilization of the package car and more effective work. Parameters and restrictions which will affect this decision are the package car size and the utilization, load capacity and stop density.

Today, new influences, restrictions, and parameters evolve affecting decisively existing parcel delivery tours within urban areas. These challenges are e.g. the entry restrictions in pedestrian zones, customer availability during delivery times in pedestrian zones, insufficient parking spaces for delivery vehicles, long walking distances for service provider and delivery helper, conflicts with city authorities—e.g. parking fines, traffic obstruction in city centers, and emission targets (i.e. standards that limit access of older vehicles).

To meet these challenges UPS tested an alternative concept known as the delivery helper aid, in combination with an e-assisted tricycle, called "cargo cruiser". This concept would allow a vehicle to be parked in a pedestrian zone during the entire day as well as to reduce walking distances with stop to stop delivery. Furthermore, the (service provider) delivery by package car in the city center would be reduced to a minimum (only large shipments would be delivered by car, before the closing of the pedestrian zone) and the service provider and helper work independently without mutual waiting times and reduced traffic obstruction. The usage of an e-tricycle is expected to reduce emissions in the city center and to support city emission targets.

The "cargo cruiser" has a volume load capacity of 2.2 m^3, a payload of 306 kg and a maximum speed of 24 km/h. For an efficient use of the "cargo cruiser", it is necessary to create additional transport and storage capacities in the delivery area, in order to load the "cargo cruiser" with the proper amount of packages. Therefore, a bigger size of package cars, e.g. the replacement of the P80 by the P100 (scenario 2*), or a storage container (scenario 3*) is needed.

Scenario 2* was implemented in the delivery area Hamburg Hammerbrook (see Fig. 3, right). Part of this quarter was covered by the routes 15C and 15E using package cars P80 with an average utilization of 60 % each.

The characteristics of this area are a high density with small packages. Obviously this is an indication for a delivery helper area. A negative aspect is that the possible walking distances are too long and a delivery helper with carry aid would not be effective. The e-tricycle "cargo cruiser" was the reasonable solution for that issue and so the tour area 15E was switched to a delivery helper area. The challenge to feed the "cargo cruiser" with packages was answered by using a bigger

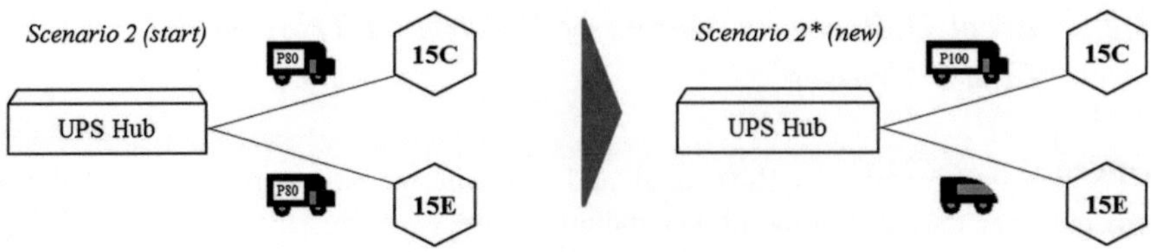

Fig. 3 Parcel delivery by means of "cargo cruiser"

package car size for the route 15C, a P100. This size is 1 m longer and has a load volume of 28, 4.5 m^3 more than that of the P80.

Using an e-tricycle in Hammerbrook, UPS is moving one package car less and a staging and load position was saved in the UPS building. The delivery operation in this area has the same effectiveness as before.

3.3 Parcel Delivery by Means of E-Assisted Tricycle "Cyclo Cargo" and Storage Container

In cooperation with the city authorities in Hamburg a package storage container was implemented in the quarter "Neuer Wall" which is a main shopping area in the city center of Hamburg (Doelling 2013). The requirements for this project were the reduction of traffic in particular in terms of package car movements, relaxation of the staging/parking situation, reduction of noise pollution and direct emissions in the quarter. In order to meet these requirements it was necessary to increase the driver helper delivery. The idea was to give driver helpers aids, such as an e-tricycle, a conventional cargo tricycle or a carry aid, to improve their efficiency and hence reduce package car movements. Therefore one package storage in the delivery area was essential.

A regular container which was used for package transportation in the UPS network was modified. Shelves were installed inside to permit a structured loading according to the loop planning during preload operation.

The storage container project allowed UPS to test an alternative to the "cargo cruiser", the e-assisted tricycle "cyclo cargo". The "cyclo" has a volume load capacity of 1.5 m^3, payload of 220 kg and a maximum speed of 24 km/h. In comparison to the "cargo cruiser", the "cyclo" has a 30 % lower capacity.

The original operational process for Scenario 3 included the P100 route 13A and two driver helpers (H3A each), one delivering with a conventional cargo tricycle and the other with a carry aid. After implementing the storage container the delivery in the area "Neuer Wall" was covered by three driver helpers. They used a carry aid, a conventional and an electrically assisted cargo tricycle (see Fig. 4). This way, the use of one package car was reduced in the city.

Because of the larger load capacity in comparison to a P100 the storage container covered a greater delivery area, so the assigned work on the other routes of loop 13 was reduced. Due to this reason additional package car movements were saved.

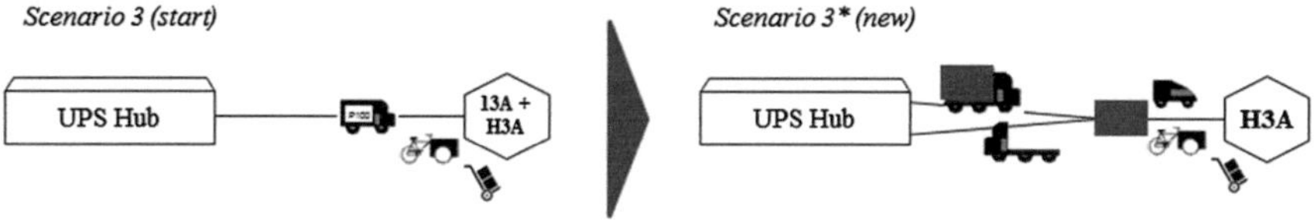

Fig. 4 Parcel delivery by means of "cyclo cargo" and storage container

One new issue in the new scenario was the storage container transport from the UPS building to the staging area in the city. This was done by a feeder which transported the container to the city, placed it there and returned to the building. In the evening the container was picked up by the feeder.

4 Methodology

4.1 Ecological Assessment Context

Already today, logistics service providers can realize eco-efficient solutions and products. However, today's methodological standards offer only insufficient guidance for calculating the respective (possible) environmental benefit of new concepts and for providing comparable results.

Life Cycle Assessment (LCA) is a method of systematic registration, analysis, and assessment of the environmental effects of production and services systems. The methodology, by which the environmental effects associated with the life cycle of a product or service are estimated (i.e. for the manufacturing, usage and disposal phases), is formulated for example in the norms (DIN ISO 14040 2006) and (DIN ISO 14044 2006). Due to its complexity and other reasons, the life cycle assessment is hardly applied in the area of logistics services. Against the backdrop of increasing discussions on climate change, the assessment of greenhouse gas emissions is rather focused when talking about environmental effects of logistics services today, as can be seen from the various recent initiatives on international level, such as the Greenhouse Gas Protocol Initiative and its standards (WRI and WBCSD 2005, 2011a, b), ISO standards 14064 (ISO 14061-1/2/3 2006a, b, c) and ISO 14605 (2007), DIN EN 16258 (2013), the International Workshop Agreement on harmonization of CO_2e emission quantification schemes of freight transport (IWA 16 2015), Global Logistics Emissions Council (GLEC 2015) etc. It is customary in published methods and the procedures of many companies at present to focus solely on greenhouse gases. Examinations conducted in the frame of the R&D-project "Green Logistics" however outlined that relevant environmental impacts of logistics services are also noise, particulate matter (PM), and other local air pollutants (Dobers et al. 2012a, b). This is because even though progress has been made in the segment of transport emissions in recent years, transport nevertheless continues to have a large influence on local air quality. According to the European Environmental Agency (EEA 2011) 58 % of all nitrogen oxide (NO_x) emissions and 20 % of the respective additional pollutant emissions are attributed to transport (utility and private vehicles). The differentiation between modes of transport also provides an interesting insight: The EEA attributes e.g. almost all sulfur dioxide (SO_2) and 10 % of the $PM_{2.5}$ transport emissions to shipping transport. Noise also

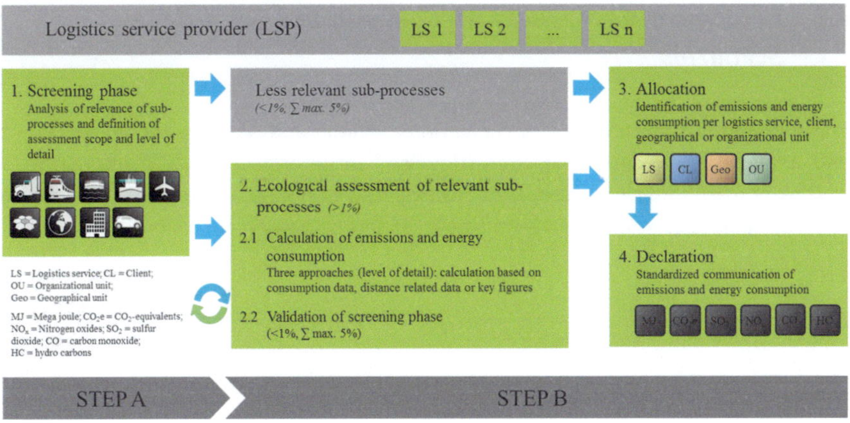

Fig. 5 Recommended procedure of "Green Logistics" method (Green Logistics 2015)

remains a relevant environmental effect in the transport sector (EEA 2011; Europäische Kommission 2011; Althaus et al. 2009a, b).

Consequently, the project consortium "Green Logistics" expanded existing assessment methods (DIN EN 16258 2013 and WRI and WBCSD 2005 in particular) to include missing processes and local environmental effects (e.g. SO_2, NO_x), and a uniform procedure was developed (see Fig. 5). Similar to the GHG Protocol, this procedure also uses a two-stage approach to identify, at the start, the relevant sub-processes of the logistics service to be assessed and define the level of further investigations. In step B, the actual assessment of these sub-processes is carried out, during which the level of input data is differentiated between detailed and rough, depending on the scope and the relevance of the sub-processes of the overall system. Step B covers the calculation of the overall environmental impact, the allocation of these impacts on the level of logistics services, clients, organizational, and geographical units of the company, and finally, if preferred, to communicate the calculated environmental impact.

The "Green Logistics" approach offers a standardized, comprehensive, and usage-related method for the ecological assessment of logistics processes and services. The method was discussed with national and international stakeholders and was finalized and published in the spring of 2015. A further description of the method and some preliminary results were published in Dobers et al. (2014) as well as in Green Logistics (2015).

In application fields where a logistics service provider wants to assess the differences of the environmental impact of various scenarios, this method can be used with focus on the varied sub-processes which is described in more detail in the following.

4.2 Ecological Assessment Method Applied for UPS Scenarios

The scenarios of the analysis cover various vehicle and delivery technologies, i.e. P80, P80e, P100, feeder, e-assisted tricycles “cargo cruiser” and “cyclo cargo” as well as conventional tricycles, hand trucks, and a storage container. The relevant items for calculating costs and emissions of the processes are listed in Table 2.

4.2.1 Equations

The costs of each delivery scenario could be calculated by using one of the equations below

$$C_{scenario} = \sum_{option} C_{fix}^{day} + C_{var}^{km} * d + C_{fuel} * fc * d + C_{lab} * h \tag{1}$$

$$C_{scenario} = \sum_{option} C_{fix}^{day} + C_{var}^{day} + C_{FC}^{day} + C_{lab} * h \tag{2}$$

where $C_{scenario}$ is the total costs of the delivery options applied in the scenario in Euro per day (€/d); C_{fix}^{day} is the fixed costs in Euro per day (€/d); C_{var}^{km} is the variable costs for maintenance in Euro per kilometer (€/km) and d the delivery distance in kilometer per day (km/d); C_{var}^{day} and C_{FC}^{day} are the variable costs for maintenance or fuels[1] in Euro per day (€/d); C_{fuel} is the costs for fuels (see Footnote 1) in Euro per fuel unit; fc is the specific amount of consumed fuel (see Footnote 1) in the relevant fuel unit, i.e. diesel in liter per kilometer (l/km) and electricity in kilowatt hours per kilometer (kWh/km); and C_{lab} is the labor costs in Euro per hour (€/h) and h is the daily working hours (h/d). Equation (1) is used for the delivery options 1 to 3, whereas Eq. (2) is used for other options 4 to 7.

The amount of emissions of each delivery scenario was calculated according to the following equation

$$E_{scenario}^{i} = \sum_{option} fc * d * EFA^{i} \tag{3}$$

where $E_{scenario}^{i}$ is the total amount of emissions (i.e. CO_2e, NO_x, SO_2, PM or CO) of the scenario in grams (g); and EFA^{i} is the emissions factor for pollutant i in liter or kilowatt hours of fuel (see Footnote 1) consumed (g/l) or (g/kWh).

[1]Fuels stands for any drive energy, i.e. here diesel and electricity.

Table 2 Scope of the ecological assessment

Delivery options			Costs			Drive energy	Labor
			Fixed	Variable[a]	Labor		
1	P80, P100		Depreciation, insurance, taxes (€/day)	Maintenance, consumption of drive energy (€/km)	Wage, nonwage (€/h)	Diesel (l/km)	Driver 1
2	Feeder						Driver 2
3	P80e					Electricity (kWh/km)	Driver 1
4	Cargo cruiser, cyclo cargo		Depreciation (€/day); no insurance or taxes	Maintenance (€/day)		Electricity (kWh/d)	Helper
5	Tricycles					–	
6	Hand trucks			–			
7	Storage container				–		–

[a]No city toll or equivalents were relevant

For reasons of sensitivity of the input data and the possibility of drawing inferences from the result about those, all results are normalized as follows

$$\left(E^i_{scenario\,j}\right)_{normalised} = \frac{E^i_{scenario\,j}}{\max\left(E^i_{scenario\,1}, E^i_{scenario\,1*}, \ldots, E^i_{scenario\,3*}\right)}. \tag{4}$$

4.2.2 Data Sources and Derived Information

With the objective of calculating costs and emissions as realistic, i.e. UPS specific, as possible, the highest degree of primary data available was used. Therefore, primary data was documented for all scenarios for at least 10 weeks to 54 weeks at the maximum (see Table 3). In doing so information on the daily total distance, paid work time, and total number of stops and parcels (both delivery and pickup) for the delivery options 1 to 3 was collected.

With respect to cost calculation, the current German UPS rates were integrated. Here, labor costs refer to those relevant in the city of Hamburg. Since no distance of any helper activity is documented today, variable costs for maintenance for option 4 and 5 have been estimated on a daily basis.

As fuel consumption of option 1 and 3, current German UPS average information was used. The current UPS average fuel consumption for option 2 (feeder) is derived on the basis of an application field which is not comparable to that of scenario 3* (road train on main carriage, hardly empty trips). Therefore, the average fuel consumption was estimated based on secondary data, i.e. Handbook of emission factors for road transport (HBEFA) (INFRAS 2010). For this, the following assumption was used: Solo trucks (20–26 t) with equal share of both, EURO 5 SCR[2]/EURO 6 and urban/extra-urban roads; with an average load factor of 99 % (container loaded) and 0 % (empty trip). By this, a specific fuel consumption, which is 80 % of the UPS average of feeder transport, was calculated.

Due to the lack of records of real electricity charging cycles for option 4, the maximum capacity of 2.25 kWh per day was assumed, which is roughly equivalent to a range of approximately 40 km. This assumption is subject to one of the detail analyses described below.

Relevant diesel calculation factors were taken from (DIN EN 16258, 2013), i.e. 3.24 kg CO_2e per liter diesel (well-to-wheel, WTW), 2.67 kg CO_2e per liter diesel (tank-to-wheel, TTW), and diesel density of 832.0 g per liter diesel.

Since so far, no internationally aligned calculation figures for air pollutants other than CO_2e from diesel exist, the emission factors per liter diesel (tank-to-wheel) were calculated on the basis of HBEFA data (INFRAS 2010). For this, the following assumption was used: Solo trucks (7.5 t) with EURO 4 SCR and urban roads; P80 with an average load factor of 85 % and P100 of 89 %.

[2]SCR—selective catalytic reduction.

Table 3 Timeframe and number of recording days[a] for primary data for scenarios

	Scenario 1		Scenario 2		Scenario 3	
	Timeframe	Recording	Timeframe	Recording	Timeframe	Recording
Old	01/02/2013–01/20/2014	~248 days in 54 weeks	10/16/2013–01/21/2014	~73 days in 13 weeks	12/24/2013–03/05/2014	40 days in 10 weeks
New	01/21/2014–07/25/2014	~122 days in 26 weeks	01/22/2014–07/22/2014	~125 days in 25 weeks	09/02/2013–07/22/2014[b]	127 days in 35 weeks

[a]Number of recording days listed for scenarios 1 and 2 is an average for all included tours, for scenario 3 it is that of the determining single tour 13A
[b]With intermediate use of scenario 3

Emission factors for well-to-tank emissions per liter diesel were taken from ecoinvent data base (version 3.01, activity “Market for diesel Europe without Switzerland”) (ecoinvent 2013). The same data base was used for the relevant emission factors for electricity: activities “Market for electricity, low voltage (DE)”, and for the detail analyses “Market for electricity, low voltage (NO)” and “Market for electricity, low voltage (FR)”.

For the ecoinvent data (ecoinvent 2013) the following life cycle impact analysis (LCIA) methods were used: CO_2e according to “IPCC 2007—climate change, GWP 100a”, and all others according to “selected LCI results—air”, namely carbon monoxide, nitrogen oxides, sulfur oxides, and particulate >10 μm.

4.2.3 Detail Analyses

In addition to the economic and ecological assessment of the scenarios at hand, two detailed analyses were performed to outline the influences of the following assumptions made: (1) influence of the maximum load capacity of e-assisted tricycles and (2) the influence of the national electricity mix on the ecological performance of scenario 1* using P80e with electric drive. In addition to Germany, data from France and Norway with a high share of nuclear and renewable (resp.) electricity were used for this comparison (see Fig. 6).

5 Results and Discussion

5.1 Parcel Delivery with Diesel and Electric Drive

The first scenario for urban parcel delivery focuses on classic UPS parcel delivery vehicles, i.e. P80 (scenario 1) and its version with electric drive P80e (scenario 1*). In both alternatives three vehicles are used simultaneously on three tours (12B, 12D, 12G),

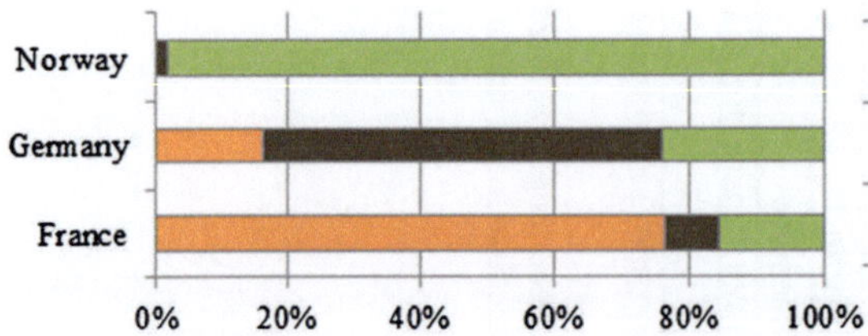

	Nuclear	Fossil	Renewables
Norway	0%	2%	98%
Germany	16%	60%	24%
France	76%	8%	16%

Fig. 6 Electricity generation mix of Norway, Germany and France in 2012 (EIA 2015)

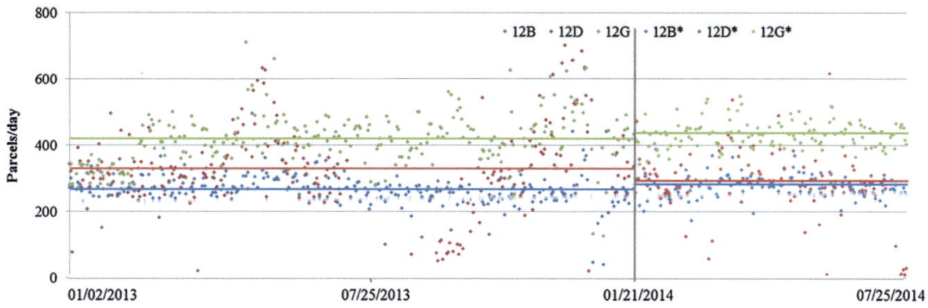

Fig. 7 Total parcel delivery/pick-up during scenario 1 and 1* recording

driving an average of about 125 km (1) and 130 km (1*) daily. Altogether more than 400 daily stops were made and about 1000 parcels were delivered or picked-up daily on these tours. In scenario 1* (after 01/21/2014) 3 % more stops, but 2 % less parcels were recorded compared to scenario 1. This difference occurs, besides others, due to a shorter period of recording, namely without the peak season in December (see Fig. 7).

Figure 8 clearly shows that, while costs are almost the same, scenario 1* (green) causes distinct more emissions than scenario 1 (brown), which is valid for all emission categories assessed.

However, it is interesting to look at the effect of the underlying electricity production mix, which is defined by about 2/3 by fossil fuels, 1/4 by renewables and 1/5 by nuclear power. The French scenario (1* FR, orange) represents an electricity mix basing mainly, i.e. 3/4, on nuclear power. The Norwegian scenario (1* NO, blue) represents an electricity mix basing almost only on renewables.

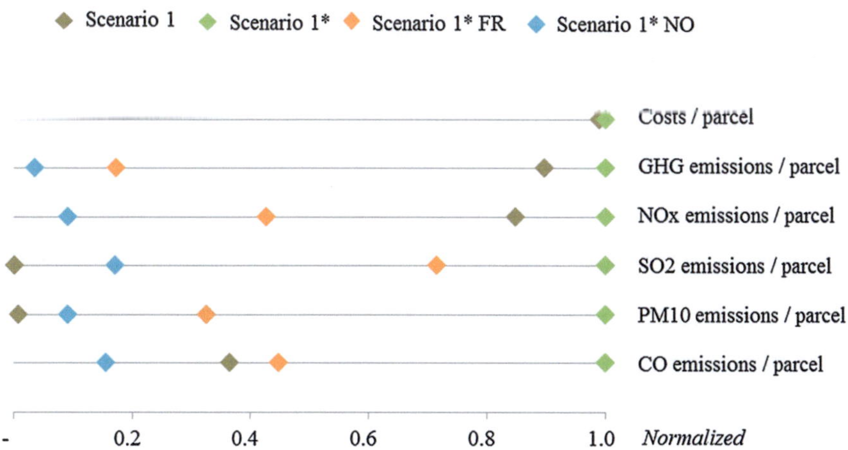

Fig. 8 Normalized key figures (costs/emissions per parcel) of the scenario 1 and 1*

These scenarios cause less greenhouse gas and NO_x emissions than scenario 1 with diesel drive: While the German mix causes 12 % more GHG emissions than scenario 1, the French causes only 19 %, the Norwegian 4 % of the overall amount of GHG emissions of scenario 1. Concerning NO_x emissions the following differences resulted: Germany +18 %, France −50 % and Norway −89 %.

Looking at CO emission, only the electricity mix basing on renewables would result in less emission than parcel delivery with diesel drive. The latter, however, results in better values for SO_2 and PM_{10} emission than any electric drive scenario. With regards to the SO_2 emissions, they clearly depend on the sulfur content of diesel, which is sulfur-free in Germany (maximum limit is 10 mg per kg Diesel since 2011, 10. BImSchV 2010).

The above description is valid for the overall amount of emissions influenced by the different scenarios. However, since electric drive uses power at a different area than where the power has been produced, it is allowed to distinguish between

- City emissions which are local air pollutants caused directly by driving the vehicle [so-called tank-to-wheel (TTW) emissions] and
- Non-city emissions: in this analysis, these are caused by the power production and distribution (scenario 1*) or by the diesel production and supply (scenario 1), which are the so-called well-to-tank (WTT) emissions.

Their respective shares to the overall emissions of each scenario and emission category are visualized in Fig. 9. From this perspective, scenario 1* is responsible for less (or even zero) city-emissions compared to scenario 1.

Nevertheless, one should consider that particulate matter has only been calculated with respect to the diesel or electric drive. In addition to this, it also can be caused by tire wear, brakes or risen dust of the street surface, and via precursors such as sulfur or nitrogen oxides (Umweltbundesamt 2015). However, the modelling and estimation of these potential relationships are beyond the scope of this study. Therefore, one may conclude that zero inner-city particulate matter emissions from P80 parcel delivery are unlikely, but this fact is similar for scenario 1 and 1*.

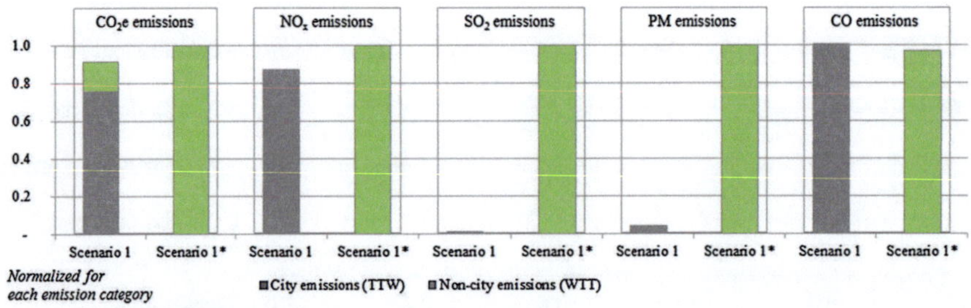

Fig. 9 City and non-city emissions of scenario 1 and 1*

5.2 *Parcel Delivery with E-Assisted Tricycle "Cargo Cruiser"*

The second scenario for urban parcel delivery compares the classic UPS parcel delivery vehicle, i.e. P80 (scenario 2), with the combination of P100 and the e-assisted tricycle "cargo cruiser" (scenario 2*). In scenario 2, two P80 drove a daily average of about 65 km, meanwhile ca. 250 stops and 875 parcels (delivery and pick-up) were recorded. Scenario 2* covered 36 km/d by P100 and for the combined delivery as an average 330 stops and almost 1000 parcels were recorded per day.

Figure 10 summarizes the normalized key figures (costs/emissions per parcel) of all scenarios. Scenario 2 is represented by triangles. With respect to costs, greenhouse gas, NO_x and CO emissions, the new delivery concept results in lower key figures than the classic P80. Again, the calculated amount of SO_2 and PM_{10} emissions is higher for the new scenario 2*, comparable to scenario 1*. The difference between 2* and 2 is strongly dependent on the electricity mix assumed, as it has been discussed above.

Again, the distinction between city and non-city emissions is examined in the following (see Fig. 11). According to that, scenario 2* is responsible for less city-emissions than scenario 2. Except for SO_2 and PM emissions, the overall emissions are lower as well.

As described in Sect. 4.2, the maximum load capacity of the e-assisted tricycles has been assumed for the daily energy use for parcel delivery, since no primary data was recorded during the pilot phase. Figure 12 (left) shows the influence on the relative amount of air pollutants of this assumption. Again, SO_2 and PM_{10}

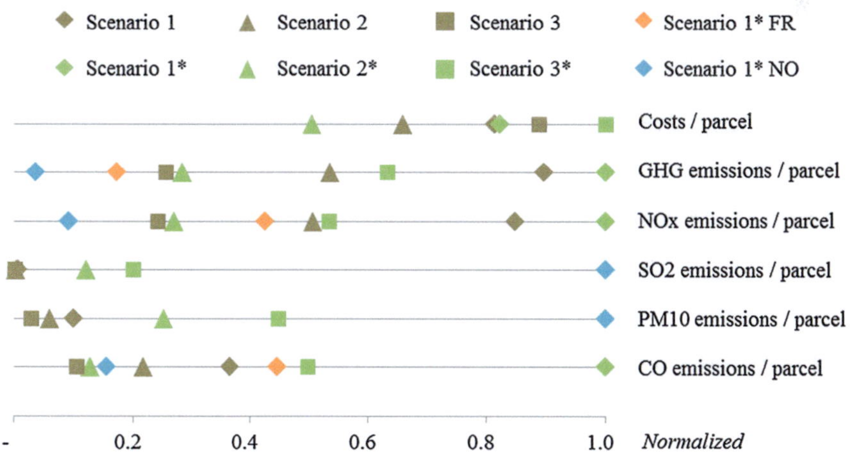

Fig. 10 Normalized key figures (costs/emissions per parcel) of all scenarios. SO_2 and PM_{10} key figures of scenario 1* and 1* FR: see relations to scenario 1* NO in Fig. 8

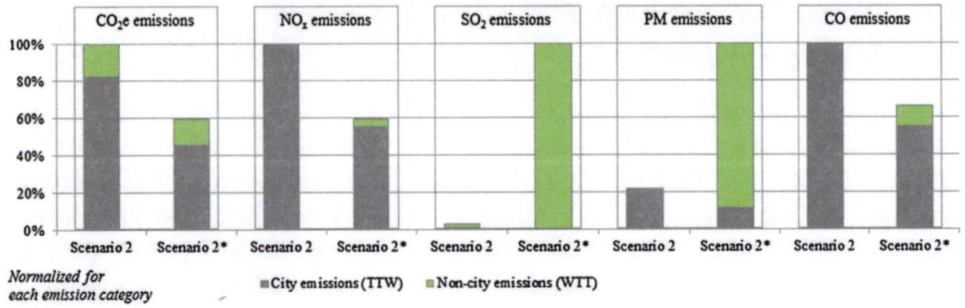

Fig. 11 City and non-city emissions of scenario 2 and 2*

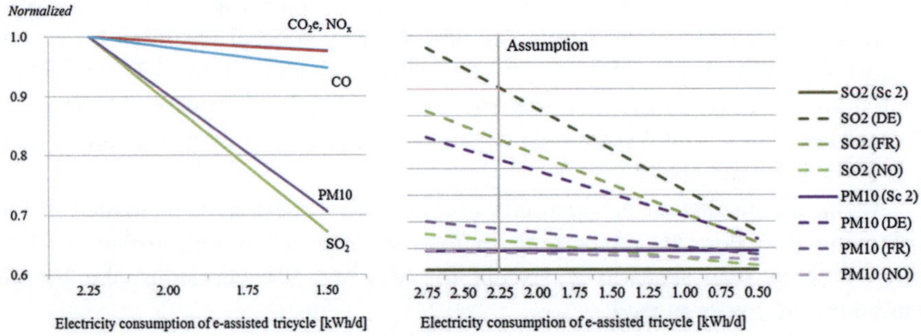

Fig. 12 Influence of the maximum load capacity of e-assisted tricycles (*left* scenario 2*; *right* focus SO_2/PM_{10} and scenario 2 and 2* with varied electricity mix)

emissions are influenced most, as those are directly dependent on the underlying German electricity mix.

A reduction of electricity consumption of the cargo cruiser by 22 % (i.e. to 1.75 kWh/d) results in reduced overall emissions by approx. 22 % for SO_2 and approx. 20 % for PM_{10} for scenario 2*. However, this reduction does not affect the rating of the scenarios with respect to those emission categories.

Figure 12 (right) shows the influence of reduced consumption in combination with the assumed electricity mix. In case of a Norwegian mix (renewables) a consumption of even 2.5 kWh/d still results in less particulate matter emission. SO_2 emissions are not affected at all.

5.3 Parcel Delivery with E-Assisted Tricycle "Cyclo Cargo" and Storage Container

The third scenario focuses on the replacement of the P80 (scenario 3) by a combination of feeder transport to a storage container and following parcel delivery by

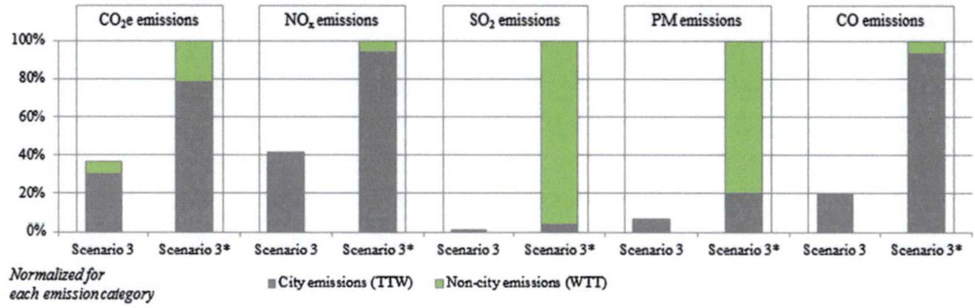

Fig. 13 City and non-city emissions of scenario 3 and 3*

means of the e-assisted tricycle "cyclo cargo" (scenario 3*). Here the relevant tour key figures recorded are:

- Scenario 3: 20 km/d (kilometers per day) of one P80 with overall 285 stops/d and 550 parcels/d (delivery and pick-up)
- Scenario 3*: 33 km/d of one feeder with overall 315 stops/d and 610 parcels/d (delivery and pick-up)

In Fig. 10 the results for the normalized key figures (costs/emissions per parcel) of scenario 3 and 3* are represented by squares. It clearly shows that, with respect to all, the costs and any emission category, scenario 3* results in higher values than scenario 3. This is valid also for the distinction between city and non-city emissions (Fig. 13).

5.4 *Further Discussion*

5.4.1 Electric Drives

In the framework of rising requirements concerning the environmental performance of urban logistics services, electric drives are an interesting alternative to diesel drives, as shown by means of the example of the P80e in Hamburg. On the one hand, they fulfill the requirements of the local authorities and operational challenges. On the other hand, e-drives offer the advantages of reduced city-emissions to the air, and noise emissions. However, as shown before, the introduction of electric drives is recommended to be combined with the purchase of renewable energies to fully unlock environmental potentials. From the operational point of view it is interesting that some communes offer privileges to vehicles with electric drives, such as the use of dedicated bus lanes. The supported extension of such privileges (e.g. for all economic traffic in cities) might enhance the use of e-driven vehicles in the future.

5.4.2 E-Assisted Cargo Tricycles

The tested e-tricycles have shown to be an excellent aid for realizing more efficient and environmentally friendly parcel delivery. Next to improved operational key figures, they have realized direct emission reductions in the model regions. Furthermore, tricycles provide extended access to areas, e.g. pedestrian zones, where parking is only allowed for a limited amount of time. The latter might be one reason for failing to meet compulsory delivery times which in turn might cause conflicts with clients. E-assisted cargo tricycles could solve these time restrictions. However, since the legal framework for pedestrian zones differs between communities (e.g. in some communities cycles do not have any access at all), the transfer of this alternative delivery concept requires detailed analysis of the individual local framework beforehand.

5.4.3 Distributed Parcel Storage Container

In cities with a high stop density, the combination of a parcel storage container with delivery helpers and tricycles can be recommended to reduce inner-city traffic. Classic parcel delivery vehicles enhance the competition of space: the driver needs to continuously identify possible places to stop or even congests traffic lanes or minor streets. Such problems could be solved by the use of a parcel storage container, which implies only the transport to and from. With respect to the city-emissions the tested scenario 3* needs to be further analyzed concerning the real daily energy demand of the tricycles as well as the effect of enhanced use of renewable electricity. Besides, the feeder transport combined with other loops, which could be realized in theoretical analyses in Dortmund, might offer further approaches to reduce emissions.

However, one possible obstacle of parcel storage containers that needs to be overcome is the access to reasonable locations for the container, as well as its permission to stay there. Here, the close co-operation with the local authorities is essential. Nevertheless, for levelling the playing-field, the authorities need to offer all CEP service providers the same opportunities. This in turn might provoke additional competition for the use of space in the already tight city centers (e.g. restricted disposability of locations, change of cityscape). The initiated pilot test in Hamburg was successful for UPS. After having gained the relevant positive experiences, the case study in Hamburg is planned to be extended by three additional UPS' parcel storage containers.

6 Conclusions

In the framework of continuous technological and conceptual further development of parcel delivery, UPS has tested various approaches. Here, the main focus was put on new options to face rising challenges and restrictions of the last mile logistics, especially concerning efficiency and sustainability. To measure and monitor the environmental performance of the tested concepts, all relevant process and consumption data have been recorded and applied in the comprehensive ecological assessment method, developed in the R&D project "Green Logistics". The ecological assessment has proven that not only economic and single ecological key figures, such as greenhouse gas emissions, need to be discussed when evaluating the advantages and disadvantages of new concepts. Moreover, while some emission categories show positive effects of the tested delivery options (e.g. CO_2e and NO_x), others are contrary (e.g. SO_2, PM) and/or others even hard to measure, such as noise, and overall PM. Besides, the alternative parcel delivery concepts need to be embedded into a comprehensive concept within an area or even a region. Since isolated pilot tests might result in more emissions or costs, the combination with other tours or even the re-planning of delivery districts might overcome this initial burden, which changes in running systems often face.

One approach is the realization of a parcel vehicle with electric drive, which offers—in comparison to UPS' classic P80—equivalent ergonomics for the driver, retained operation processes in combination with reasonable investment costs and resource savings. All this could be realized with the development of P80e. Therefore, UPS has extended its implementation successively in several cities, such as Dortmund, Frankfurt a. M., Düsseldorf, Hamburg, and Karlsruhe.

In addition, the concept with e-assisted tricycles, such as "cargo cruiser" and "cyclo cargo", was successful as well. In the case of appropriate framework conditions, they result in efficient and sustainable procedures, especially in city centers with relevant access restrictions to delivery areas. The successful test in the model region Dortmund and the subsequent transfer to Hamburg convinced UPS to persist with the delivery concept "cargo tricycles": For the city center of Hamburg, three additional tricycles have been ordered (Kopp 2015), and the concept is planned to be transferred to the Rhine-Main area. Besides, other UPS operations of main European cities have shown interest in this applied concept, e.g. UPS France has started to analyze options for a future realization of e-assisted tricycles in Paris.

Acknowledgments The detailed examinations have been performed in the frame of the research project "Green Logistics" (Green Logistics 2015) and its case study "ecoPlan". The project is part of the LogistikRuhr efficiency cluster (EffizienzCluster LogistikRuhr 2015), which is funded by the German Federal Ministry of Education and Research (BMBF) (Funding code 01IC10L06).

References

Althaus et al (2009a) Traffic noise in LCA. Part 1: State-of-science and requirement profile for consistent context-sensitive integration of traffic noise in LCA. Int J Life Cycle Assess 14:560–570

Althaus et al (2009b) Traffic noise in LCA. Part 2: Analysis of existing methods and proposition of a new framework for consistent, context-sensitive LCI modeling of road transport noise emission. Int J Life Cycle Assess 14:676–686

Arnold et al (2008) Handbuch Logistik, 3rd edn. Springer-Verlag, Berlin
BIEK Bundesverband Internationaler Express- und Kurierdienste e.V. (2014) Wirtschaftliche Bedeutung der KEP-Branche. Die Kurier-, Express- und Paketbranche in Deutschland. KEP-Studie 2014 - Marktanalyse. http://www.biek.de/index.php/studien.html [03/15/2015]
DIN EN 16258 (2013) Methode zur Berechnung und Deklaration des Energieverbrauchs und der Treibhausgasemissionen bei Transportdienstleistungen (Güter- und Personenverkehr)
DIN ISO 14040 (2006) Umweltmanagement - Ökobilanz - Grundsätze und Rahmenbedingungen
DIN ISO 14044 (2006) Umweltmanagement - Ökobilanz - Anforderungen und Anleitungen
Dobers et al (2012a) Ökologische Bewertung von Logistikprozessen und -systemen: Systemdefinition Green Logistics (Langdokumentation). http://www.green-logistics-network.de/images/stories/Download/systemdefinition_green_gogistics_v3_5.pdf [03/15/2015]
Dobers et al (2012b) Ecological assessment of logistics processes and systems: System definition Green Logistics (brief documentation). http://www.green-logistics-network.de/images/stories/Download/systemdefinition_summary_en.pdf [03/15/2015]
Dobers et al (2014) Comparability of the Environmental Effects of Logistics Services. Sector Guidance for Ecological Assessment. TRA 2014 Proceedings. Paris, 14.-17.4.2014
Doelling (2013) Neue Containerlösung soll innerstädtischen Lieferverkehr und Emissionen reduzieren. http://www.energie-experten.org/experte/meldung-anzeigen/news/neue-containerloesung-soll-innerstaedtischen-lieferverkehr-und-emissionen-reduzieren-4150.html [03/15/2015]
EEA European Environment Agency (2011) Laying the foundations for greener transport. TERM 2011: transport indicators tracking progress towards environmental targets in Europe. Copenhagen
ecoinvent (2013) v3.01 Master Data. http://www.ecoinvent.org [09/19/2014]
EFA-S (2015) EFA-S Elektrofahrzeuge Schwaben GmbH. http://www.efa-s.de/ [03/15/2015]
Effizienzcluster LogistikRuhr (2015) http://www.effizienzcluster.de/ [03/15/2015]
EIA (2015) U.S. Energy Information Administration. International Energy Statistics. http://www.eia.gov/cfapps/ipdbproject/iedindex3.cfm?tid=2&pid=alltypes&aid=12&cid=r3,&syid=2010&eyid=2012&unit=BKWH [03/15/2015]
ELMO (2014) http://www.projekt-elmo.de [03/15/2015]
Europäische Kommission (2011) Roadmap to a Single European Transport Area – Towards a competitive and resource efficient transport system. White Paper. COM(2011) 144 final
GLEC Global Logistics Emission Council (2015). http://www.theglec.org/ [03/15/2015]
Green Logistics (2015) http://www.green-logistics-network.de/ [03/15/2015]
INFRAS (2010) Handbuch Emissionsfaktoren des Straßenverkehrs (HBEFA). Version 3.1. Bern. http://www.hbefa.net [03/15/2015]
ISO 14064-1 (2006) Greenhouse gases - Part 1 - Specification with guidance at the organization level for quantification and reporting of greenhouse gas emissions and removals
ISO 14064-2 (2006) Greenhouse gases - Part 2 - Specification with guidance at the project level for quantification, monitoring and reporting of greenhouse gas emission reductions or removal enhancements
ISO 14064-3 (2006) Greenhouse gases - Part 3 - Specification with guidance for the validation and verification of greenhouse gas assertions
ISO 14065 (2007) Greenhouse gases - Requirements for greenhouse gas validation and verification bodies for use in accreditation or other forms of recognition
IWA International Workshop Agreement 16 (2015) International harmonized method(s) for a coherent quantification of CO_2e emissions of freight transport
Kopp (2015) Paketdienst UPS testet die Paketzustellung zu Fuß. Neue City-Logistik. http://www.abendblatt.de/hamburg/article136896611/Paketdienst-UPS-testet-die-Paketzustellung-zu-Fuss.html# [03/15/2015]
Rüdiger and Clausen (Ed.) (2014). Studie zu alternativen Antriebsformen im Straßengüterverkehr. Status Quo und Entwicklungsperspektiven - 2014. Stuttgart: Fraunhofer Verlag
Umweltbundesamt (2015) Feinstaub. http://www.umweltbundesamt.de/themen/luft/luftschadstoffe/feinstaub [03/15/2015]

UPS (2013) UPS Corporate Sustainability Report 2013. http://sustainability.ups.com/media/UPS-2013-Corporate-Sustainability-Report.pdf [03/15/2015]
WRI and WBCSD (2005) The Greenhouse Gas Protocol. The GHG Protocol for Project Accounting. USA, Switzerland. http://www.ghgprotocol.org/files/ghgp/ghg_project_protocol.pdf [03/15/2015]
WRI and WBCSD (2011a) The Greenhouse gas protocol. Product life cycle accounting and reporting standard. USA, Switzerland. http://www.ghgprotocol.org/files/ghgp/public/Product-Life-Cycle-Accounting-Reporting-Standard_041613.pdf [03/15/2015]
WRI and WBCSD (2011b) The Greenhouse gas protocol. Corporate value chain (Scope 3) accounting and reporting standard. USA, Switzerland. http://www.ghgprotocol.org/files/ghgp/public/Corporate-Value-Chain-Accounting-Reporing-Standard_041613.pdf [03/15/2015]
10. BImSchV (2010) ZehnteVerordnung zur Durchführung des Bundes-Immissions-schutzgesetzes: Verordnung über die Beschaffenheit und die Auszeichnung der Qualitäten von Kraft- und Brennstoffen. http://www.gesetze-im-internet.de/bundesrecht/bimschv_10_2010/gesamt.pdf [03/15/2015]

Order Fulfillment and Logistics Considerations for Multichannel Retailers

Kees Jan Roodbergen and Inger B. Kolman

Abstract This paper addresses the challenge of making multichannel decisions for order fulfillment and logistics. We present a framework for multichannel strategies consisting of seven elements. Some channel decisions are part of the marketing mix, with the ultimate choice left to the customer. Other channel decisions concern logistics activities and can be made, for example, based on efficiency measures. We study the situation of a Dutch retailer that sells large household items (e.g. appliances or furniture). Inventories of products are available at the stores as well as at the central warehouse. Customers can select and order the products in the store; the store then delivers the products to the customers' homes, since the items are typically considered too big to be transported by the customers themselves. Recently, a distribution network was added to deliver products from the central warehouse to customers in response to online orders. The two distribution networks are operated separately, as is common at many retailers. We model the company's logistics operation by means of a variant on the Multi-Depot Vehicle Routing Problem to make dynamic channel assignments on a per customer basis. Our study aims at creating awareness of the wider portfolio of multichannel decisions, at stimulating dynamic assignment of orders to available channels, and at providing case-based evidence of the potential gains of such a strategy.

Keywords Distribution · Vehicle routing · E-commerce · Multi-channel

K.J. Roodbergen (✉)
University of Groningen, PO Box 800, 9700 AV Groningen, The Netherlands
e-mail: k.j.roodbergen@rug.nl

I.B. Kolman
Districon, Raadhuisstraat 32–34, 3603 AW Maarssen, The Netherlands
e-mail: i.kolman@districon.nl

H. Zijm et al. (eds.), *Logistics and Supply Chain Innovation*,
Lecture Notes in Logistics, DOI 10.1007/978-3-319-22288-2_11

1 Introduction

Retail organizations increasingly consider the internet a cornerstone for their sales strategy. According to the Thuiswinkel Markt Monitor (Homeshop Market Monitor), the share of online sales in total sales of products in The Netherlands accounted for 7.6 % in 2014, adding a full percent compared to the 6.6 % share in 2013 (Jongen 2015). Already before the advent of internet sales, retail operations were considered to be among the most important, dynamic and difficult operations to manage (Hill et al. 2002). Adding an internet sales channel only increases that challenge for several reasons. As was noted in Roodbergen et al. (2012), the consumer loyalty is decreasing while at the same time product and price awareness of consumers is increasing, which results in higher fluctuations of demand. The channel preferences (online or physical store) of consumers are continuously changing depending on product type, weather conditions, price levels, and numerous other reasons. “Moreover, during the decision phase, consumers tend to frequently switch channels, gathering information on the web, and then buying the product in a store, or the reverse. Retailers are evidently aware of consumers’ behavior and are rapidly adapting their sales strategies. Many brick-and-mortar companies have introduced an online store or upgraded the existing. And online players are opening physical retail outlets. All of this, however, only concerns the front-end. The online and in-store operations are typically organized and run as separate businesses.” (Roodbergen et al. 2012, p. 13).

It is in this space we seek to make a contribution. Or as Maxwell and Hudson (2011, p. 2) state “One of the biggest overall conclusions is that consumers are leading the way in multichannel shopping, with many retailers lagging behind in terms of meeting their needs. Today’s global retailers have a huge opportunity to enhance the mechanisms necessary to keep up with shoppers who are demanding more customization in terms of delivery and returns, product choice, and number of channels from which to choose.” To this end, we first define seven elements of multichannel strategies, expanding on a framework from marketing research (Neslin et al. 2006). Our framework adds elements specific to the logistics operations of retail and internet sales organizations, which can support the choices a multichannel (online) retailer faces in this respect. In the second part of the chapter we select one prominent logistics element from the framework and show how a dynamic channel selection mechanism in the logistics process execution can be exploited to enhance efficiency of the order fulfillment.

2 Multichannel Strategies for Product Sales to Consumers

In literature, the term *multichannel* is used for a variety of activities. For example, Frazier (1999) describes a multichannel strategy as a situation in which two or more distinct channels for goods flows sell the same product line to the same target market. Quite differently, in marketing research, a multichannel strategy is considered the use of multiple information channels "through which firms and customers interact, with the goal of enhancing customer value through effective customer acquisition, retention, and development" (Neslin et al. 2006). Such channels may include a website, social media, a call center, or a store employee. Zhang (2009) states that "the term multichannel retailing is increasingly used to refer to the practice of retailers using both traditional bricks-and-mortar retail stores and the internet to sell merchandise (i.e. the so-called 'bricks-and-clicks' format)".

Based on interviews and meetings with numerous stakeholders, we define a *multichannel strategy* for product sales to consumers more broadly, as consisting of seven elements: (1) search channels, (2) purchase channels, (3) forward fulfillment channels, (4) forward logistics channels, (5) return logistics channels, (6) return fulfillment channels, and (7) after sales channels. Various options for these channels are depicted in Fig. 1. Elements 1, 2, and 7 originate from Neslin et al. (2006). The choice of companies to offer multiple channels may have several reasons. As Neslin et al. (2006, p. 100) state: "Businesses could grow through multichannel strategies by acquiring more customers or selling more to each customer. However, a multichannel environment might erode loyalty, whether to firms or to the brands offered by these firms, because it encourages extensive search, which may lead to purchases from different firms. […] However, more channels also suggest better service, which often leads to greater loyalty".

The *search channels* refer to the options consumers have to become informed about a product's properties, price, and accompanying services such as warranties. The *purchase channels* provide the options for consumers to buy products. The purchase may be made in a store or showroom, or via communication platforms, including phone, e-mail, and internet websites. After the order has been placed, the product needs to be retrieved from some storage location, for example, from a warehouse, from a store shelf, or it may still need to be obtained from a supplier. These choices form the *forward fulfillment channels*. Often there is a one-to-one correspondence between the purchase channel and the fulfillment channel.

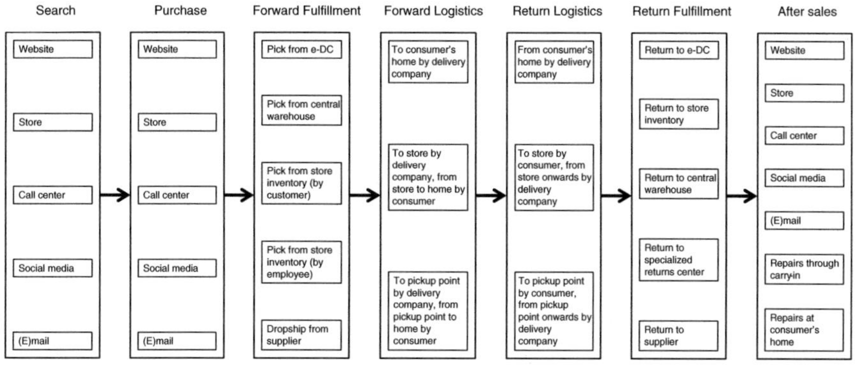

Fig. 1 Overview of channels for seven elements of physical goods sales to consumers

For example, in traditional bricks-and-mortar stores, the only fulfillment option is for the consumer to retrieve the product from the store's shelves. However, when buying furniture or appliances in a store, the forward fulfillment may be performed from a storage room by a store employee in response to an in-store order.

After placing the order and retrieving the product from storage, it must be shipped to the consumer's home. For this, several *forward logistics channels* exist. In traditional bricks-and-mortar shopping, transportation would be performed by the consumer. In traditional e-commerce channels, a parcel delivery company brings the product to the consumer's home. However, almost any combination of forward fulfillment channel and forward logistics channel can be observed. For example, supermarkets exist where consumers retrieve products from the store themselves, but the store delivers the products to the consumers' homes (e.g. Metro, Canada). Likewise, supermarkets exist where employees retrieve the products for online orders from storage, and the consumers come to pick up the products with their own cars (e.g. Carrefour, France). An intermediate form occurs when the company transports the products to a pickup point near the consumer's home, and the consumer retrieves the products from there.

Note that for a forward logistics channel consisting of a consumer's pick up from the store, the company still has the choice in the forward fulfillment channel to pack and ship those products from a central warehouse to that store (e.g. HEMA, The Netherlands), or to retrieve the requested products from the stores' inventories (e.g. Stop&Shop, United States).

Recent European Union regulations state that for contracts concluded as of June 13, 2014, consumers have the right to withdraw from online purchases within 14 days (European Commission 2014). In e-commerce a significant percentage of goods is already being returned, varying from 1 % (e.g. printer cartridges) to 70 % (e.g. in shoe sales). The *return logistics channels* and *return fulfillment channels* are therefore of importance to many companies. The options for these elements are largely similar to those for the forward fulfillment and logistics channels. The *after sales channels* refer to all forms of communication after the purchase, as well as to warranty and repair issues.

3 Efficiency Gains by Dynamic Channel Assignment

Some channel decisions are driven by marketing considerations, legacy infrastructure, or (local) expectations of consumers. For example, in The Netherlands standard delivery of e-commerce orders is within 24 h, which prevents many webshops from invoking dropshipments, since the supplier's processes are not designed for such response speed. In other cases, channel decisions could be made based on cost considerations. Here we investigate the potential for efficiency gains by introducing a dynamic scheme for assigning consumers' orders to existing fulfillment and logistics channels.

We find inspiration in the distribution network of a Dutch retailer that sells large household items (e.g. appliances or furniture). The company has two distribution networks in place. The first network has historically grown from the bricks-and-mortar sales. Suppliers deliver products to a warehouse, which subsequently replenishes the stores using a fleet of company owned trucks. Customers buy products based on the available selection in the store. Due to size and weight, products are delivered to the consumers' homes by means of trucks owned by the store. Some of the smaller products are directly taken from the store by the consumers, but these purchases are considered outside the scope of this study.

The second network was implemented for consumers' online orders placed via the company's website. Fulfillment of these orders occurs in the dedicated e-commerce distribution center (e-DC). Trucks deliver the products from the e-DC to consumers throughout the country. The e-DC itself is replenished from the same warehouse that supplies the stores. In the remainder of this chapter we will have a closer look at the integration of these two networks. At this point, we emphasize that the current situation of this company is in the authors' experience quite common. Separated networks are easier to manage, activities are easier to schedule, and agreements on profit margins need not be made with franchise stores in the network. The networks are depicted in Fig. 2.

What we proposed to the company is the following. When an order arrives, either in a store or online, we determine from which location it is best served. This means that an order placed online, may be fulfilled from a store. Or an order placed in a store may be fulfilled from the e-DC. Or an order placed at one store, may be fulfilled from another store. All of these options are feasible. Evidently, an order placed online can be delivered from a local store with a truck owned by that store. But also consumers who buy a product in the store, are indifferent to the fulfillment and logistics channels used. This holds, because such consumers will only see the

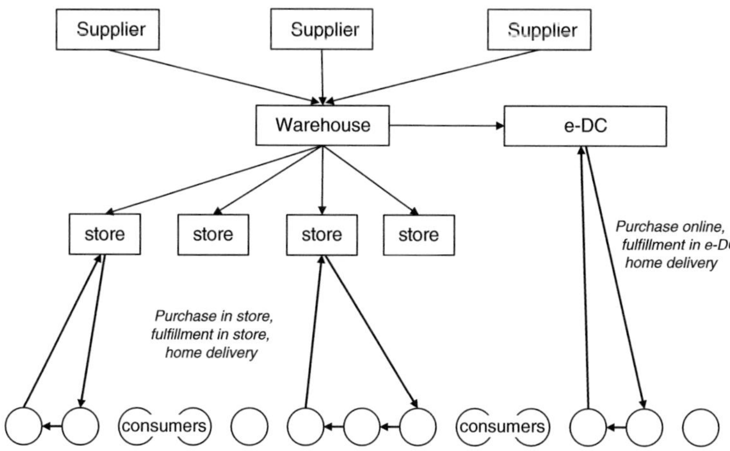

Fig. 2 Structure of current forward fulfillment and logistics channels

(branded) truck delivering the products. The origin of that truck is unknown and irrelevant to the consumer. This indifference is not true in general, for example, if consumers carry home their own products from the store. Also in other cases, it may not be desirable to retrieve online orders from a store's inventory. For example, several supermarkets started their online operations by picking online orders from within the store, but switched to a dedicated e-DC when volumes reached the point that store customers started complaining about the professional pickers in the store. Since appliance and furniture stores typically have a storage room separated from the showroom floor, this is not an issue here.

Cost savings from network integration can be achieved in the following manner. When stores are provided with their usual store supply from the warehouse, the available empty space in the truck is used to further increase store inventories for expected or known online orders. Thus, this transport from the warehouse to the store is virtually free of charge. Additional handling costs do arise. Cost advantages are the result of the fact that when orders occur, some products are already closer to the customer and need less transportation. Potentially additional inventory risks, like obsolescence, may occur. It is therefore important to limit additional stock keeping to products that are ordered frequently, are also sold in the store itself, and will remain in the assortment for the immediate future. This chapter, however, does not treat inventory decisions, but rather focuses on channel decisions. Desirability of stock depletion can be incorporated relatively easily by means of virtual cost incentives.

We present a hybrid genetic heuristic for solving this variant of the Multi-Depot Vehicle Routing Problem (MDVRP). The MDVRP is an extension of the classical Vehicle Routing Problem (VRP). The VRP involves determining a set of routes that minimize total costs or travel distances, while satisfying certain constraints. The MDVRP adds to this the option of sourcing from multiple locations, in our case stores and the DC. A difference between the regular MDVRP and our problem is that sources are unrelated in the regular MDVRP. The stores in our situation, however, use stock supplied by the DC, while at the same time the DC can serve as a source for deliveries. This gives additional handling costs at the stores that result in a more complex cost structure and objective function.

4 Literature on Vehicle Routing

A significant body of VRP literature exists, as well as on various extensions, such as the VRP with time windows, the VRP with backhauls, the VRP with pick-ups and deliveries, and the VRP with multiple use of vehicles (Gendreau et al. 2008). Exact algorithms are time consuming and usually not applicable to problems with over 50 customers (Liong et al. 2008). Therefore a majority of papers present heuristics. Tabu search algorithms on the MDVRP were introduced by Renaud et al. (1996) and Cordeau et al. (1997). Pisinger and Ropke (2007) develop a unified heuristic to solve five extensions of the VRP, including the MDVRP. This heuristic uses an adaptive large neighborhood search framework. Ho et al. (2008) propose two hybrid genetic algorithms for the MDVRP. They first generate the initial population randomly and subsequently use the savings algorithm and the nearest neighbor routing heuristic. Mirabi et al. (2010) present three hybrid heuristics for the MDVRP that combine constructive heuristic search and improvement methods.

The heuristic presented in this chapter is based on Vidal et al. (2012). Their Hybrid Genetic Search with Adaptive Diversity Control (HGSADC) addresses three extensions of the VRP including the MDVRP. One notable difference between regular genetic algorithms and the algorithm developed by Vidal et al. (2012) is in the control of population diversity. No mutation phase is included, but instead 'advanced population diversity management schemes' are used with both feasible and infeasible solutions, while the population is refreshed if a prespecified maximum number of iterations has been made without finding improvements. Also, a diversity contribution is included in the fitness measure used to evaluate the chromosomes. This enhances the evolution of the population while avoiding premature convergence. Vidal et al. (2012) demonstrate that their algorithm is the best performing method to approach the MDVRP, which makes it a natural candidate for our work.

5 Hybrid Genetic Algorithm

The general scheme of our meta-heuristic for assigning consumers' orders to the best fulfillment and logistics channels is presented in Fig. 3. The heuristic generates an initial population of feasible and infeasible chromosomes, where each chromosome represents a routing scheme corresponding to a randomly determined assignment scheme. Next, the chromosomes undergo some route improvement techniques during the education step. Thereafter, roulette wheel selection is applied to select parents. The parents' assignment schemes are randomly combined to produce offspring, and corresponding routing schemes for this offspring are determined. The offspring undergoes the education procedure and is included in the subpopulations based on its (in)feasibility. Below, the various steps of the algorithm will be explained. Our algorithm follows the lines of the algorithm of Vidal et al. (2012) with modifications as presented by Kolman (2012).

5.1 Generation of Initial Population

In the first step of the algorithm, an initial population is generated containing 4μ chromosomes. The set P_{op} represents the chromosomes. The first information for the chromosomes is created by randomly assigning the n customer locations to a

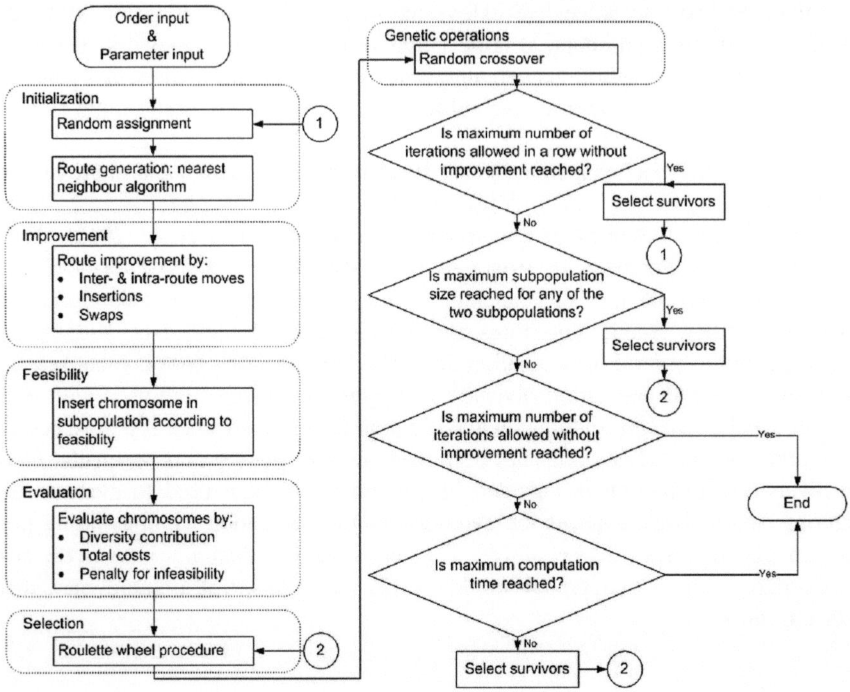

Fig. 3 Flowchart representation of the hybrid genetic algorithm

source i where i represents the DC ($i = 1$) and the stores otherwise ($i = 2,\ldots, s$). We define the sets of locations $V_D = \{1,\ldots, s\}$, $V_C = \{s + 1,\ldots, s + n\}$ and $V = V_D \cup V_C$. The assignment of a customer location to a source can be defined by matrix $Z(r) = z_{ij}(r)$ where $z_{ij}(r) = 1$ if customer location $j \in V_C$ is assigned to source $i \in V_D$, and zero otherwise. Besides the source assignments, the chromosomes also contain the information on the routes and vehicle assignments. We use a nearest neighbor routing method with vehicle capacity constraints to generate routes for the initial population. The matrix $X(r) = x_{ijk}(r)$ represents the routing scheme of chromosome r where $x_{ijk}(r) = 1$ if arc (i, j) is contained in route $k \in R(r)$ and zero otherwise, with the set $R(r)$ representing the routes of chromosome r.

Chromosomes may be infeasible since the maximum route duration T was not yet taken into account. Let t_{ij} denote the time for traveling from location i to location j, and let τ_j denote the required unloading time at customer j, then the maximum route duration constraint is given by $\sum_{i \in V} \sum_{j \in V} x_{ijk}(r)(t_{ij} + \tau_j) \leq T$. Chromosomes are inserted in the feasible or infeasible subpopulation, based on this constraint. At the end of the initialization step, at least one of the subpopulations contains μ chromosomes.

5.2 Evaluation

The chromosomes are evaluated by their fitness as represented by a function, which combines the rank of the costs and the rank of the diversity contribution of the chromosome. According to Vidal et al. (2012), "empirical studies show that these new mechanisms not only avoid premature population convergence efficiently, but also lead to higher-quality solutions in reduced computation time when compared to traditional approaches". The costs are composed of two components: total costs (i.e. transport and handling costs) and a penalty cost corresponding to the feasibility of the chromosome.

Transport costs are calculated as a linear function of distance

$$c_{tr}(r) = \sum_{i \in V} \sum_{j \in V} x_{ijk}(r) \cdot d_{ij} \cdot c_1,$$

where c_1 represents a cost per kilometer, including fuel, driver, insurance, and fleet costs, and where d_{ij} represents the distance of driving from i to j.

Handling costs are counted at all intermediate points as

$$c_h(r) = \sum_{i \in V_D \setminus \{1\}} \sum_{j \in V_C} z_{ij}(r) \cdot q_{ij} \cdot c_2,$$

where the handling costs are represented by c_2, a cost per unit of volume, and where q_j gives demand at customer location $j \in V_C$ expressed in the same unit of volume. Total cost are then given by $c_t(r) = c_{tr}(r) + c_h(r)$.

Penalty costs $\psi(r)$ and the *diversity contribution* $\Delta(r)$ of chromosome r, and the evaluation function are calculated as in Vidal et al. (2012) and given by

$$EV(r) = \xi_c \cdot RANK\left(\frac{1}{c_t(r) + \psi(r)}\right) + \xi_{dc} \cdot RANK(\Delta(r)).$$

The function RANK(.) here presents the rank of chromosome r relative to the other chromosomes, based on the criterion given between brackets. The best chromosome is the one that maximizes the evaluation function. Ranks are weighted by ξ_c and ξ_{dc} which are dynamically adjustable.

5.3 Education

When the evaluation process is complete, all chromosomes are educated to find improvements over the initial nearest neighbor routing scheme. The procedure uses the nine route improvement techniques from Vidal et al. (2012) which mainly consist of various ways for swapping vertices or partial routes. The nine moves are randomly investigated, and the first move yielding an improvement of the evaluation value $EV(r)$ is taken. This continues until none of the nine moves results in an improvement. Then the feasibility of the chromosome is investigated and the new evaluation value of the chromosome, $EV(r)$ is obtained. When the feasibility of the chromosome has changed, it also changes between subpopulations.

5.4 Generating Offspring

The first step in creating offspring is the selection of parents. We apply the roulette wheel selection operation (Goldberg 1989) which takes the evaluation criterion of the chromosomes into account to select the best chromosome with the highest probability and the worst with the lowest. Let the probability of selecting chromosome r as parent be defined by $(r) = EV(r)/\sum_{\rho \in P_{op}} EV(\rho)$.

To create offspring the assignment schemes of two parents, each selected by means of the roulette wheel, are randomly combined. This random crossover results in a new assignment scheme for a new chromosome. From this point onwards the newly created offspring follows the same procedure as the chromosomes do in the process of generating the initial population, i.e. routing of vehicles, feasibility check, calculating the evaluation value, and education. Offspring is created as long as the following two conditions are satisfied: (1) the two subpopulations have not reached their maximum size $\mu + \lambda$, and (2) the prespecified number of iterations allowed in a row without improving over the best solution so far has not been reached.

When condition (1) is no longer satisfied, only the μ best chromosomes are kept of each subpopulation and the process re-enters the stage of generating offspring. When condition (2) is no longer satisfied, only the $\mu/3$ best chromosomes are kept from both subpopulations and the process re-enters the stage of creating an initial population to diversify the chromosomes.

5.5 Stopping Criterion

The process evolves generation by generation until either one of the two stopping criteria is met. The two stopping criteria are a maximum computation time, T_{max}, and a maximum number of iterations the algorithm is allowed to make without improving over the best solution, $IT_{allowed}$. When the computation time or the number of iterations reaches its maximum value, the algorithm stops and the best chromosome, i.e. the best found routing scheme is returned as the solution.

6 Results

To represent the wide diversity of existing practical situations, we use three descriptors. The first descriptor concerns the number of locations (source locations plus customer locations). We consider a high value of 350 zip code areas including one DC, nine stores and 340 consumer locations; and a low value of 50 zip code areas including one DC, four stores and 40 consumer locations. Zip code areas are randomly drawn from the total of 4039 possible four-digit zip codes in The Netherlands. The program XCargo is used to obtain both the distance matrix and the time matrix for each pair of locations.

The second descriptor concerns the consumers' demand, expressed in square meters. Note we do not use cubic meters, since normally in home delivery the products are not stacked in the truck to make unloading faster and safer. Random draws for consumers' demand are taken from either the set {1.5; 2; 2.5} or the set {4.5; 5; 5.5}. The third descriptor concerns the required service time at the customer locations, in minutes, where high values are randomly drawn from {30; 60; 90} and low values from {10; 30; 60}. Maximum route duration is set at 10 h. The values for these factors and constraints are based on data and expert opinions from the company.

The program NEA was used to obtain a transport cost per kilometer of €1.53, which includes fuel costs, driver costs, insurances, fleet costs and maintenance costs. Expert knowledge was used to define the average time required to load and unload a vehicle and the cost per hour for an employee to do this work. This information is combined to acquire handling costs of €4.25 per m^2. Truck capacity is set at 15 m^2. Parameter tuning was performed before the hybrid genetic algorithm was put into action, as described in Kolman (2012). As a baseline for comparisons, we use a heuristic that assigns customers to the nearest store and determines truck routes using a nearest neighbor routing heuristic with capacity constraints. We performed a series of experiments with 25 replications per situation, which allows for determining confidence intervals for the results.

When comparing the hybrid genetic algorithm to the baseline heuristic, we observed improvements ranging from 1.3 to 9 % on average for the various scenarios. This improvement actually is lower than we anticipated and may be caused by several reasons. Our initial thought was that the routing scheme in the genetic algorithm was not adequate. However, experiments could not confirm this hypothesis. On the contrary, on our search for explanations we found that even switching off the entire education step seemed to have only a limited effect in many situations. The explanation for this is found in the number of customers that are served per route. For the high demand volume scenario, two or three locations are visited per route; for the low demand volume scenario, an average of seven locations. These numbers match the actual number of stops in the company. However, from an Operations Research perspective on the Traveling Salesman Problem, visiting fewer locations implies fewer possible routes and hence an easier problem to solve. In the extreme case, for routes visiting two locations, there is only one possible route, so all routing heuristics will provide the optimal solution. The improvements we found are therefore largely due to the assignment of customers to locations, which is directly linked to our prime research goal of channel assignment.

Next, we investigated efficiency gains from choosing the best fulfillment and logistics channel on a per customer basis. The confidence interval for the paired differences between the genetic algorithm with dynamic channel assignment and the genetic algorithm with only delivery from the DC is [18.3 %; 27.7 %]. This implies an average improvement of 23.0 % when switching from pure centralized distribution to a setup with partly centralized and partly localized fulfillment and distribution for all online orders. Note that additional efforts for the extra handling at the stores have been taken into account in these experiments.

7 Conclusions

In this chapter we have introduced a framework for multichannel order fulfillment and logistics for online and store sales of physical goods. Seven elements of multi-channeling were identified: the search channel, the purchase channel, the forward fulfillment channel, the forward logistics channel, the return logistics channel, the return fulfillment channel, and the after sales channel. Often companies offer multiple channels and allow customers to select the channel of their preference. From a marketing perspective this process is widely studied for the search, purchase and after sales channel. The choice of the fulfillment and logistics channels, however, in many cases can be made by the company itself based on efficiency, as was investigated in this chapter.

We studied the online and in-store sales of large household items, such as appliances and furniture, as inspired by a Dutch retailer. In this company, and in the authors' experience in many similar companies, fulfillment and logistics channels are assigned based on the consumer's choice for the purchase channel. Thus, online orders are fulfilled from the central warehouse, and store orders from the store. However, since products are home delivered and all trucks of the company look similar, customers cannot tell the difference between the two channels. Using the algorithm we developed, it is shown that a dynamic assignment of fulfillment and logistics channels can result in an average efficiency gain of 23 %, thus providing a compelling signal for retailers to investigate the possibilities of more dynamically organizing multichannel decisions within their network.

8 Limitations and Future Work

The work presented in this paper provides research results based upon a blend of practical and theoretical aspects for multichannel decisions. Based on academic literature, the authors' experience, and discussions with companies, a framework for channel selection for physical goods sales to consumers was presented. Inspired by the challenges faced by a Dutch retailer selling large household items a new heuristic was developed for managing the forward fulfillment and forward logistics channels, which builds upon the academic work of Vidal et al. (2012). As a consequence of taking both practice and theory into consideration in the research, limitations of the work and possible future work also arise from both viewpoints.

From a practical point of view, we note the following limitations to the forward channel selection heuristic. The heuristic assumes a setting that is described primarily in terms of the physical locations of the distribution points (typically: shops). In practice, not all shop managers may be willing to participate, or only to a limited extent. This may hold in particular when the company uses a franchise structure, where individual stores each have their individual (local) owners. Furthermore, depending on the increased product flows at the various locations due to the proposed new configuration, additional storage capacity, material handling equipment and employees may be required. Or alternatively, product flows may need to be restricted to comply with current limitations at the various locations. This causes additional aspects that may not in all cases be reducible to the cost parameters of the current model. From a theoretical point of view, the channel assignment decisions of the heuristic are shown to be effective. However, since the instances cannot easily be solved to optimality, it is unclear as to whether further solution quality improvements may or may not be possible.

For further research, several aspects can be considered. Firstly, the seven elements of multichanneling as depicted in Fig. 1, though thoroughly based on literature and observations in practice, could potentially benefit from further refinements that may be obtained from a test of this framework in a number of practical settings. Secondly, the heuristic for forward channel decisions can be extended in several ways. A first and important step in the advancement of the heuristic would require to allow for multiple different products in the model, such that trade-offs in transport capacities and costs between product categories can be made explicit. Subsequently, extensions may include the addition of capacity restrictions or other restrictions at locations or the inclusion of return flows.

Acknowledgment This study was supported by a grant of the Dutch Institute for Advanced Logistics (Dinalog).

References

Cordeau JF, Gendreau M, Laporte G (1997) A tabu search heuristic for periodic and multi-depot vehicle routing problems. Networks 30:105–119

European Commission (2014) Shopping: returning unwanted goods. http://europa.eu/youreurope/citizens/shopping/shopping-abroad/returning-unwanted-goods/index_en.htm. Accessed 15 Sept 2014

Frazier GL (1999) Organizing and managing channels of distribution. J Acad Mark Sci 27 (2):226–240

Gendreau M, Potvin JY, Bräysy O, Hasle G, Løkketangen A (2008) Metaheuristics for the vehicle routing problem and its extensions: a categorized bibliography. In: Golden B, Raghavan S, Wasil E (eds) The vehicle routing problem: latest advances and challenges. Springer, Heidelberg, pp 143–169

Goldberg DE (1989) Genetic algorithms in search optimization and machine learning. Addison-Wesley Longman Publishing Co., Inc., Boston

Hill A, Collier D, Froehle C, Goodale J, Metters R, Verma R (2002) Research opportunities in service process design. J Oper Manag 20:189–202

Ho W, Ho GTS, Ji P, Lau HCW (2008) A hybrid genetic algorithm for the multi-depot vehicle routing problem. Eng Appl Artif Intell 21(4):548–557

Jongen W (2015) Aandeel online productverkoop in NL detailhandel naar circa 7,6 %. http://wijnandjongen.nl/aandeel-online-productverkoop-in-nl-detailhandel-naar-circa-76/. Accessed 4 May 2015

Kolman IB (2012) A hybrid genetic algorithm to solve the multi-depot vehicle routing problem. Master's thesis, University of Groningen, The Netherlands

Liong CY, Wan Rosmanira I, Khairuddin O, Zirour M (2008) Vehicle routing problem: models and solutions. J Qual Meas Anal 4(1):205–218

Maxwell JG, Hudson M (2011) Preface. In: Customers take control, how the multi-channel shopper is changing the global retail landscape. http://www.pwc.com/gx/en/retail-consumer/retail-consumer-publications/global-multi-channel-consumer-survey/assets/2011-multi-channel-survey.pdf. Accessed 15 Sept 2014

Mirabi M, Fatemi Ghomi SMT, Jolai F (2010) Efficient stochastic hybrid heuristics for the multi-depot vehicle routing problem. Robot Comput Integr Manuf 26(6):564–569

Neslin SA, Grewal D, Leghorn R, Shankar V, Teerling ML, Thomas JS, Verhoef PC (2006) Challenges and opportunities in multichannel customer management. J Serv Res 9(2):95–112

Pisinger D, Ropke S (2007) A general heuristic for vehicle routing problems. Comput Oper Res 34 (8):2403–2435

Renaud J, Laporte G, Boctor FF (1996) A tabu search heuristic for the multi-depot vehicle routing problem. Comput Oper Res 23:229–235

Roodbergen KJ, Vis IFA, Boter J (2012) Cross-chain order fulfillment for Internet sales. Dinalog, The Netherlands. http://www.dinalog.nl/cms_file.php?fromDB=2832&forceDownload. Accessed 15 Sept 2014

Vidal T, Crainic TG, Gendreau M, Lahrichi N, Rei W (2012) A hybrid genetic algorithm for multidepot and periodic vehicle routing problems. Oper Res 60(3):611–624

Zhang X (2009) Retailers' multichannel and price advertising strategies. Market Sci 28 (6):1080–1094

Part III
Value Chain Management

Connecting Inventory and Repair Shop Control for Repairable Items

Martin A. Driessen, Jan Willem Rustenburg, Geert-Jan van Houtum and Vincent C.S. Wiers

Abstract This paper presents control structures for integrating decisions on spare parts inventories and repair shops. A different control structure is proposed for the various repair shop types recognized in practice. An overview of related literature is presented and open research topics are identified. The applicability of the control structures is verified in case studies. This paper furthermore aims to provide a basis for further research on designing control structures for the integrated control of spare parts inventory and repair shops. An agenda for further research is presented.

Keywords Repairable inventory systems · Repair shops · Spare parts planning and control · Case studies

1 Introduction

Companies in many industries rely on the dependability, availability and safety of high value capital assets in their primary processes. Unplanned unavailability of the assets results in lost revenues, customer dissatisfaction and high claims or down-time costs. To obtain a high availability of the assets against acceptable maintenance logistics costs is of increasing importance, as companies operate in increasingly competitive markets (e.g. aerospace) or face decreasing budgets (e.g. military organizations).

A significant part of the companies in these industries both use and maintain their high value capital assets (i.e. a "user network" according to Basten and van Houtum (2014)). During the maintenance of the assets, components are removed

M.A. Driessen (✉) · G.-J. van Houtum · V.C.S. Wiers
School of Industrial Engineering, Eindhoven University of Technology,
P.O. Box 513, 5600 MB Eindhoven, The Netherlands
e-mail: m.driessen@gordian.nl

M.A. Driessen · J.W. Rustenburg
Gordian Logistic Experts, Groenewoudsedijk 63, 3528 BG Utrecht, The Netherlands

H. Zijm et al. (eds.), *Logistics and Supply Chain Innovation*,
Lecture Notes in Logistics, DOI 10.1007/978-3-319-22288-2_12

and directly repaired or replaced by spare parts. In the latter case, availability of the required spare parts is critical to timely complete asset maintenance. Spare parts can be repairable or non-repairable (consumable); the first are inspected and subsequently repaired (or refurbished) in a dedicated shop.

Repairable inventory systems are systems in which failed parts are repaired and returned to service, rather than scrapped (Hausman and Scudder 1982). Decisions in repairable inventory systems are interrelated; e.g. the required number of repairable spare parts depends on the rate at and sequence in which the repairs are conducted. This requires a certain coordination between an inventory department, responsible for setting spare parts stock levels, and the repair shop, responsible for conducting repairs.

This turns out to be hard in practice, as we observed in in-depth case studies at six companies. We observed serious lacks of consistency between the departments of inventory management and the repair shops. For example, an inventory department often assumes a fixed repair lead time when determining stock levels. The repair shop may face serious problems in realizing these lead times. Risks for unanticipated waiting times are multiple, due to e.g. specialized repair shops with a small number of repair men and a reasonably high utilization rate, where waiting times may vary from long to short. In such cases, the assumption of 100 % reliable lead times is far from realistic and needs to be adjusted to the specific characteristics of repair shops.

A huge body of literature exists both on repair shop (and production) control and spare parts management (see e.g. the review papers Guide Jr. and Srivastava 1997; Kennedy et al. 2002; Basten and van Houtum 2014). Nevertheless, the literature is limited to models that assume a single type of repair capacity and deny the material uncertainty encountered by repair shops in practice (e.g. Sleptchenko et al. 2003; Caggiano et al. 2006). Literature on control structures for repairable inventory systems is, to the best of our knowledge, lacking. In this paper, we study such control structures, and we hope that it stimulates more research in this area.

Literature on designing control structures for production systems (Bertrand et al. 1990) could provide relevant starting points for setting up the control structures. However, applying those starting points is far from straightforward, because repair shops are not all one and the same (as will be discussed in the sequel) and in addition show characteristics different from classical production shops. Both arrival patterns and repair times may be stochastic, while also material requirements may depend on the nature of the failure (and hence are less predictable in advance). The fact that a repairable inventory system is a closed-loop system (turn-around stock), rather than an open system (customer orders from outside), marks another difference. Our control structures are based on results from the spare parts literature, insights from case studies at six companies, the experience of the authors in the field of service logistics and logical reasoning. We present our design principles and if possible we refer to the literature in case it supports our design choices.

Driessen et al. (2013) indicate that a different control structure is required for the four repair shop types recognized in practice. This paper presents control structures for these four repair shop types. We describe the designs in detail and we (i) map the various decision functions on the interface between the inventory department and the repair shop, (ii) assign each decision function to either the inventory department or the repair shop and (iii) review the available literature and give suggestions for further research.

The control structures presented in this paper serve as a useful starting point for creating control structures for repairable inventory systems. For companies with existing control structures, the proposed control structures have a mirror function. That is, they serve as a useful reference point to redesign existing control structures of repairable inventory systems. The use of these control structures, as a reference point for redesigning existing control structures of repairable inventory systems, is verified in workshops at two companies.

This paper is organized as follows. In Sect. 2, we describe the repairable inventory system. Section 3 briefly describes the different repair shop types presented by Driessen et al. (2013). In Sect. 4, we present differentiated control structures for all repair shop types, refer to related literature and identify open research topics. In Sect. 5 we report on findings from two workshops, in which we used our control structures as a reference point for redesigning existing control structures of repairable inventory systems. Section 6 concludes the paper and suggests further research on developing and improving control structures for repairable inventory systems.

2 Repairable Inventory Systems

2.1 System Description and Objective

We consider a company that uses and maintains high value capital assets. The assets are maintained according to a "repair-by-replacement" strategy: parts that require maintenance are removed from the asset and, if available, replaced by ready-for-use (RFU) spare parts. A request for a spare part that cannot be met from the RFU stock, is backordered and fulfilled as soon as the requested part becomes available.

The parts are removed either (i) upon failure (corrective maintenance), (ii) to prevent future failure (preventive maintenance) or (iii) on exceeding a certain condition threshold value (known via inspection or condition monitoring). Consequently, the demand for spare parts is planned in advance or not (i.e. unplanned). Table 1 summarizes the different maintenance types, drivers for spare parts demand and their characteristics.[1]

Parts removed from the asset are called *Line Replaceable Units* (LRU's) and are sent to the "Failed LRU's" stock. Failed LRU's that are (economically) repairable

[1]Note that modificative maintenance (to improve the performance of the asset) in principal does not lead to the repair of the removed LRU's.

Table 1 Different maintenance types including drivers and characteristics

Maintenance type	Driver(s)	Characteristics
Corrective	Failure	Unplanned, random
Preventive (age/usage based)	Time, running hours	Planned, usually known way ahead (e.g. one year)
Preventive (inspection, condition monitoring)	Condition	Planned, usually known some time ahead

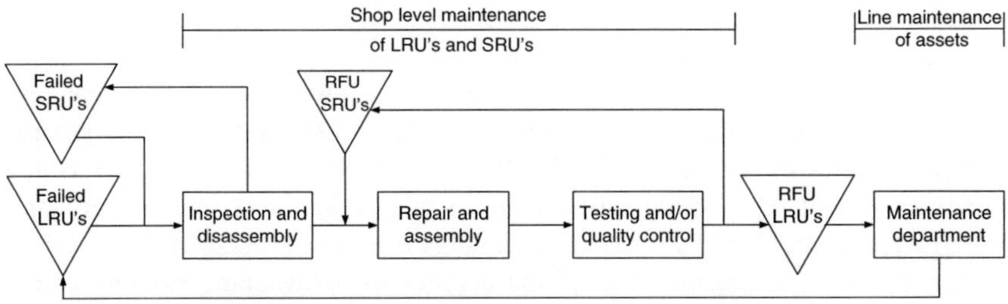

Fig. 1 Supply chain of repairable spare parts including the repair process

are sent to a repair shop for repair. The repair process consists of three phases: (1) inspection and disassembly, (2) repair and assembly and (3) testing and/or quality control. After the inspection phase, it is known which parts need to be removed from the LRU and replaced by spare parts. Parts removed at the repair shop are called *Shop Replaceable Units* (SRU's). SRU's can be either repairable or consumable and hence there are multiple levels of repair (indenture levels). The process described above is illustrated in Fig. 1.

In the repair process, repair capacity and often materials are required. The exact requirements are known after the inspection phase. This implies that the inspection phase is of importance, because the part failure and hence the requirements are not known beforehand.

Each repair job consists of multiple repair steps. A repair man is required during each step of the repair job, and each repair step requires a certain skill. We make a distinction between specialized and basic skills. Each repair man has a subset of these skills and is hence able to perform certain repair steps. The availability of specialized skills is often limited in the repair shops.

The repair shops have the possibility to subcontract repair jobs and the possibility to apply overtime policies for all repair men, though with a certain maximum threshold value per period. Furthermore, the repair shops have the possibility to conduct emergency repairs. This is relatively easy for repair jobs that require only basic skills; in case specialized skills and capacities are required, this may result in extra set up or changeover times.

2.2 Decisions and Coordination in Repairable Inventory Systems

The company that we consider has an existing supply chain and multiple repair shops. The layout (and costs) of the repair shops (i.e. facilities, machines, tools and manuals that are shared by multiple repair men in a repair shop) is assumed to be given and fixed for the long term, whereas the repair men capacity (including tools that are permanently allocated to a repair man) may vary.

The objective of a repairable inventory system is realizing a certain availability of the high-value capital assets (from a spare parts availability point of view) against acceptable maintenance logistics costs. The boundaries of a repairable inventory system are determined by its repair limitations: a limited set of capacities (typically grouped in one repair shop) is only able to conduct repairs for a limited set of repairable parts. This implies that a company generally has multiple repair shops and hence multiple repairable inventory systems (one per repair shop).

According to Hausman and Scudder (1982), the relevant decisions in a repairable inventory system are: (1) the inventory policy (parameters) of the LRU's and SRU's, (2) the repair capacities available to repair failed parts and (3) the scheduling system used to control the flow of work in the repair shop. Since decisions in repairable inventory systems are interrelated, this requires a certain coordination (set of agreements) between the inventory department and the repair shop. The agreements between the repair shop and inventory department are arranged in an interface agreement. The structure of the interface agreement, i.e. the type of agreements that are made, is part of the control structure design. The interface structure prescribes how the repair shop and the inventory department cooperate.

Figure 2 outlines decision functions related to both repair capacity planning and spare parts planning on a strategic, tactical and operational level. We concentrate on decisions that are directly interrelated; these decision functions are highlighted in

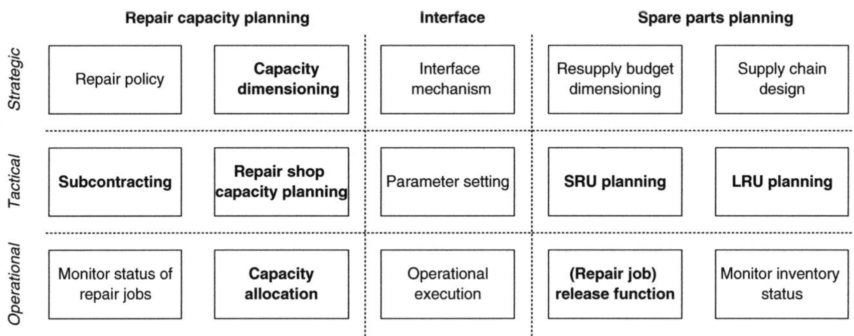

Fig. 2 Decisions on different levels related to the repair and planning and control of repairable spare parts

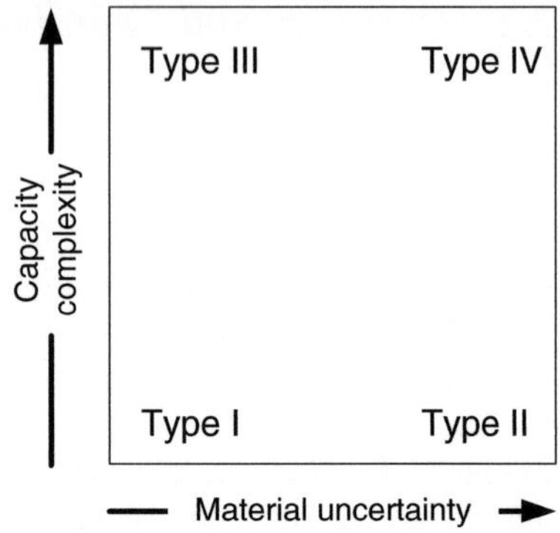

Fig. 3 Different types of repair shops for maintenance spare parts (Driessen et al. 2013)

bold. Strategic decisions regarding supply chain design and layout and location of repair shops are beyond the scope of this paper. We explicitly note here that decisions regarding spare parts (repair capacity) planning are not automatically assigned to the inventory department (repair shop), rather it depends on the design of each specific control structure.

3 Repair Shop Typology

Empirical research by Driessen et al. (2013) shows that different types of repair shops can be found in practice.[2] Figure 3 presents the repair shop typology in which the variables on the two axis are the distinctive criteria for the classification. The *capacity complexity* concerns the extent to which specialized capacities are required in the repair process. The *material uncertainty* concerns the extent to which the set of required materials varies for repair jobs of the same LRU. Table 2 summarizes the characteristics of the capacity complexity and material uncertainty.

In repair shops with a high material uncertainty, repair jobs are often interrupted due to missing materials. The repair job lead time mainly consists of waiting time for materials. In repair shops with a high capacity complexity, the repair job lead time mainly consists of waiting time for the required (specialized) resources. In repair shops with a high capacity complexity, the parts that require repair are sent from workstation to workstation (i.e. the part goes "on a journey"), whereas the part needs only one or two workstations in case of low capacity complexity. Shops with a high capacity complexity experience shifting bottlenecks, especially for repair men with specialized skills.

[2]For a classification of the repair shops in our case study companies, see Sects. 5.2 and 5.3.

Table 2 Characteristics of a low/high capacity complexity and material uncertainty (Driessen et al. 2013)

	Low	High
Capacity complexity	Homogenous repair capacity, low number of specialized skills required	Heterogeneous repair capacity, high number of specialized skills required
	Narrow product mix, i.e. low diversity in the type of items to repair	Wide product mix, i.e. high diversity in the type of items to repair
Material uncertainty	Often the same SRU's required for repair jobs of the same LRU	The sets of SRU's that are required for repair jobs of the same LRU vary strongly
	Small set of SRU's that has to be kept on stock	Large set of SRU's that has to be kept on stock

4 Control Structures for Repairable Inventory Systems

As mentioned in the introduction, the control structure designs that we present in this paper follow from an empirical approach and not from a theoretical methodology. The designs should in the first place contribute to the aim of a repairable inventory system: realize a certain availability of the high-value capital assets (from a spare parts availability point of view) against acceptable maintenance logistics costs.

We use the following guiding design principles:

1. The interface agreement should be part of the design.
2. The coordination between (or integration of) interrelated decision functions should be part of the design.
3. The repair process should be part of the design.
4. The degree of coordination between individual decision takers should be minimal.
5. The design should be as simple as possible.

In this paper, we limit ourselves to control structure designs for the four repair shop types of the typology, presented in Sect. 3. This typology describes the characteristics of the repairable inventory system regarding the capacities and materials required in the repair process.[3]

In the remainder of this section, we outline the design of the control structures for all repair shop types. We combine Figs. 1 and 2 and assign the decision functions in Fig. 2 to the repair shop or to the inventory department. For each repair

[3]Repair shops with a high 'no fault found' or scrap rate are, from a control perspective, comparable to repair shops with a high material uncertainty. Therefore we do not account for scrap of parts and 'no fault found' in the design of the various control structures.

shop type, we outline (i) the characteristics and design principles, (ii) the control structure design, and (iii) a review of available literature, including suggestions for further research.

4.1 Repair Shop Type I

4.1.1 Characteristics and Design Principles

Repair shop type I is characterized by a low capacity complexity and a low material uncertainty. Usually this type of repair shop repairs a small number of different parts with a low diversity. The added value of the repairs is typically low because no expensive materials and specialized skills are required, and utilization rates will be high. Typically a relatively small number of parts is responsible for a large portion of the repair work of the repair shop. Examples are the emergency bottles shop or the wheels and tires shop at one of our case study companies.

A low material uncertainty means that no SRU's or often the same set of SRU's is required. The number of SRU's to stock is low and it is easy and relatively cheap to arrange for a high availability of the SRU's. The benefits of integrating SRU and LRU inventory control are low. The repair capacity is homogeneous and the number of specialized skills is low. This means that the repair job release is only influenced by the aggregate release plan and not by specific capacity aspects, and repair jobs can be released as soon as (repair man) capacity becomes available.

The choice of how to set up the interface between the inventory department and the repair shop depends on how the repair process is organized. Typically this shop type has multiple identical servers (repair men) that are able to conduct all types of repairs, generally without interruptions and with short setup and changeover times. For this reason, the WIP is usually low and repair lead times can be kept short (equal to the effective repair time).

The inventory department has the best knowledge about which repair jobs to prioritize from an inventory control perspective. Moreover, the inventory department is best able to balance the costs of internal and external repairs. Since the capacity complexity is low, the priority from a repair capacity perspective (i.e. to optimize repair capacity utilization) is less important. For these reasons, the inventory department should be able to set the repair job priorities. In this control structure, the inventory department 'pushes' the repair jobs into the repair shop.

4.1.2 Control Structure

We propose to set a target level on the workload (in hours of work) that is cleared by the repair shop each period (e.g. one week), given the constraint that sufficient work is available for the shop. This target level is jointly determined with the capacity decision and the LRU base stock levels at the aggregate level. It is the

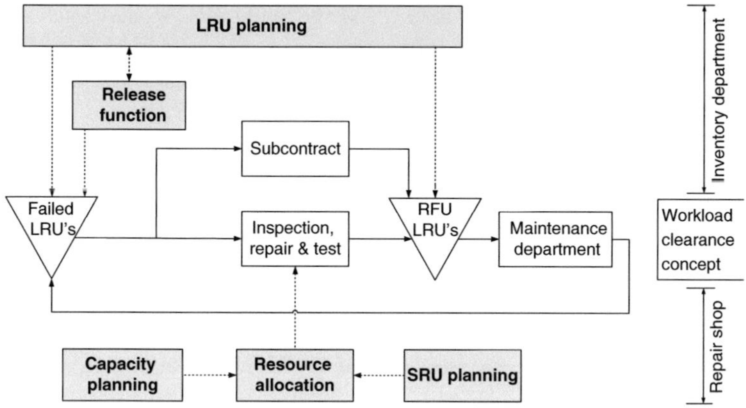

Fig. 4 Process and control structure design for repair shop type I

responsibility of the repair shop to timely clear the shop, i.e. to make sure that the (cumulative) backlog based on the target levels is below a certain threshold.

The repair shop and the inventory department agree on (i) the maximum amount of cumulative backlog and (ii) the maximum amount of work that is released to the repair shop per period. The process and control structure design of repair shop type I can be found in Fig. 4.

The low number of repair job interruptions, combined with a restriction on the maximum amount of work released to the repair shop, enables the repair shop to ensure a high probability that the backlog is below the maximum threshold value. We propose to use a constraint (input rule) for the weekly offered workload (in hours). The workload constraint is a trade-off between inventory capacity, repair capacity and subcontracting options.

As the repair shop is responsible for timely clearing the workload, and the benefits of integrating SRU and LRU inventory control are low, it is rather obvious that the repair shop is responsible for the inventory control of the SRU's. In this way, the repair shop has full responsibility for timely clearing the workload.

At the aggregate control level, a queuing type analysis can be used to determine the repair capacities, the LRU base stock levels and the budget for subcontracting repairs. In this analysis, one should take into account a scheduling rule that accounts for the differences between inventory holding costs of various LRU's. This analysis implicitly also determines the threshold value for subcontracting repair jobs, that is, repair jobs are subcontracted as long as the amount of repair work in the "Failed LRU's" stock exceeds this threshold value.

Detailed scheduling of repair jobs is not required, and the repair jobs can be released by the inventory department as soon as repair man capacity becomes available. Repair jobs are released based on the run-out times of the LRU's (i.e. dynamic priority rules). Note that in this design, emergency repair jobs do not really affect the repair shop planning. Resource allocation is easy as the repair capacity is homogenous and both repair man capacity and SRU's may be allocated first-come-first-serve to the repair jobs.

Alternatively, sometimes we see that type I repair shops face relatively high setup or changeover times for repair. An example is the emergency bottles shop at one of our case study companies. For this shop, it is important to account for the changeover times between repair of different LRU parts, in order for the repair shop to efficiently utilize its resources. In such cases, it is sometimes preferred to work with static priority rules so that the repair shop itself can determine when to switch the repair from one LRU to another.

4.1.3 Literature Review and Suggestions for Further Research

The interface between the inventory department and the repair shop is based on a workload clearance concept, in which the workload constraint is a trade-off between the number of LRU's to stock, the repair shop capacity and subcontracting options. This trade-off is an integrated (long-term) decision made by the inventory department and the repair shop.

Table 3 provides references to literature on decision support models for repair shop type I. We describe how to make/support the decision, how the available literature fits the need for decision support and provide suggestions for further research.[4]

4.2 *Repair Shop Type II*

The control structure design of repair shop type II is not much unlike the control structure of repair shop type I. For this reason, we only describe the main differences between both types, including the implications for the design of the control structure.

4.2.1 Characteristics and Design Principles

Repair shop type II differs from repair shop type I in that it is characterized by a high material uncertainty. The added value of repairs mainly consists of the value of the materials used in the repair process. Repair jobs are conducted one by one; batching of repair jobs of the same LRU parts is more difficult because different SRU's are required in different repair jobs. An example is the electronics repair shop in one of our case study companies.

[4]Research to apply batching policies (of e.g. repairs, returns or transshipments) is beyond the scope of this paper.

Table 3 Literature review and suggestions for further research for repair shop type I

Decision function	Responsibility	Suggestions and available decision support	Open research topics
Workload constraint	Inventory department and repair shop	Trade-off repair and inventory capacity: Sleptchenko et al. (2003)	Extension of Sleptchenko et al. (2003) with the option of subcontracting repairs
Workload clearance levels	Inventory department and repair shop	Set equal to workload constraint	
Repair shop capacity planning	Repair shop	Follows directly from workload constraint	
LRU planning	Inventory department	(Variants of) METRIC with infinite repair capacity: Sherbrooke (1968, 1986)	Model with expediting and subcontracting repairs
		Expediting: Verrijdt et al. (1998), Arts (2013)	
		Regular and subcontracting options: Keizers et al. (2001)	
SRU planning	Repair shop	Single-item, single location inventory policies, see e.g. Silver et al. (1998)	
Release function	Inventory department	Real-time rolling horizon model: Caggiano et al. (2006)	Include option of subcontracting and working overtime
Resource allocation	Repair shop	Apply "first-come-first-serve" rule to allocate the repair capacity. Apply reactive overtime policy, see e.g. Scudder and Chua (1987)	

A high material uncertainty implies that often different sets of SRU's are required for repair jobs of the same LRU and the number of SRU's to stock is high. The number of repair job interruptions is moderate and mainly due to missing SRU's. The benefits of integrating SRU and LRU inventory control can be substantial (Muckstadt 1973).

Before the inspection phase of a repair job is completed, there is a large uncertainty on which SRU's are required in the repair process. In general, decoupling points are introduced at points in a repair (or a manufacturing) process where the predictability changes significantly (Bertrand et al. 1990). For this reason, we decouple the inspection phase and the repair phase of the repair jobs via a stocking point of parts that are "Ready to repair".

To minimize the waiting time for SRU's, it is important that the inspection phase of a repair job starts as quickly as possible after the part enters the "Failed LRU's" stock. The waiting time for parts that are ready to be repaired will increase though, if inspection of parts has a higher priority than the repair of parts. For repair shop type II, it is important to be able to find the optimal balance between inspection of parts (to decrease the waiting time of SRU's) and repair of parts (to decrease the waiting time of LRU's). It is the task of the inventory department to dynamically set the right priorities to inspection or repair of the right parts (based on actual stock levels of both LRU's and SRU's).

4.2.2 Control Structure

Similar to the design for repair shop type I, we propose to use a target level on the total workload that is cleared by the repair shop each period. In contrast to the design of repair shop type I, the cleared workload may consist of jobs in which only the inspection of a part is conducted. The inspection and repair phase of a repair job are decoupled, in order to be able to dynamically set the right priority to the inspection or the repair of parts. In this control structure, the inventory department 'pushes' the repair jobs into the repair shop. The control structure design for repair shop type II is visualized in Fig. 5.

The inventory department is responsible for the SRU and LRU inventory control. With an integrated approach the inventory department is able to find the optimal balance between the (average) waiting time for SRU's, that directly influences the waiting time to repair a part and hence inventory levels of the LRU's, and the inventory levels of the SRU's. The inventory department can influence the time a part spends in the "Failed LRU's" stock and the "Ready to repair" stock, by assigning the desired repair job priorities.

At the aggregate control level, a queuing type analysis can be used to determine the repair capacities, the LRU and SRU base stock levels and the budget for subcontracting repairs. In this analysis, one should take into account a scheduling

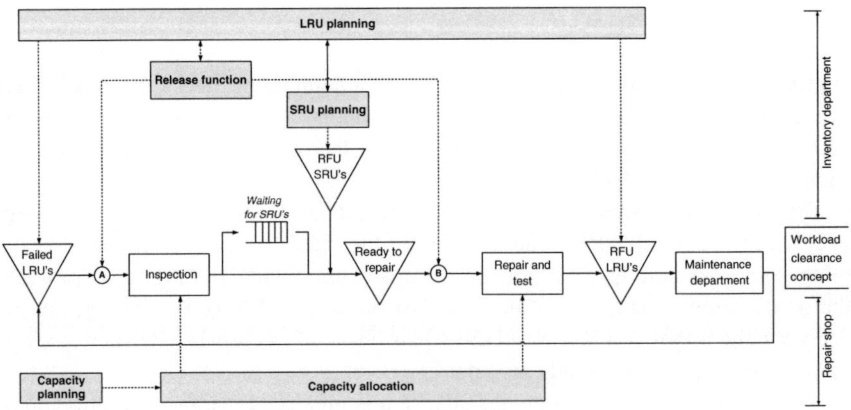

Fig. 5 Process and control structure design for repair shop type II

Table 4 Literature review and suggestions for further research for repair shop type II

Decision function	Responsibility	Suggestions and available decision support	Open research topics
Workload constraint	Inventory department and repair shop	Trade-off between LRU's and SRU's in MOD-METRIC model: Muckstadt (1973)	Include option that multiple SRU's can fail upon failure of a LRU
		Trade-off between inventory and repair capacity: Sleptchenko et al. (2003)	Account for priorities on operational level (inspection vs. repair)
Release function	Inventory department	Priority rules: Hausman and Scudder (1982)	Priority rules that incorporate distinction between inspection and repair phase

rule that accounts for the differences in inventory holding costs of each LRU and SRU, the decoupling of the inspection and repair phase of a repair job, and the inefficiency caused by the interruption of repair jobs.

Repair jobs can be released by the inventory department as soon as repair man capacity becomes available. Repair jobs are released based on the run-out times of the LRU's and availability of the SRU's (i.e. dynamic priority rules). Capacity allocation is easy as the repair capacity is homogenous, however allocating SRU's to repair jobs may be more difficult especially when SRU's are required for the repair process of multiple LRU's.

4.2.3 Literature Review and Suggestions for Further Research

Table 4 provides references to literature on decision support models for the decisions in the control structure of repair shop type II, in case it differs from repair shop type I.

4.3 Repair Shop Type III

4.3.1 Characteristics and Design Principles

Repair shop type III is characterized by a high capacity complexity and a low material uncertainty. The added value mainly consists of the use of specialized resources in the repair process. The number of repair job interruptions is moderate and mainly due to unavailable (specialized) capacities. For this reason, the WIP in repair shop type III is usually high and a repair man may be working on multiple repair jobs at a time. An example is the mechanics repair shop in one of our case study companies.

The repair capacity is heterogeneous and the number of specialized skills is high. Therefore it is important to effectively utilize the capacity of specialized skills, i.e. repair men with specialized skills should conduct specialized repair steps rather than steps for which only basic skills are required.

Since the material uncertainty is low, like in repair shop type I, the benefits of integrating SRU and LRU inventory control and decoupling the inspection and repair phase are low. Hence the repair shop is responsible for the complete repair job, and it is rather obvious that the repair shop is also responsible for the inventory control of SRU's.

The repair shop has the best knowledge about which repair jobs to prioritize, in order to effectively use the available (specialized) repair capacity. Since the capacity complexity is high, the priority from a repair capacity perspective is very important. Therefore the repair shop is responsible for setting the repair job priorities.

4.3.2 Control Structure

The inventory department has the best knowledge about which repair jobs to prioritize from an inventory control perspective. However, the repair shop is responsible for setting the (dynamic) repair job priorities. Therefore we propose that the inventory department is able to set static repair job priorities. The repair shop is able to dynamically set its own repair job priorities, though respecting these static priorities. The proposed control structure of repair shop type III can be found in Fig. 6. In this control structured design, the repair shop 'pulls' the repair jobs into the repair shop.

We propose to use a control mechanism that decouples the capacity planning of specialized skills and the LRU planning via these static repair priorities. The inventory department and the repair shop agree on static priorities, e.g. minimum and maximum stock levels for LRU's, based on which the repair shop has to schedule the repair jobs. With these constraints, the inventory department is able to

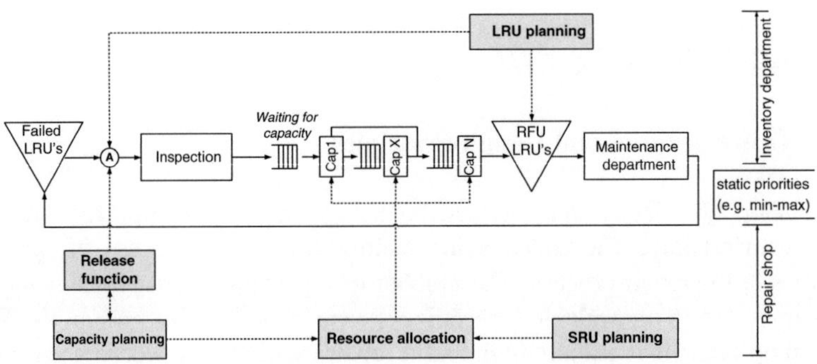

Fig. 6 Process and control structure design for repair shop type III

determine the number of parts required per LRU. These constraints also ensure that parts are repaired in approximately the right order from an inventory control perspective.

Within the constraints, the repair shop is able to optimize the utilization of the specialists skills, by combining repair steps of different repair jobs. It is important to timely anticipate on shifting bottleneck resources, e.g. by effectively scheduling ahead for the capacity of specialized skills. Proper overtime policies should be applied to make extra use of specialized skills in busy periods and let them idle in quiet periods. For the latter to work out, the static repair priorities should not be too tight.

Repair priorities are in general tight for slow moving LRU's, because the number of parts per LRU is low (typically one). If the repair shop has to repair a relatively high number of slow movers, then an alternative design may be to work with planned lead times instead of static repair priorities (lead time coordination concept). These planned lead times are usually longer than the effective repair time, because of the waiting times for capacities.

In such a concept, the repair shop is responsible for meeting the planned lead times with a high probability. Secondly, the repair shop is responsible to respect a certain repair job release pattern that is agreed upon with the inventory department, e.g. a minimum amount of repair jobs should be released by the repair shop per period. In this alternative control structure design, the inventory department should to some extent have the possibility to interrupt the repair shop schedule with emergency repair jobs.

4.3.3 Literature Review and Suggestions for Further Research

Table 5 provides references to literature on decision support models for the decisions in the control structure of repair shop type III, in case it differs from other repair shop types.

Table 5 Literature review and suggestions for further research for repair shop type III

Decision function	Responsibility	Suggestions and available decision support	Open research topics
Integrated repair capacity and LRU planning with static repair priorities	Inventory department and repair shop	Inventory planning with static priorities: Adan et al. (2009)	Model that integrates repair capacity planning (with a set of specialized skills per repair man) and inventory planning
Integrated release function and capacity allocation	Repair shop	Caggiano et al. (2006)	Extend to multiple capacity types (specializations)

4.4 Repair Shop Type IV

4.4.1 Characteristics and Design Principles

Repair shop type IV is characterized by a high capacity complexity and a high material uncertainty. Usually this type of repair shop repairs a large number of different parts with a large diversity and high value. The added value of the repairs is typically high because expensive materials and specialized skills are required. Utilization rates will be alternating high and low. Examples include the avionics and composites repair shop in one of our case study companies.

The number of repair job interruptions is high and due to unavailable (specialized) repair capacities and/or materials. A high material uncertainty calls for an integrated planning of SRU's and LRU's and the inventory department to be responsible for repair job release. A high capacity complexity calls for an optimized utilization of the (specialized) repair capacities and hence the repair shop to be responsible for repair job release. The design of the control structure should cope the fact that the repair job release function is a joint responsibility of the inventory department and the repair shop.

4.4.2 Control Structure

Like in the design for repair shop type II, we propose to decouple the inspection phase and the repair phase of the repair jobs via a stocking point of parts that are "Ready to repair". This decoupling point enables us to combine the control structure design of repair shop type II and the alternative design (lead time coordination concept) of repair shop type III. Hence the inventory department and the repair shop need to agree on repair lead times for the repair and test phase of a repair job. The control structure design for repair shop type IV can be found in Fig. 7.

We propose a workload clearance concept for the inspection of failed parts, that is, the inventory department and the repair shop agree on: (1) the maximum backlog of inspection work and (2) the maximum amount of work (in hours) related to the inspection of parts, released to the repair shop per period. The inventory department hence 'pushes' the repair jobs into the repair shop. The workload constraint is only a fraction of the total available repair capacity, as inspection requires only a small fraction of the total repair time. Via the workload constraint, the WIP in the repair shop can be kept at an acceptable level.

Furthermore, the inventory department is able to balance between inspection of parts with a high expected waiting time for SRU's (to decrease SRU waiting time) and parts that should be RFU in a short time period (to decrease LRU waiting time). Like in the design for repair shop type II, the inventory department is responsible for both SRU and LRU inventory control in order to be able to find the optimal balance between SRU and LRU stock investments.

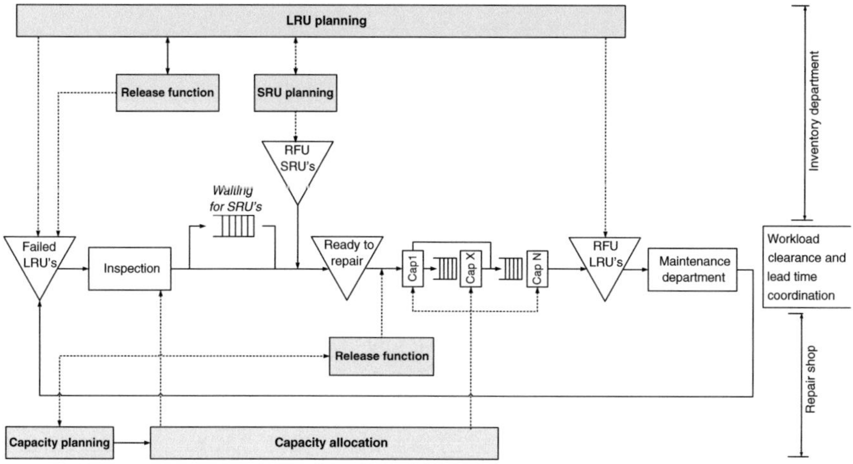

Fig. 7 Process and control structure design for repair shop type IV

We propose that the repair shop is able to release ('pull') the repair jobs from the "Ready-to-repair" stock, in order to effectively utilize the available (specialized) repair capacity. Released jobs from the "Ready-to-repair" stock should be completed within the planned lead time with a high probability, and the jobs should be released according to a certain release pattern agreed upon with the inventory department. We propose to use the lead time coordination concept as discussed in the alternative design of repair shop type III.

4.4.3 Literature Review and Suggestions for Further Research

Table 6 provides references to literature on decision support models for the decisions in the control structure of repair shop type IV, in case it differs from other repair shop types.

Table 6 Literature review and suggestions for further research for repair shop type IV

Decision function	Responsibility	Suggestions and available decision support	Open research topics
Coordination between repair lead times and repair capacity	Inventory department and repair shop		
Integrated release function and capacity allocation	Inventory department and repair shop separately	Caggiano et al. (2006)	Extend to multiple capacity types (specializations), include SRU planning

5 Case Studies

This paper presents control structures for repairable inventory systems. These control structures contribute to the following goal: to realize a high availability of high-value capital assets (from a spare parts availability point of view) against acceptable maintenance logistics costs and at an acceptable degree of coordination between decision makers.

Our hypothesis is that these control structures serve as a useful reference point for companies to redesign the control structure of existing repairable inventory systems. That is, they can be used for a comparison to their current design. This comparison provides insights in how to find an improved balance in the spare parts availability, maintenance logistics costs and degree of coordination.

In this section, we report on case studies at two companies to verify the hypothesis. This section is organized as follows. In Sect. 5.1, we describe the case study methodology. In Sects. 5.2 and 5.3 we describe the case studies at KLM Engineering & Maintenance (KLM E&M) and Royal Netherlands Air Force (RNLAF) in more detail. Section 5.4 summarizes the case study conclusions, and we reflect on the hypothesis.

5.1 Case Study Methodology

We selected two companies, that both use and maintain high value capital assets and that operate in different markets: an airline company (KLM E&M) and a military organization (RNLAF). At each of the two companies, we organized a separate workshop. Each company has multiple repair shops and therefore we were able to generate multiple cases per company.

Participants of the workshops represented either the inventory department or the repair shop. The workshops were led by a moderator, who provided a short introduction to repairable inventory systems and the repair shop typology. We first verified whether all types of repair shops are recognized by the participants. Repair shop types II, III, and IV occur at both case study companies; repair shop type I occurs at KLM E&M but not at RNLAF. The division of the repair shops over the various types can be found in Table 7 for KLM E&M and in Table 8 for RNLAF. At each company, we selected one repair shop per repair shop type. Hence, we studied in depth 4 cases at KLM E&M and 3 cases at RNLAF to test our hypothesis.

Per case, the moderator outlined the appropriate control structure presented in Sect. 4 while each design choice was discussed in detail. The control structure was compared to the current control structure and next the main differences between the current and the proposed control structure were identified. Subsequently, advantages and disadvantages of changing from the current to the proposed control structure were discussed. The discussion stopped when consensus among the participants was reached. The main findings were summarized by the moderator and the summary was completed based on individual feedback of the participants.

5.2 *Case Studies at KLM Engineering and Maintenance*

KLM E&M is an aircraft maintenance organization. KLM E&M carries out maintenance, repairs and modifications on aircraft, engines, and components for their fleet and various other airlines worldwide. KLM E&M has relatively homogenous parts repair streams, because the total volume of parts repair work is relatively high. The parts repair work is distributed among 15 independently operating components repair shops, each responsible for the repair of an individual set of repairable parts.

The participants of the workshop at KLM E&M hold the following positions: one supply chain pool manager, responsible for (LRU) inventory stocking decisions and two managers cell development, together responsible for the coordination of the repair shop decision functions and the design of the repair process in all repair shops within KLM E&M.

One of the participants presented the current control structure at KLM E&M, which is in fact the same for all repair shops and thus also for all repair shop types. The department responsible for LRU inventory control has made an agreement on a fixed repair lead time for each LRU with each repair shop. Based on the fixed repair lead times, the LRU inventory levels are determined. A target is set on the percentage of repair jobs that has to be completed within the agreed lead time of an LRU. The repair shops are responsible for the inventory control of the SRU's. Hence, the repair shops have full responsibility for achieving the repair lead times.

In Table 7 we present the final results of the workshop at KLM E&M. For each repair shop type, we list the number of repair shops classified per type, the repair shop that served as a case, its characteristics and finally the mapping of the current on the presented control structure.

As can be concluded from Table 7, (elements of) the presented control structures provided useful insights for the participants. The main difference in the control structures is that currently all repair shops are responsible for the inventory control of SRU's, while we propose to integrate the LRU and SRU stocking decisions for repair shop types II and IV. The participants admitted that the current control structures may result in far from optimal stock levels. The participants pointed out, however, that the choice for changing the current control structure would depend on the expected cost reductions (stock reduction versus increase in coordination).

We ended the workshop with formulating final conclusions. The participants believe that their current 'one-size-fits-all' approach is not optimal, and both the typology and the presented control structures generated important insights for designing a more differentiated approach.

Table 7 Mapping of the control structure in four cases at KLM E&M

Shop type	Number of shops	Case study repair shop	Characteristics	Control structure mapping
I	7	Wheels and tires	High volume repair shop with one-for-one repair and very short lead times	The current control structure matches with the proposed control structure
II	4	Toilet repairs	Large number and high diversity of required SRU's with limited availability at suppliers. This leads to many interrupted repair jobs with long and variable lead times	Integrating the LRU and SRU stocking decision may be useful to improve LRU availability and reduce stock levels. Decoupling inspection and repair is probably not desired for this specific repair shop because it is very inefficient to structurally interrupt the repair jobs in this repair shop
III	2	Mechanical repairs	Due to long (inefficient) repair set ups, one cannot follow a pure earliest due date rule for the scheduling of the repair jobs, while such a rule would be desirable because of the lead time targets	Control structure with static repair priorities and repair job priorities set by the repair shop are an alternative to align repair job priority setting and capacity utilization targets
IV	2	Composites repairs	Because of the high diversity in required SRU's and their limited shelf life, often SRU's are too expensive to stock	Introduction of a decoupling point between inspection and repair, in combination with dedicated repair job priority setting, will probably result in shorter and more reliable lead times

5.3 *Case Studies at Royal Netherlands Air Force*

RNLAF is a high-tech armed forces service that contributes to peace and security on a global basis. For this purpose, it uses and maintains aircraft, helicopters and other weapon systems. At RNLAF, the heterogeneity of parts repair streams is higher than at KLM E&M, because the total volume of parts repair work is small compared to KLM E&M. The total parts repair work is distributed among four independently operating repair shops.

The participants of the workshop at RNLAF hold the following positions: one logistics manager, responsible for the coordination between the inventory department and the repair shops, the manager of all repair shops, and a planning employee, responsible for the planning of the repair jobs. Within the RNLAF, the inventory department is responsible for LRU and SRU inventory control and the release of repair jobs to the repair shops. The repair shops are responsible for timely executing the repair jobs. Currently agreements are made on the repair lead time for each individual repair job. This results in a high degree of coordination between the repair shops and the inventory department.

In Table 8, we present the final results of the workshop at RNLAF. For each of the repair shop types II, III, and IV, we again list the number of repair shops per type, the repair shop that served as a case, its characteristics and finally the mapping of the current on our presented control structure. No repair shop was classified as type I.

The main bottleneck in the current control structure is the fact that the repair shops are responsible for meeting targets on the repair lead times, but are not responsible for the availability of the required SRU's. Introduction of a decoupling point between the inspection and repair phase of the repair job is an interesting option, because in that way the repair shops are able to provide the inventory department with more reliable lead times (after the inspection phase). This however may further increase the already high degree of coordination between the inventory department and the repair shops. Therefore, the costs reductions and increased

Table 8 Mapping of the control structure in three cases at RNLAF

Shop type	Number of shops	Case study repair shop	Characteristics	Control structure mapping
II	1	Avionic repairs	Many repair job interruptions and low repair lead time reliability due to missing SRU's	Decoupling inspection and repair might enable the repair shop to provide more reliable completion dates of the repair jobs to the inventory department
III	2	Mechanical repairs	Much coordination as reliability of the repair shop depends on SRU availability and repair priorities arranged by the inventory department	The high degree of coordination between repair shop and inventory department can be reduced by assigning the responsibility for SRU inventory control and repair job priorities to the repair shop
IV	1	Engines and modules repairs	Low lead time reliability and many changes in capacity allocation due to missing SRU's	Decoupling inspection and repair (lead times) enables the repair shop to create a robust capacity plan and provide more reliable completion dates of the repair jobs

coordination should be quantified first before this option can be considered as a serious option.

The workshop was ended with the conclusion that the presented control structures gave interesting insights for RNLAF to redesign their current control structure, and that a more differentiated approach is required.

5.4 Case Study Conclusions

In each of the two case study companies, we see that the same control structure is used for all repair shops. Within the workshops, practitioners agreed that differentiated control structures are required for the various repair shop types. The typology presented in Sect. 3 can be used to classify the repair shops.

The presented control structures served as a reference point. In workshops we made clear how the companies deviate from the presented control structures and how they can (re)design their control structures, in order to decrease the degree of coordination or to decrease the cost level. However, the workshops made clear that, besides a qualitative motivation for the design choices, also quantitative support is required for the companies to redesign their control structures.

6 Development and Improvement of the Designs

This paper presents control structures for repairable inventory systems. A differentiated control structure is proposed for the various repair shop types recognized in practice. The use of the control structures as a reference point for existing control structures is verified in case study research. The main goal of this paper is to create a foundation for further research on control structures for repairable inventory systems.

In this paper a qualitative motivation for the design choices is given. Quantifying the impact of the design choices is a suggestion for further research and helps to further improve the designs. One open research question is whether LRU and SRU stocking decisions should be integrated. Second, quantitative analysis on the impact of decoupling the inspection and repair phase of a repair job on the total maintenance logistics costs is a relevant topic for further research.

References

Adan IJBF, Sleptchenko A, Van Houtum GJ (2009) Reducing costs of spare parts supply systems via static priorities. Asia Pac J Oper Res 260(4):559–585

Arts JJ (2013) Spare parts planning and control for maintenance operations. PhD thesis, Eindhoven University of Technology, School of Industrial Engineering

Basten RJI, van Houtum GJ (2014) System-oriented inventory models for spare parts. Surv Oper Res Manage Sci 19:34–55
Bertrand JWM, Wortman JC, Wijngaard J (1990) Production control: a structural and design oriented approach. Elsevier, Amsterdam
Caggiano KE, Muckstadt JA, Rappold JA (2006) Integrated real-time capacity and inventory allocation for repairable service parts in a two-echelon supply system. Manuf Serv Oper Manag 80(3):292–319
Driessen MA, Wiers VCS, van Houtum GJ, Rustenburg WD (2013) Typology of repair shops for maintenance spare parts. Beta working paper 427
Guide VDR Jr, Srivastava R (1997) Repairable inventory theory: models and applications. Eur J Oper Res 102:1–20
Hausman WH, Scudder GD (1982) Priority scheduling rules for repairable inventory systems. Manag Sci 280(11):1215–1232
Keizers J, Adan I, van der Wal J (2001) A queueing model for due date control in a multiserver repair shop. Naval Res Logisitics 48:281–292
Kennedy WJ, Patterson JD, Fredendall LD (2002) An overview of recent literature on spare parts inventories. Int J Prod Econ 76:201–215
Muckstadt JA (1973) A model for a multi-item, multi-echelon, multi-indenture inventory system. Manag Sci 200(4):472–481
Scudder GD, Chua RCH (1987) Determining overtime policies for a repair shop. Omega 150 (3):197–206
Sherbrooke CC (1968) Metric: a multi-echelon technique for recoverable item control. Oper Res 160(1):122–141
Sherbrooke CC (1986) Vari-metric: improved approximations for multi-indenture, multi-echelon availability models. Oper Res 340(2):311–319
Silver EA, Pyke DF, Peterson R (1998) Inventory management and production planning and scheduling. Wiley, New York
Sleptchenko A, van der Heijden MC, van Harten A (2003) Trade-off between inventory and repair capacity in spare part networks. J Oper Res Soc 540(3):263–272
Verrijdt J, Adan I, de Kok T (1998) A trade off between emergency repair and inventory investment. IIE Trans 30:119–132

Knowledge Lost in Data: Organizational Impediments to Condition-Based Maintenance in the Process Industry

Robert M. van de Kerkhof, Henk A. Akkermans and Nils G. Noorderhaven

Abstract In this chapter we report on a pilot study of Condition-Based Maintenance (CBM) in the process industries, such as the steelmaking- and chemical industries, and propose a diagnostic instrument for systematically analysing the challenges to the successful implementation and execution of a CBM program. Although the field of predictive maintenance is growing and considerable research effort has been targeted to the technical aspects of CBM, we observe that many firms in the process industry do not yet systematically use advanced CBM approaches. This research therefore aims to contribute to our understanding of the contextual barriers (beyond the technical issues) that hinder organisations from employing condition-based maintenance programs to improve their asset management.

Keywords Condition-based maintenance · Inter-organizational collaboration · Organizational design · Maintenance strategy

1 Introduction

For many complex capital goods, the costs of maintenance represent a large fraction of the Total Cost of Ownership. Additionally, with the rapid development of modern technology, products have become more and more complex while better quality and higher reliability are required. Given this it is not surprising that in the Netherlands and in Europe the field of maintenance, repair and overhaul (MRO) is growing in importance (Verbraeken 2015). In the field of maintenance it is essential to develop strategies that minimize cost whilst maximizing the availability and safety of assets (Garg and Deshmukh 2006). One promising approach to boost

R.M. van de Kerkhof (✉) · H.A. Akkermans · N.G. Noorderhaven
Tilburg School of Economics and Management, Tilburg University, Tilburg 5037 AB, The Netherlands
e-mail: r.m.vdkerkhof@tilburguniversity.edu

H. Zijm et al. (eds.), *Logistics and Supply Chain Innovation*,
Lecture Notes in Logistics, DOI 10.1007/978-3-319-22288-2_13

maintenance productivity is Condition-Based Maintenance (CBM); a pro-active maintenance program that aims at predicting the time of a future malfunction by monitoring several conditions (e.g., temperature, vibrations), so the maintenance can be executed at 'just the right time' (Jardine et al. 2006). Multiple studies have empirically found that CBM results in a substantial reduction of the equipment's downtime and total maintenance costs compared to other maintenance strategies (see Jardine et al. 2006; Veldman et al. 2011), particularly when CBM activities can be clustered with scheduled maintenance activities.

Considerable research effort has been expended on the technical aspects of CBM, driven by the development of hardware technology, remote condition monitoring instrumentation, data mining ('big data') and data analysis techniques (e.g., Aggarwal 2013; Peng et al. 2010; Panagiotidou and Tagaras 2010; Heng et al. 2009). At the same time, firms in the process industry work with high capital investments, large expenses for downtime and high safety requirements, which puts pressure on the maintenance function and causes a need for advanced maintenance technology and practice (Arts et al. 1998; Tan and Kramer 1997). However, in contrast to our expectations, we observe that many firms in the process industry apply CBM approaches in general, and data-driven prognostics in particular, only to a very limited extent.

Perhaps even more surprising is how little attention this gap between state of the art and general practice has yet received in the literature. This raises the question: why do many asset owners within these industries struggle when trying to successfully set up and execute systematic CBM approaches, either by themselves or in collaboration with third parties? Apparently, studying technical factors alone is insufficient to answer this question, as technical solutions can rarely solve an organizational or a cultural problem. As yet, however, hardly any empirical evidence on the organizational aspects of designing and implementing condition-based maintenance approaches has been published (Veldman et al. 2011).

Hence we propose that there is a need to increase our understanding of the contextual barriers (beyond the technical issues) that hinder organisations from employing condition-based maintenance programs to improve their asset management. In time this understanding will be transformed into practical guidelines for the implementation of intra- and inter-organisational collaborative CBM activities and data sharing. Given that the asset owners do not have sufficient resources to set up the CBM programs by themselves, the aim is to contribute to the optimal design of condition-based maintenance relations, based on both general knowledge on inter-organisational relations and experiences acquired in the research.

The chapter begins with a brief description of condition-based maintenance and its applications in the process industry. Then we provide a diagnostic framework for analysing the contextual barriers to the implementation and execution of a CBM program, after which we discuss the initial findings of a pilot study we have conducted in the Dutch process industry. Here we have included several empirical examples (in grey boxes), in order to make the discussed concepts more tangible. The pilot study involved interviews with Maintenance-, Process-, and Reliability engineers, Inspectors and Operators, as well as plant tours and observations of maintenance-related meetings. This pilot study is part of the larger research program CAMPI, a collaborative effort of the University of Groningen, Tilburg University, Eindhoven University of Technology, and Dinalog.

2 Condition-Based Maintenance

2.1 Maintenance Strategies

A prevalent taxonomy of maintenance strategies (Fig. 1) distinguishes between reactive and proactive maintenance, in which the asset is replaced or repaired either after or before breakdown, respectively (Kothamasu et al. 2006). In the proactive maintenance category we can further distinguish between preventive maintenance, which sets a predetermined periodic interval to perform maintenance (regardless of the physical status of the asset), and predictive maintenance, which recommends maintenance actions based on the actual status of the asset. Since reactive maintenance is (by definition) unplanned and frequently involves collateral damage, the costs of breakdown (especially of critical equipment) are expected to be higher than in the case of proactive maintenance. Ideally, one would want to conduct maintenance shortly before the asset will fail, not too early, but not too late either. By monitoring the state of the asset and basing the maintenance decision on its current condition, the number of (unplanned) breakdowns and the number of unnecessary scheduled preventive maintenance operations can be reduced (Tsang 1995).

Jardine et al. (2006) describe CBM as "a maintenance program that recommends maintenance actions based on the information collected through condition monitoring." According to this definition, CBM contains a large range of predictive maintenance activities, such as a visual inspection of the tires of a car, vibration

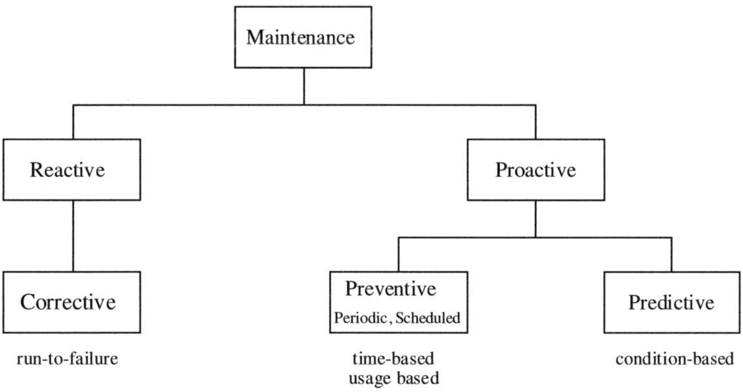

Fig. 1 Categorization of maintenance strategies

analyses of rotating equipment (tool-driven techniques) and fully automated approaches for hard-to-reach assets, in which many conditions are remotely and continuously monitored, stored in large data bases (for data mining purposes) and used in one or multiple prognostic modelling techniques (data-driven techniques). Over the years many specialized tools and sensors have been developed to monitor specific degradation processes (see for example Moubray 1997) and technological progress has enabled the more automated data-driven techniques. Each of these approaches varies in their Technology Readiness Level (TRL) and requirements for the IT-infrastructure, but also (and most importantly) in their monitoring sensitivity and predictive accuracy.

To illustrate the general rationale behind CBM, the degradation process of an asset is depicted in Fig. 2, a so-called P-F curve (Moubray 1997). Within this curve two states are prevalent: (1) a functional failure, entailing the inability of the equipment to meet a specified performance standard, and (2) a potential failure, entailing an identifiable condition that indicates that a functional failure is imminent (early signal). Note that the several early signals indicating the degradation of the asset are likely to be received by people within different departments. In this fictional example, the reduced pressure (P_1) can for instance be noted by an operations engineer, the decreased quality of output (P_2) by a quality inspector and the audible noise (P_3) by a maintenance engineer. Whenever proactive maintenance is desired, the P-F interval (i.e. the interval between the first signal indicating a possible degradation and the functional failure, see Fig. 2) can be used to determine the optimal inspection interval for periodically measured conditions.

Three observations are important here. First, the earlier one detects the asset's degradation (which requires more sensitive monitoring equipment), the more time is available to schedule and prepare the maintenance. Second, in reality the degradation process is often much less 'smooth' than the function depicted in Fig. 2

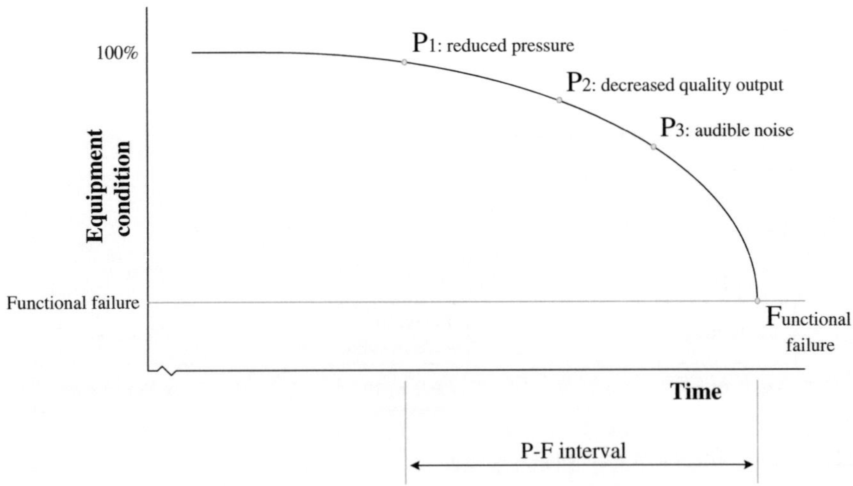

Fig. 2 P-F curve

and the actual timing of the functional failure might be earlier or (much) later than predicted (partly due to the fact that most machines have multiple failure modes). Third, accurately predicting the downtime of a complete unit (i.e., a system of connected machines) requires predictions about each (non-redundant) equipment's failure timing and is therefore in many cases (very) hard.

In essence every CBM program consists of three key steps (Jardine et al. 2006): (1) data acquisition, (2) data processing, and (3) maintenance decision-making. Data acquisition is the process of collecting and storing useful information, such as process and event data, preferably in a centrally accessible system. Before the data can be analysed with a variety of models and algorithms (for a review, see Jardine et al. 2006), the data should be cleaned, involving the removal of human entry errors and accounting for irregularities arising by start-ups and sensor faults. Finally, the techniques for maintenance decision support can be divided in either of two categories: diagnosis and prognosis. Fault diagnostics focuses on detection (i.e., whether something goes wrong), isolation (i.e., where it goes wrong) and identification (i.e., what goes wrong) of faults when they occur. Prognostics on the other hand deals with fault prediction before it occurs, providing an estimation of either the remaining useful life time or the probability that the equipment operates faultlessly until the next inspection (thereby determining the condition monitoring interval).

2.2 *CBM in the Process Industry*

The process industries are industries in which the primary production processes are either continuous or occur on repeated batches, such as the food, chemical and pharmaceutical industries. In the pilot study we have mainly focused on the chemical-, refinery and steelmaking industries, which are characterized by large multinational asset owners, high capital investments, complex processes within complex installations and high financial and safety risks connected with breakdown. Therefore the maintenance function is of crucial importance.

At the same time, most of the processes in these industries are continuous, which limits the ability to inspect the equipment, and dynamic, which makes it harder to analyse trends. Since a large portion of the equipment is custom-made and every context is slightly different, data gathered from one unit is not directly transferable to other units. Also, because most assets have a relatively long life-span and are replaced preventively before breakdown, limited failure data is available. Furthermore, a trend of outsourcing of specialized maintenance activities has led to the dispersion of data and expertise (Garg and Deshmukh 2006; Tsang et al. 2006) and increased the need for collaboration and coordination to minimize downtime.

Some trends relevant to maintenance are worth mentioning as well. Over the years the governmental and media attention on these industries has risen, fueled by several incidents all around the world (in which people were fatally wounded and major damage was inflicted to the environment). As a result the level of governmental and internal regulation has steadily increased, leading to the establishment of new procedures to 'guarantee' a quality of work and more bureaucracy. At the same time, the '*don't fix it when it ain't broken*' culture of the '90s turned gradually into a safety culture (Guldenmund 2000), in which new procedures had to safeguard against mistakes previously made. Although the once prevalent reactive culture is slowly diminishing, many of the asset owners are still primarily trained in performing corrective maintenance. A real "data culture" has not yet been established, although reliability engineers and asset management advocates are pushing in this direction.

Through conversations with several disciplines, we realized however that (somewhat to our surprise) a lot of data is gathered and stored already, though for different *purposes*. For example, Optimization continually monitors and stores the process data (input data, output data, operational data), Inspection periodically assesses the quality of the products and the integrity of the assets, and Maintenance generally records the failure data (failure mode). When combined these data contain a lot of information about the current functioning of the asset, common failure modes and trends, but to date this potential has remained largely untapped.

2.3 Adoption of CBM

In the current state the asset owners neither possess sufficient resources nor have sufficient knowledge to execute advanced prognostic CBM activities by themselves. Therefore asset owners have to make use of third parties for specialized condition-based maintenance tasks (Veldman et al. 2011). Three general stakeholders can be identified for the use and provision of CBM, which together have the capabilities to successfully set up a CBM program: the asset owner, the original equipment manufacturer (OEM), and maintenance contractors specialized in condition monitoring and/or data analysis. The potential values each of these stakeholders can derive from engaging in CBM are listed in Table 1. Recently OEM's have started exploiting remote monitoring tasks in their service offering and explored possibilities for 'leasing' their equipment, a trend called 'servitization' (Persona et al. 2007; Veldman et al. 2011).

Table 1 is based on Zaki and Neely's (2014) overview of values and benefits for condition monitoring services in asset-heavy organizations.

Although these potential values seem promising, it is important to realize that each of them can only be realised if the right processes are in place. Minimizing downtime for example requires the asset owner to accurately and timely predict the timing of an impending failure, schedule the maintenance (preferably during an already planned periodic maintenance stop), prepare the maintenance and execute the maintenance as planned. In the same line of reasoning, reducing safety hazards requires the asset's health assessment to be translated into adjusted operational guidelines and enhancing the product's quality requires the OEM to access and gather the data (in the right format), and to analyse and translate them into an improved product design.

Table 1 Value of using/providing CBM for the main stakeholders

User	Asset owner	–	Decrease total cost of ownership
		–	Maximize asset productivity
		–	Minimize downtime
		–	Extend effective life of asset
		–	Reduce long-term cost of maintenance
		–	Reduce safety hazards
		–	Facilitate certification and reports for legislative bodies
Provider	OEM	–	Grow spare parts and repairs business
		–	Strengthen relationship with asset owners
		–	Understand equipment behaviour
		–	Build dataset of equipment in different contexts
		–	Improve product quality
		–	Identify and leverage best practices
	Maintenance contractor	–	Increase the value of the contract
		–	Improve and leverage algorithms
		–	Gain access to data; pooling
		–	Collaborate with and learn from engineers

3 A Diagnostic Instrument

The factors explaining the relative absence of advanced CBM programs in the process industry can be grossly subdivided in one of three categories: technical, organizational and cultural. Given the current technological capabilities, we expect that a large fraction of the absence of CBM can be explained from (1) organizational and cultural factors, and (2) the interaction between the three categories (e.g., different technological designs synergize better with specific organizations and cultures).

The waterfall diagram, as depicted in Fig. 3, provides a diagnostic framework for systematically analysing the organisational and cultural barriers to CBM. The purpose is to uncover *why* a certain critical equipment is *not* maintained predictively; what contextual barriers hinder the successful execution of the CBM program? Only if all contextual barriers are dealt with, the CBM program can and will be executed successfully. Since we are primarily interested in the organisational impediments, we will not consider equipment that is impossible to maintain through CBM due to technological or legal impediments.

Reading from left to right, the first block contains all critical equipment for which CBM is technologically and legally possible. The block in the middle contains all critical equipment for which CBM is selected as the maintenance strategy (which is a subset of the requipment referred to in the first block) and for which the required CBM technology such as the monitoring equipment, a data platform and specialized prognostic software is in place. The final block on the right

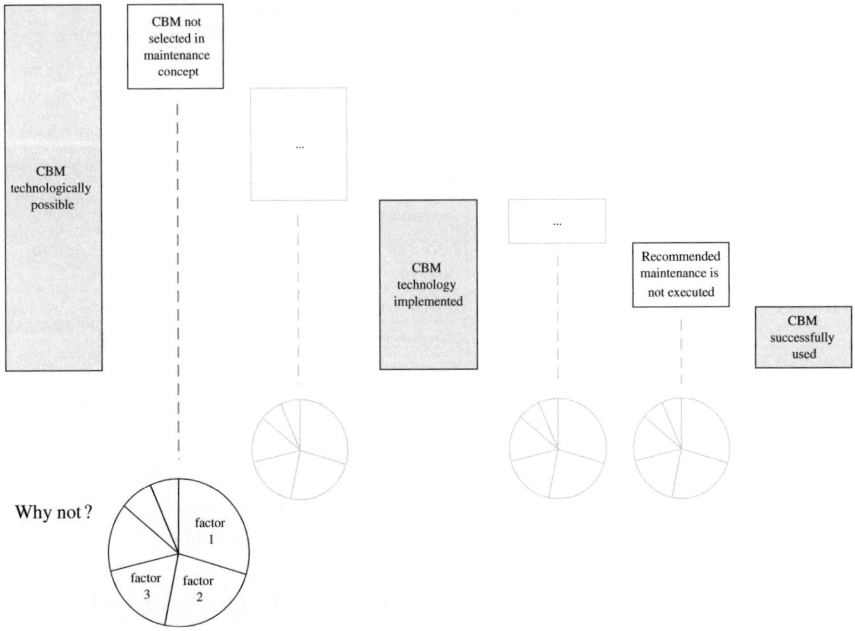

Fig. 3 Systematic (and truncated) presentation of organizational impediments to CBM

contains all critical equipment for which CBM is successfully implemented and used. (Note: although not shown in Fig. 3, we also keep a list of equipment for which CBM is unsuccessfully used, or delivers unsatisfactory results). Thus, some equipment can be found in all three blocks, while others ‘do not make it’ to the second or third block; these are stopped in between by a contextual barrier.

In between the three blocks are smaller rectangles, containing broad categories of contextual barriers to either the implementation or the execution of CBM programmes. These are for example “CBM is not in the maintenance concept” or “the recommended maintenance is not executed”. The final components of the waterfall diagram are the pie diagrams at the bottom, one for each broad category of contextual barriers. This is the part in which we are particularly interested; the pie diagram contains the arguments *why* a certain critical equipment could not pass the barrier. For example, CBM is not in the maintenance concept, because (1) this year’s budget did not allow it, (2) the equipment was identified as not critical, (3) the engineers responsible for the maintenance concept were not aware of the technological monitoring possibilities for this type of equipment, (4) the engineers responsible for the maintenance concept had a bad experience with previous CBM activities, etc.

4 Organizational Impediments

Lastly we report on several common organisational barriers, combining a theoretical and empirical perspective. These are selected based on their frequent occurrence in interviews in the pilot study. The aim of this section is to provide a first exploration of important contextual barriers to collaborative CBM activities and data sharing within the chemical-, refinery and steelmaking industry (and is not intended to be exhaustive). An effective (collaborative) condition-based maintenance program needs to be designed and managed in such a way that these barriers can be overcome.

4.1 Barriers to the Implementation of CBM

Most organizations in the process industry work with a so-called 'maintenance concept' (that specifies for each equipment what type of maintenance activities will be performed, as well as its frequency) and a 'maintenance plan' (that specifies how these activities will be performed; with what tools and by whom the data will be collected, processed and analysed). Thus, a CBM program is considered to be 'implemented' when CBM is selected in the maintenance concept, specified in the maintenance plan, when the tools are purchased and installed and when the engineers are trained to work with these tools. Within this context, multiple barriers have been identified that refrain CBM from being implemented.

The first draft of the maintenance concept is often constructed with a multi-disciplinary team that consists of one or multiple reliability engineers, maintenance engineers and operators. Together this team decides on the maintenance strategy and control measures for each type of equipment, based on the criticality and the known failure modes and effects (FMEA) and (sometimes) an estimation of the Life-Cycle Costs (LCC). A potential *lack of knowledge of this team* is thus the first barrier and limiting factor. From observations and interviews we derive that its decisions are mainly driven by (1) expertise and experience of the engineers and (2) the attached manual of the OEM. However, we've also observed that (a) most of the engineers in the team have been working in the industry for a long time and are not familiar with state of the art condition monitoring techniques, (b) it requires a lot of time to comprehend the manual of the OEM, which often recommends more preventive maintenance than actually needed, (c) none of the teams contained condition monitoring specialists or data experts, and (d) the teams didn't have access to an internal library of best practices or condition monitoring techniques.

Then, the initial draft of the maintenance concept is discussed with the maintenance managers (who are responsible for the maintenance budget), after which the maintenance manager decides what maintenance activities will be executed and how they will be performed (including what activities will be outsourced). These decisions are mainly based on the *yearly maintenance budget*, in which corrective and legally obliged preventive maintenance activities get priority. Since many of the advanced CBM techniques require a significant initial investment (for the purchasing and installation of sensors, the IT-infrastructure and data-processing software, the training of engineers and the validation of the prognostic models) and are only profitable in the long run, the maintenance manager is more likely to remove these activities from the maintenance concept (Marginson and McAulay 2008). Note that the initial investment is even larger for techniques which are still under development (TRL of 5, 6 or 7; such as big data analytics) and the returns are more uncertain. Unfortunately, most of the advanced CBM techniques to date are not yet readily applicable in practice, as the models lack specificity to the assets in question and many algorithms require a large amount of historic data, resulting in a long learning time (Peng et al. 2010). This is especially true for companies in the process industry that deal with dynamic processes (which makes it harder to develop accurate prognostic models) and custom-made assets, each operating in its own context (which limits the possibilities for leveraging prognostic models to similar assets).

Although these first organisational barriers are not exhaustive, they contain an important lesson: *people* decide what maintenance strategy will be adopted (and consequently design the CBM process, buy and install the monitoring equipment, etc.). We have limited mental capabilities, limited time and are susceptible to local incentives. Thus, any organisation that aims to adopt CBM on a larger scale should attempt to enable the involved people to make well-informed decisions and encourage them to make the decisions that are most profitable for the entire organisation.

A positive note: enablers. While discussing the barriers that hinder the implementation of CBM, the interviewees identified multiple enablers of the implementation process. The absence of these factors doesn't make it impossible to implement CBM, but the presence of these factors certainly speeds up the process. Some examples are:

- visionary managers that prioritize reliability over short-term profits and free up resources for innovation (e.g., time, money);
- engaged plant managers that believe in the potential of CBM;
- an internal knowledge team with condition monitoring specialists and data experts;
- an internal knowledge base that captures best practices with proven technologies and contains an overview of the existing condition monitoring techniques (and for which type of equipment they are suited);
- standards for data collection and storage (based on for example the ISO 14.224 standard);
- a wall of fame, that spreads success stories of previous CBM activities, and;
- strategic partnerships with innovative OEMs or maintenance contractors (who share their knowledge, sell and use their tools, have built a trustful relationship over time and don't charge high initial investments).

4.2 Barriers to a Successful Execution of the CBM Program

The next stage is the execution of the CBM program, which can be organized in many different ways. In essence three sub processes can be identified: (1) the condition monitoring process, in which the data requirements are determined, after which the data is (continuously or periodically) monitored and stored, (2) the condition-based maintenance process, in which the data is processed and analysed, the equipment's remaining life-span is predicted, proactive maintenance decisions are made and the maintenance is scheduled and executed, and (3) the failure elimination process, in which recurring failures are eliminated through root-cause analyses (RCAs) and consequent changes in procedures or the equipment's design. How these processes are designed depends on the requirements of the CBM program, the in-house knowledge and capabilities and the availability of capable partners. Note that the value and predictive power of the CBM program diminishes significantly if the effort into one of the process' steps is abandoned at any moment in time (Braaksma et al. 2011).

Due to the specificity of the techniques and knowledge needed for effective condition-based maintenance in the process industry, effective condition-based maintenance activities often require the asset owner, the original equipment manufacturer (OEM) and specialized maintenance contractors to work together closely and share their data (Veldman et al. 2011), even though their *organisational incentives* to do so are initially *misaligned*. Within these complex, multi-party and multi-phased transactions, formal agreements as such are likely to be insufficient to solve the contracting problems (Henisz et al. 2012), as can be seen in Box 1.1.

Box 1.1 Asset owner and Maintenance contractor

Consider the case in which an asset owner has decided to outsource the condition monitoring and maintenance execution to a specific maintenance contractor. This scenario typically leads to multiple conflicts of interests. By nature, an asset owner benefits from a maximized uptime of the asset and minimized maintenance costs, while the maintenance contractor generally benefits from more maintenance actions. Also the question arises who is responsible for breakdowns and additional costs: the party performing the maintenance or the party operating the asset?

Clever design of the formal agreements can to some extent align the interests of both parties, but the asset owner will always remain legally responsible for safety and environmental incidents.

Failure to recognise and acknowledge these differences of interests is likely to ultimately result in abandonment of the CBM program. Therefore the organisation of any successful CBM program has to be designed in such a way that the interests of these direct stakeholders are either (1) aligned, or (2) met separately (Parmigiani and Rivera-Santos 2011). In case the interests cannot be aligned perfectly, bonding, collaborative problem solving, open communication and trust (relational contracting; Ring and Van de Ven 1994) may complement the formal agreements, such that the collaboration (and the CBM program) is sustained and strengthened over time (see e.g., Autry and Golicic 2010; Krishnan et al. 2006; Narayandas and Rangan 2007). For improving collaboration under these difficult conditions the concept of 'partnering' has been developed and implemented, in particular in the construction industry (Gadde and Dubois 2010). Partnering emphasises trust building, transparency, and the construction of shared goals, and has been proposed also in the context of maintenance relationships (Olsson and Espling 2004). The relationship is strengthened by building value together, by engaging in inter-organisational learning processes (Akkermans et al. 2004; Laan et al. 2011) or by integrating the supply chain (Prajogo and Olhager 2012). Also building and maintaining understanding for each other's processes (with each new successor) enhances mutual trust and reduces conflicts (Akkermans et al. 2004; Johnston et al. 2004). That trust is essential for a successful CBM collaboration is underscored in the example in Box 1.2.

Box 1.2 Maintenance scoping by a maintenance contractor
Instead of hiring and training an internal vibration monitoring specialist, some asset owners have decided to outsource the periodic vibration analyses to a specialized maintenance contractor. In discussion with the vibration expert, he identified that he adds value for his clients if and only if:

- he has access to the client's data (process data, machine archive, etc.) and is allowed to do a visual inspection upfront
- he has the opportunity to use high-end tools (i.e., no minimal budget) and is allocated the time to perform extensive analyses
- the internal knowledge and experience (of the engineers from the maintenance contractor) is up to date and they maintain a customer-oriented attitude
- (most importantly,) the client trust his recommendations. According to him, this trust is a product of the client's trust in the tool, the maintenance contractor and (mainly) in the vibration expert himself. Without trust, the client is very unlikely to perform the recommended actions.

Additional challenges with the successful execution of a CBM program are (1) the continuous collection and sharing of high-quality data, which requires time from engineers who do not themselves directly benefit from the data collection, (2) the recognition of weak and ambiguous signals, which requires sensitive monitoring equipment and well-trained engineers, (3) the processing and analysis of large piles of data, which requires fast computers, decision-support tools and well-trained engineers (data-overload impending), (4) proactive and collaborative decision-making, which requires well-trained engineers to get together and invest time (without a direct sense of emergency), (5) the increased variability in the plant's maintenance (and production) planning, which is particularly bothersome if the maintenance is executed by external maintenance contractors, etc. All of these activities require time from the engineers and well-structured processes which are challenging when spanning multiple departments and even more challenging when spanning multiple organisations (Paulraj et al. 2008). Moreover, many asset owners have indicated that they struggle with measuring and proving that the CBM program is actually an improvement over the previous maintenance policy, since "we'll never know what would have happened to the equipment if we hadn't preventively maintained it" (Carnero 2006). This puts even more pressure on inter-organisational relationships, as can be seen in the final example in Box 1.3.

Box 1.3 External monitoring

The developments in remote monitoring technology gave rise to contractors that offer to 'monitor the health of your asset' and predict impending failures with clever algorithms. Although the services of these specialized monitoring organisations seems promising, it turned out to be hard to establish such functions in the past. In one representative example, the contractor was challenged with:

1. *providing valid alarms*. Although the contractor was specialized in data processing, they lacked the specific engineering and situational knowledge to determine all failure mechanisms and relevant indicators (a list of 'usual suspects' was the best possible result);
2. *providing valid alarms, in time*. Many of the alarms given were 'too late' (i.e., the issues were already spotted by an operator or a maintenance engineer), limiting the direct added value of the contractor, and;
3. *proving your worth, continuously*. The contractor had to prove his added value year after year, particularly when previous stakeholders of the collaboration changed positions within the organisation;

Recently, the contract was terminated after 5 years of collaboration, right after the former manager of the contract moved to a different position.

5 Conclusion

In this chapter we have discussed the condition-based maintenance strategy and its implication in the process industry and proposed a diagnostic instrument for identifying the organizational impediments to CBM. Although considerable research effort has been expended on developing the technical aspects of CBM programs, our pilot study indicates that many firms in the process industry struggle with systematically employing CBM activities in general and prognostic CBM approaches in particular. Therefore this research aims to contribute to our understanding of the contextual barriers that hinder organisations to adopt condition-based maintenance programs to improve their asset management. We hope to contribute to the establishment of a safe, more sustainable future, in which advanced maintenance approaches enhance the machines' uptime and reduce unnecessary waste. We believe that this is only possible by taking the evident organisational challenges into account and therefore encourage both scholars and engineers to further explore the organisational conditions that enable the successful execution of collaborative CBM activities.

As a final note; while data gathering can be done by machines, analytics is about people (Zaki and Neely 2014). No matter how advanced the technologies, transforming data into information and information into knowledge and action requires an equal investment in the people handling these technologies.

References

Aggarwal CC (2013) Managing and mining sensor data. Springer, Berlin

Akkermans H, Bogerd P, van Doremalen J (2004) Travail, transparency and trust: a case study of computer supported collaborative supply chain planning in high-tech electronics. Eur J Oper Res 153:445–456

Arts R, Knapp G, Mann L (1998) Some aspects of measuring maintenance performance in the process industry. J Qual Maint Eng 4(1):6–11

Autry CW, Golicic SL (2010) Evaluating buyer–supplier relationship—performance spirals: a longitudinal study. J Oper Manag 28:87–100

Braaksma AJJ, Klingenberg W, van Exel PWHM (2011) A review of the use of asset information standards for collaboration in the process industry. Comput Ind 62:337–350

Carnero M (2006) An evaluation system of the setting up of predictive maintenance programmes. Reliab Eng Syst Safety 91(8):945–963

Gadde L-E, Dubois A (2010) Partnering in the construction industry—problems and opportunities. J Purch Supply Manag 16:254–263

Garg A, Deshmukh S (2006) Maintenance management: literature review and directions. J Qual Maint Eng 12(3):205–238

Guldenmund FW (2000) The nature of safety culture: a review of theory and research. Saf Sci 34 (1):215–257

Heng A, Zhang S, Tan AC, Mathew J (2009) Rotating machinery prognostics: state of the art, challenges and opportunities. Mech Syst Signal Process 23(3):724–739

Henisz WJ, Levitt RE, Scott WR (2012) Toward a unified theory of project governance: Economic, sociological and psychological supports for relational contracting. Eng Proj Organ J 2:37–55

Jardine AKS, Lin D, Banjevic D (2006) A review on machinery diagnostics and prognostics implementing condition-based maintenance. Mech Syst Signal Process 20(7):1483–1510

Johnston DA, McCutcheon DM, Stuart FI, Kerwood H (2004) Effects of supplier trust on performance of cooperative supplier relationships. J Oper Manag 22(1):23–38

Kothamasu R, Huang S, VerDuin W (2006) System health monitoring and prognostics: a review of current paradigms and practices. Int J Adv Manuf Technol 28(9):1012–1024

Krishnan R, Martin X, Noorderhaven NG (2006) When does interorganizational trust matter to strategic alliance performance? Acad Manag J 49(5):894–917

Laan A, Noorderhaven NG, Voordijk H, Dewulf G (2011) Building trust in construction partnering projects: an exploratory case study. J Purch Supply Manag 17(2):98–108

Marginson D, McAulay L (2008) Exploring the debate on short-termism: a theoretical and empirical analysis. Strateg Manag J 29(3):273–292
Moubray J (1997) Reliability-centered maintenance. Industrial Press Inc., New York
Narayandas D, Rangan VK (2007) Building and sustaining buyer-seller relationships in mature industrial markets. J Market 68:63–77
Olsson U, Espling U (2004) A framework of partnering for infrastructure maintenance. J Qual Maint Eng 10(4):234–247
Panagiotidou S, Tagaras G (2010) Statistical process control and condition-based maintenance: a meaningful relationship through data sharing. Prod Oper Manag 19(2):156–171
Parmigiani A, Rivera-Santos M (2011) Clearing a path through the forest: a meta-review of interorganizational relationships. J Manag 37(4):1108–1136
Paulraj A, Lado AA, Chen IJ (2008) Inter-organizational communication as a relational competency: antecedents and performance outcomes in collaborative buyer-supplier relationships. J Oper Manag 26(1):45–64
Peng Y, Dong M, Zuo MJ (2010) Current status of machine prognostics in condition-based maintenance: a review. Int J Adv Manuf Technol 50(1–4):297–313
Persona A, Regattieri A, Pham H, Battini D (2007) Remote control and maintenance outsourcing networks and its applications in supply chain management. J Oper Manag 25(6):1275–1291
Prajogo D, Olhager J (2012) Supply chain integration and performance: the effects of long-term relationships, information technology and sharing, and logistics integration. Int J Prod Econ 135(1):514–522
Ring PS, Van de Ven AH (1994) Developmental processes of cooperative interorganizational relationships. Acad Manag Rev 19(1):90–118
Tan J, Kramer M (1997) A general framework for preventive maintenance optimization in chemical process operations. Comput Chem Eng 21(12):1451–1469
Tsang AHC (1995) Condition-based maintenance: tools and decision making. J Qual Maint Eng 1 (3):3–17
Tsang A, Yeung W, Jardine A, Leung B (2006) Data management for CBM optimization. J Qual Maint Eng 12(1):37–51
Veldman J, Klingenberg W, Wortmann H (2011) Managing condition-based maintenance technology. J Qual Maint Eng 17(1):40–62
Verbraeken H (2015) Onderhoud wordt nog te vaak gezien als kostenpost. Het Financieele Dagblad donderdag 14 mei
Zaki M, Neely A (2014) Optimising asset management within complex service networks: the role of data. Cambridge Service Alliance

Planning Services: A Control Tower Solution for Managing Spare Parts

Jan Willem Rustenburg

Abstract Planning Services is a new and promising business model for managing spare parts using a control tower concept. In this paper, we thoroughly explain the main ingredients of this new business model, such as the definition of the services and the pricing models. In addition, we deal with the aspects of the control tower concept used, such as the decision functions allocated to the control tower, the mathematical model, the remote solution including the required ICT infrastructure, and the required employee competences. Furthermore, we discuss three case studies of Planning Services at NedTrain, the Naval Maintenance Company and Marel Stork. This paper concludes with recommendations for further innovation from both a business and a scientific point of view.

Keywords Spare parts management · Control tower · Parts pooling · Business models

1 Introduction

Many industries depend on the availability of high-value capital assets to provide their services or to manufacture their products. Often a lack of spare parts is the reason for asset downtime or hampered maintenance operations. See e.g. Driessen et al. (2014) and its underlying studies for a comprehensive analysis of the impact of spare parts on maintenance.

Spare parts management can be defined as the process to ensure the right availability of spare parts against minimal integral costs. This process becomes more and more complex due to the following reasons:

J.W. Rustenburg (✉)
Gordian Logistic Experts, Groenewoudsedijk 63, 3528 BG Utrecht, The Netherlands
e-mail: jw.rustenburg@gordian.nl

H. Zijm et al. (eds.), *Logistics and Supply Chain Innovation*,
Lecture Notes in Logistics, DOI 10.1007/978-3-319-22288-2_14

- an ever increasing technical complexity of the assets, resulting in not always comprehensible failure patterns;
- smaller installed bases of production systems, resulting in lower demands and thus more complicated demand patterns;
- increased obsolescence due to fast product improvement;
- customers demanding higher and more differentiated performance levels;
- a much stronger focus on working capital;
- reduced knowledge and experience in the field of spare parts management.

For an in-depth explanation of these problems including their context we refer to Van Houtum (2010), Cohen et al. (1996, 2006), Deloitte (2006), Rustenburg (2010) and De Rooy and Rustenburg (2010).

Strongly motivated by the Dutch Institute for Advanced Logistics (Dinalog), several strategic initiatives were taken to counter the inevitable negative consequences of the complexities mentioned above, including the start of several R&D projects explicitly focusing on the relation between maintenance and spare parts logistics, and the development of control tower solutions for structural (often remote) support. See also www.dinalog.nl.

This paper describes the concept of Planning Services in detail as a way to fill in the control tower solutions approach,[1] and reports extensively on experiences with this new business model for managing spare parts.

This paper is structured as follows. Section 2 provides a high-level description of Planning Services, and positions this solution clearly in a broader range of spare parts management solutions. In Sect. 3 we explain the innovative value of Planning Services. Then we explain Planning Services in detail, by elaborating on the two key components: the control tower solution (Sect. 4) and the business solution (Sect. 5). Section 6 deals with case studies at 3 different companies. Finally in Sect. 7 we conclude the paper and suggest directions for further research.

2 The Concept of Planning Services

This section first discusses the essential ingredients of Planning Services, as well as the difference with traditional solutions. Further the key objectives of (the experiments with) Planning Services are discussed.

2.1 The Essence of Planning Services

To enhance the effectiveness of their spare parts management, companies usually buy specific software such as the SPM (Service Parts Management) module of PTC

[1]That being said this initiative is strongly connected with the R&D projects.

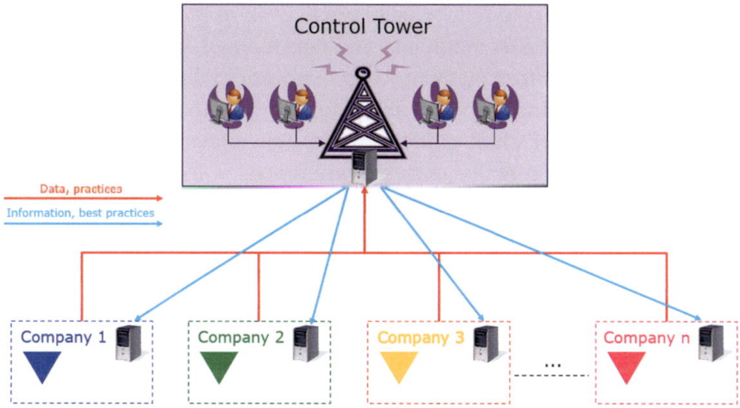

Fig. 1 Spare parts planning control tower service solution

or Synchron.[2] In the concept of Planning Services, companies do not buy the software; *they buy the planning information* such as forecasts, re-order points, re-order quantities and planning exceptions.

The planning information is produced by an outsourced control tower. The physical location of the control tower is irrelevant because the solution is "footloose". The control tower taps into the ERP system of the customers extracting a range of transactional data. In the control tower highly educated and experienced planners—using state-of-the-art algorithms—create adequate planning information, which is then returned to the ERP system of the customer.

In addition the cross-company learning process is key in Planning Services. Although assortments often differ, typical issues or trends are the same across different companies. Ultimately, the goal is to find sharing opportunities such as a centralised (outsourced) shared planning department, inventory pooling, sharing warehouses and improving purchasing agreements through aggregation.

So Planning Services is much more than just another spare parts management add-on. It is also about new processes, different people and other commercial agreements, making it a true new business model. Figure 1 provides a schematic overview of the control tower concept.

2.2 *The Objectives of Planning Services*

The first objective is to increase the planning effectiveness, i.e. creating a better balance between spare parts availability and the required working capital. As mentioned earlier the skilled planners, the use of state-of-the art algorithms and the cross-company learning process are key to realize this objective.

[2]See also www.ptc.com and www.synchron.com.

The second objective is to reduce operational costs at the customer organization. Often the planning process is inefficient and non-focused, resulting in the required workload for spare parts management to be unnecessary high. An important cause is the insufficient educational level of the planners. By providing planners with the right information and the right guidance, they will focus on the assortments for which human intervention really makes a difference.

The third objective is to develop a true outsourced and independent solution enabling entrepreneurs to provide planning services on a global level. This solution may also be considered as a Planning as a Service solution (PaaS) preferably in the cloud or otherwise on the customer's premises.

3 The Innovative Value of Planning Services

As discussed earlier, this paper does not focus on innovative mathematical models for spare parts management; merely we concentrate on the innovativeness of the business model. This section explains the innovative value from different angles.

3.1 Qualification of the Innovative Level

Van Laarhoven (2007) views a cross chain control center, aiming to coordinate the flows across various supply chains, as a revolutionary innovation with high impact. In particular they highly value the cost influence potential of outsourced planning functions like forecasting and inventory planning. At the same time, they signal that the likelihood of outsourced inventory planning is supposed to be low whereas the likelihood to outsource forecasting seems to be high.

A detailed explanation is not given, but we interpret this difference as follows. Determining re-order points and—quantities as well as the actual parts ordering are the core of inventory planning. Hence, with inventory planning key metrics such as spare parts availability and capital employed are directly influenced. This probably explains the hesitance when it comes to outsourcing inventory planning. Forecast information is important, but it constitutes an *input parameter*, just like the supplier lead time and the purchasing price. Therefore this type of information is less directly related to spare parts availability and capital employed. Hence, we need to be aware of some aversion to outsourcing spare parts inventory planning.

3.2 Experiences Documented in Literature

As indicated earlier, and also fitting the scope of this volume, the objective of Planning Services is not so much to develop new mathematical models, but to carefully select, implement, test and materialize these models. For background material, we refer to the following well-documented surveys: Guide and Srivastava (1997), Kennedy et al. (2002) and Basten and Van Houtum (2014). The difference in sophistication between models available in literature and those being used in practice (often driven by MRP logic in ERP systems), is still large. Hence, we expect significant potential by using well-tested scientific models in practice.

From an organizational point of view, Planning Services can be considered as a special form of a Shared Services Center (SSC). Strikwerda (2014) defines an SSC as: an accountable entity in the internal organization of a form or institution, tasked to deliver special services to operational units (business unit, divisions) on the basis of a service level agreement against set transfer prices.

The key difference is that Planning Services is not a part of the internal organization, instead it is an outsourced operation. Nevertheless, this definition is of use as well as the principles of an SSC as listed by Soalheira and Timbrell (2014):

- Manage the SSC like an independent business unit
- Standardize processes and establish end-to-end business owners
- Employ service levels in the delivery to end-customers
- Create customer focus
- Price the services
- Create competiveness

On internet several companies report on Planning as a Service. An example is Déhora providing workforce planning as a service (www.dehora.nl). Another example is Slimstock, an inventory management software provider, also delivering "Coordination as a Service", see www.slimstock.nl. However, to the best of our knowledge, no public reports regarding Planning Services are available.

In Sect. 4 we describe Planning Services in a systematic way, using the principles outlined above as important ingredients of the total solution.

4 The Control Tower Solution

The term "control tower" is somewhat abstract in nature. In order to avoid one-dimensional associations like "just another ICT/cloud solution" or "something remote", we make it more tangible by using the so-called PCOI model (Processes, Control, Organization and Information), because that model provides a systematic approach to describe a (new) way of working. As earlier mentioned, when describing the PBOI we clearly take into account the principles of an SSC defined by Soalheira and Timbrell (2014).

4.1 Processes

In Planning Services, several processes are outsourced to a third party. To do that properly, we have to identify all relevant processes and subsequently determine a clear demarcation of the processes.

To do so, Driessen et al. (2014) defined a framework in which all relevant decision functions are listed in a systematic way. In practice this framework proves to be an excellent aid to:

- define all relevant processes
- determine which processes can be outsourced
- determine the interface between the in- and outsourced processes

Figure 2 provides an adaptation of the framework developed by Driessen et al., complemented with a characterization of the decision functions (strategic, tactical and operational). It also indicates by the red line which decision functions are assumed eligible for outsourcing.

Let us explain this process scope in some more detail. Spare parts management starts with the main function *Assortment management*, primarily devoted to phasing in, sustaining and phasing out assortments. The criteria based on which assortments are phased in and phased out, are always determined by the customer because they strongly relate to the (service) bill of material of the equipment and the defined maintenance policies. The operational part of Assortment management such as changing and cleaning master data is eligible for outsourcing.

Forecasting is the process devoted to produce forecasts, forecast band widths (using e.g. the MSE, Mean Square Error) and the accuracy check of the forecast. In principle, all elements of forecasting are eligible for outsourcing. However, if a pro-active forecasting strategy is most suitable, i.e. a forecast is not or partly based on historical data but merely on other data such as Mean Time Between Failures

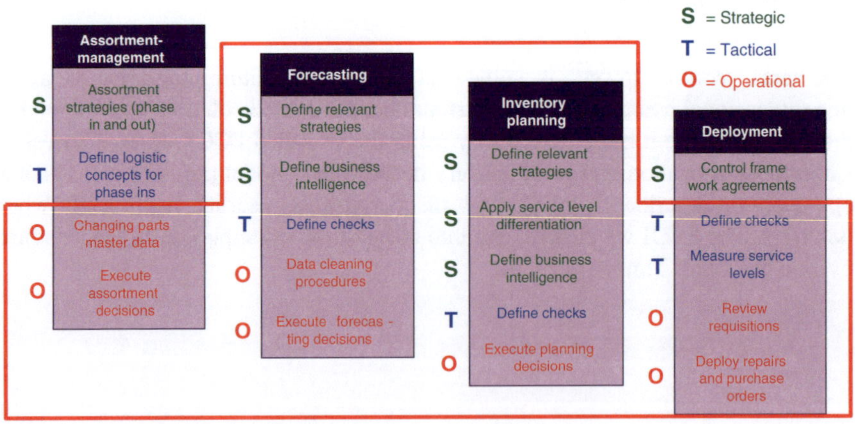

Fig. 2 The process scope of planning services

(MTBF) or maintenance plans, then the demarcation is not "black and white". Instead, a close co-operation between control tower planners and the engineers of the company is required to make accurate forecasts.

This example illustrates that Planning Services is not a 100 % remote solution. For certain assortments a close co-operation is required to be successful, leading to a hybrid solution: remote when it can, local when it must.

Inventory planning is the process to define re-order levels and re-order quantities, based on many inputs such as the forecast and required service levels. We encounter the same issues as in the Forecasting process. In principle all elements are eligible for outsourcing, however for certain assortments a close cooperation is required, e.g. when dealing with expensive and critical slow movers. By applying advanced models such as system approaches to spare parts management (see e.g. Sherbrooke 2004; Muckstadt 2005), we provide the best possible inventory parameters from a modeling perspective. Nevertheless the high costs involved make a customer validation of assumptions necessary.

Deployment is the process of actually resupplying by dispatching repairs or new buys. This process has a high potential for automation, by so-called auto-order assessments. An efficient process is only possible when proper framework agreements with suppliers are available. The establishment of framework agreements is primarily a company-commercial business and not in the scope of Planning Services.

One more remark is in place. We have assumed that Planning Services particularly provides value on the decision functions that are labeled "strategic" or "tactical". The operational parts are often not knowledge-intensive and moreover quite labor intensive. The case studies revealed though that companies might have strategic reasons to outsource the operational part as well, "focus on core business" being an important one of them.

4.2 Controls

As indicated in the introduction, the spare parts supply chains are far from stable. Demand predictability is quite low due to the fuzzy failure behavior of assets and their components, as well as the dominance of corrective maintenance. Supply predictability is also quite low, e.g. due to high risks of obsolescence and variable repair lead times.

Hence, providing planning parameters like forecast and inventory levels is not sufficient, the customer also needs information on when and how to adjust these parameters.

Basically two types of controls exist. The first one is a *pro-active* control. Typical examples are a sudden high demand for a slow mover, increasing lead times and purchase order delays at the customer. The primary objective for providing this type of information is to avoid stock-outs and to prevent rush actions.

The second one is *re-active* control. On a monthly basis we measure metrics such as the spare parts availability and the working capital. A very important characteristic is the extent to which these metrics can be influenced. For example we also provide a list of the parts that caused stock-outs in a decreasing order of occurrences (Pareto-wise or "first things first"). Each month, the customer is expected to explain and act upon 80 % of the stock-outs.

By introducing a control process like this, spare parts management becomes a natural learning process which is highly important given its unpredictable nature.

4.3 Organization

The organization depends on the type of Planning Services. Here we take the most commonly used form as a starting point, i.e. the form in which Planning Services delivers the strategic and tactical planning information, and the customer takes care of the operational planning functions.

In that case the organizational complexity is quite low. The delivery of planning services more or less follows a monthly rhythm. This is done by planners with a MSc degree in the control tower. On a weekly or monthly basis (depending on the required agility and amount of work) a Planning Services account manager confers with his counterpart at the customer, regularly complemented with representatives of the maintenance or purchasing department. This meeting is primarily about the controls and how to act upon them.

The right skills and competences of the counterpart at the customer side appear to be key to success. The concept of Planning Services may is generally not familiar and the spare parts planning information may be difficult to interpret. The absence of a skilled counterpart in the customer organisation may easily lead to putting the planning information aside, and resuming the traditional manual planning again.

An important conclusion for customers of Planning Services is that there definitely are merits to outsourcing the difficult part of spare parts management, as long as it does not lead to losing all spare parts management knowledge.

4.4 ICT

Although Planning Services is not an ICT solution, an adequate ICT infrastructure is key to success. The following requirements are imposed on the ICT infrastructure:

- *Secure*. The solution supports the use of industry standard[3] security protocols.
- *Scalable*. The multi-user solution supports hundreds to millions of SKU's, including parts pooling across multiple sites and/or customers;

[3]Our long term aim is to withstand ISO27000 audits, which also include defining and implementing extensive ICT information management processes.

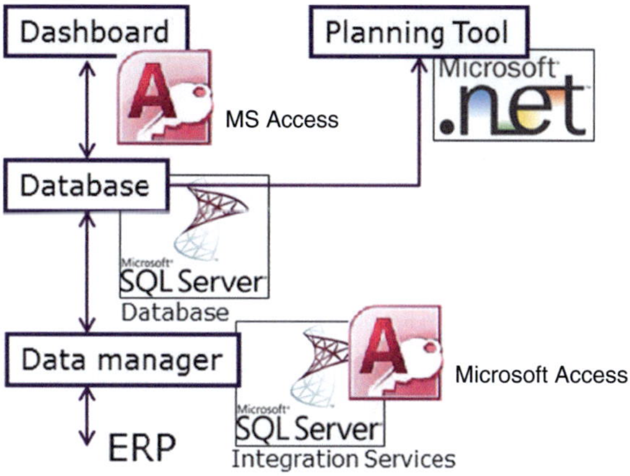

Fig. 3 Control tower ICT infrastructure. The central database is implemented as a Microsoft SQL server. Dashboards can be provided by SQL Server reporting services, but so far Microsoft access dashboard has been utilized. For ETL/data management, a combination of Microsoft SSIS packages and Microsoft access is applied. The planning tool is implemented using C#.Net

- *Future proof*. The solution supports web based and/or mobile app development without having to completely rebuild a backend;
- *In-house maintainable*. Avoid as much as possible initial dependence on external software development;
- *Universal*. The solution has to support a multitude of import and export formats from various ERP systems via a so-called ETL (Extract, Transform and Load) layer.

An ISO25010 quality assurance framework was used. Within this framework we defined requirements concerning functional suitability, performance, efficiency, compatibility, usability, reliability, security, maintainability and portability.

Given the aforementioned requirements, we created the following ICT infrastructure development schematic, see Fig. 3.

4.5 Scientific Models Used

The central goal of this paper on Planning Services is to prove that existing models can work effectively in practice. Table 1 provides an overview of models that constitute the core planning engine.

During the executions of the case studies, we did encounter one issue that was insufficiently covered by literature, i.e. the forecasting of slow moving items with a low variability in demand size (as a typical result of corrective maintenance). It appeared that existing methods such as Croston's method for intermittent demand

Table 1 Mathematical models used in the spare parts planning tool

Process		Model characteristics
1	Forecasting	• Time series based forecasting • Bootstrapping • Typical demand quantity • Advance demand information (per part and per system) • Condemnation
2	Inventory control	• Single-location, multi-item models: normal distribution, (compound) Poisson • Multi-echelon, multi-item, single indenture • Multi-echelon, multi-item, multi-indenture, budget constrained • Integration of forecast and advance demand information • Criticality, commonality • Service level differentiation
3	Other	• Initial provisioning • Last time buy • Advance order list (forecast and delivery schedules)

forecasting and Bootstrapping did not perform well. For that reason Van Wingerden et al. (2014) developed a so-called Empirical method, clearly outperforming existing methods for this specific assortment.

5 The Business Solution

In this section we discuss the business aspects of Planning Services, consisting of the definition of services, pricing model and factors that determine commercial success.

5.1 Definition of Services

Our key service is delivering tactical and strategic planning information as depicted in Fig. 2. Both from a customer as well as a commercial point of view it is important to carefully define and possibly refine the services. Without pretending to be complete, Table 2 lists a number of most commonly used services.

5.2 Pricing Models

Based on several experiments, we developed 3 types of pricing models.

Pricing model 1: "efforts based"
In this model the customer pays for the hours that the planners and the account manager actually spend. This is the most traditional model, and resembles the

Table 2 Definition of offered services

Service		Explanation
1	Tactical planning	Periodic demand forecasting and planning, determine order levels end reorder quantities, possibly supplemented with KPI-reporting
2	Operational planning	Evaluating and chasing purchase requisitions/orders. This service generally follows the customer's desire to outsource the full scope of planning for a (non-strategic) assortment
3	Monthly KPI-reporting	Providing monthly information on key metrics including "bad actors", i.e. parts that contribute most to malperformance, working capital and operational costs
4	Top 100 management	Tactical planning extended with detailed checks (e.g. checking each purchase requisition) on that part of the assortment responsible for 80 % of the budget spent and of the order lines
5	Feeding collaborative supply chains	Suppliers can often radically reduce their lead times when they are periodically provided with forecast information. Planning Services can deliver the required forecasting information
6	Logistics maintenance preparation	This service may consist of various activities • Setting up and evaluating parts lists for pre-defined (preventive) maintenance activities. Often there is a probability of usage, hence the name "percentage parts" • Definition and evaluation of kits • Standardization of parts assortment
7	Pool management	Monitor and manage inventories that are part of a pooling concept between multiple sites and/or multiple companies

pricing model for the majority of business services such as consultancy, legal and financial advice, etc.

However, from a marketing point of view this model is not attractive because in the eyes of the customer there is no real difference with consultancy. On the other hand this model seems most appropriate when the customer regularly wants additional services that cannot be specified in advance.

The reality is also that some purchasing departments at customer sites only accept "efforts based contracts".

Pricing model 2: "fixed fee"

In this model the customer periodically—generally monthly—pays a fixed fee for the services. The fee is based on a price per service. We apply a cost + model for the price per service. The most important cost drivers are the required planning capacity and the depreciation costs of the ICT development.

Because of its simplicity, this model is very attractive for both parties. In addition this model is flexible, because the customer is allowed to frequently review the services. Based on these review moments he may decide to change the mix of services.

Pricing model 3: "value based pricing"
This is the most interesting but also the most complex pricing model. From a marketing point of view this is very appealing: the customer pays for a direct contribution to his key metrics such as spares availability and working capital.

Albeit appealing, this pricing model requires direct influence of Planning Services on the key metrics. When only tactical planning decision functions are outsourced, the customer can overrule this information and incorrectly adjust purchase requisitions. In the case Planning Services is responsible for the full scope, i.e. strategic, tactical and operational planning, then a value based pricing model should be possible.

This model is still being explored. Kim et al. (2007) suggest to apply a mixture of price components: a part is based on added value but another part remains based on required efforts for planning. This idea is a leading thought in the exploration. Customers react positively, but experiments have yet to be set up.

6 Case Studies

We experimented with Planning Services at a number of companies. This section reports on 3 case studies that were quite distinct in nature.

6.1 NedTrain Haarlem

The first company is NedTrain Haarlem (NTH), part of NedTrain, which is a subsidiary of the Netherlands Railways Group. NTH performs overhaul, conversion, and upgrading of trains/trams/metro cars. We concentrate on the activities of the overhauling of bogies and wheel sets, including age based maintenance, repair, and modernization. Annually some 1100 bogies and some 3000 wheel sets are overhauled.

Experiment
Gordian ran an inventory reduction program, resulting in 40 % inventory reduction in one year while maintaining the levels of spare parts availability. During the program, NTH recruited a well-qualified tactical planner. NTH requested a solution to sustain these results taking into account the presence of this tactical planner and 2 operational planners at NedTrain.

The business solution
Given the planner capabilities at NTH, we decided to experiment with the following services:

- Tactical planning
- Top 100 management

- Feeding collaborative supply chains
- Logistics maintenance preparation

In terms of pricing NTH prefers an efforts-based contract, primarily to create mutual trust in each other regarding hours spent. We note though that for another NedTrain company, we used a fixed fee model.

The control tower solution

Processes

Following the mentioned services, the processes in scope are Forecasting, Inventory Planning, and Deployment. The demarcation is as follows. Forecasting and Inventory planning is mainly done in the control tower. Pro-active forecasting, i.e. forecasting based on maintenance plans, is carried out by the tactical planner of NTH. Deployment is primarily done by the NTH operational planners. Following the Top 100 service, the control tower reviews all main requisitions (cf. Fig. 2).

Controls

Pro-active and reactive controls are in place, just as defined in Sect. 4.2.

Organization

The delivery of planning services follows a monthly rhythm. This is done by planners with a MSc degree in the control tower. On a biweekly basis a Planning Services account manager confers with his counterpart at the customer, regularly complemented with representatives of the maintenance or purchasing department. This meeting is primarily about the controls and how to act upon them.

ICT

The ICT solution is a prototype version of the one presented in Sect. 4.4. This year it will be upgraded to the latest technology, as presented. During the experiment, the control tower is physically located at Dinalog, Breda, The Netherlands.

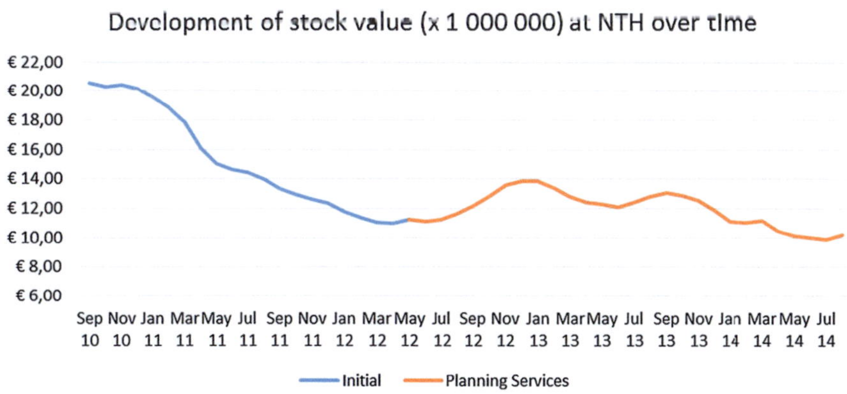

Fig. 4 Development of stock value at NTH during the initial project (*blue*) and planning services (*orange*)

Results
Figure 4 shows the results we obtained. Shortly after the project phase finished, the inventory value quite strongly increased. Several factors were responsible for this. A critical evaluation revealed though that the communication between Planning Services and NTH had room for improvement. Once achieved, the continuous learning process led to a further steady decrease in inventory while maintaining the inventory levels above 98 %.

From a business point of view the experiment was successful as well. For 2 consecutive years now, Gordian delivers Planning Services to NTH on a commercial basis.

6.2 The Naval Maintenance Company

As a part of the Royal Netherlands Navy, the Naval Maintenance Company (NMC) is responsible for the availability of all ships in use at the Navy. For that purpose, the NMC maintains, repairs, supplies, and modifies a diversity of systems. These can be propulsion systems, electronic and weapon systems of naval ships, but also guided weapons, night-vision goggles and many other equipment for the defense organization.

Experiment
The Dutch Defense is currently re-allocating the management of assortments over the various Defense maintenance companies. The reason is that for the same spare parts, Defense maintenance companies made different agreements with various suppliers. The NMC requested Planning Services to *completely* manage a certain non-strategic assortment, awaiting a transition to another Defense maintenance company. For the NMC this is a trial. Once successful, Planning Services can demonstrate their value to more complex assortments such as repairables.

The business solution
Given the full scope objective of the NMC, we decided to experiment with the following services:

- Tactical planning
- Operational planning
- Monthly KPI-reporting

In terms of pricing the NMC prefers an efforts-based contract. The primary reason is that planning resources are also involved in other projects at the NMC. From a contractual point of view, a separate fixed fee contract would then become too complex to operate and to manage.

Although a value based pricing model is clearly a bridge too far, we did make agreements upon targets with respect to spare parts availability and inventory value.

The control tower solution

Processes

Following the mentioned services, the processes in scope are Assortment Management, Forecasting, Inventory Planning, and Deployment. There is no demarcation because all activities are conducted in the control tower.

Controls

Pro-active and reactive controls are in place, just as defined in Sect. 4.2.

Organization

The delivery of planning services follows a monthly rhythm. This is done by planners with a MSc degree in the control tower. On a biweekly basis a Planning Services account manager confers with the material manager at the NMC. This meeting is primarily about the controls and how to act upon them.

ICT

The ICT solution is exactly the same as the one presented in Sect. 4.4. Following strict Defense safety regulations, the control tower is physically located at the NMC, Den Helder, the Netherlands. According to a monthly rhythm, the planners calculate and assess the parameters. The operational planning process more or less follows a weekly rhythm.

Results

In a timeframe of 10 months the inventory value decreased by 11 % from M€2.7 to M€2.4, as shown in Fig. 5. The service level decreased 91 % down to 87 %. At first sight, the combination of these figures does not present a success story. To make the picture complete, the staff was reduced with more than 50 % (!) due to cost cutting measures. Being more in control and thus being able to take painful measures without uncontrolled negative results was highly valued by the NMC.

From a business point of view the experiment was successful as well. For 3 consecutive years now, Gordian delivers Planning Services to the NMC on a commercial basis. After the transition of the assortment to other maintenance companies, we will start supporting more complex assortments such as repairables.

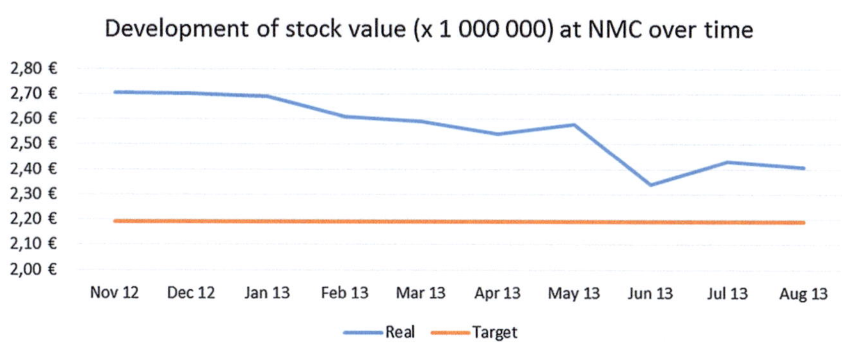

Fig. 5 Development of stock value at NMC (for a certain assortment) using planning services

Moreover we intend to use the control tower located in Den Helder, to provide Planning Services to other (maritime) companies in the Den Helder area as well.

6.3 Marel Stork Poultry Processing

Marel Stork Poultry Processing produces machines in the USA (Gainesville) and the Netherlands (Boxmeer), and also supplies (spare) parts for the maintenance of these machines. Due to an overlap in the range of (spare) parts on these two sites, there was a presumable opportunity for cost savings by pooling stock.

Experiment
To determine the benefits of pooling, Planning Services and Marel executed a project composed of four steps:

- Optimize the separate locations individually;
- Calculate the resulting amount of stock and service level when both locations are managed as a single entity;
- Redistribute the virtual stock over both locations;
- Calculate the business case: inventory reduction vs. extra emergency shipments.

The individual stock optimizations resulted in a reduction of inventory in Boxmeer of 12 % and an increase in the availability of spare parts from 97.3 to 98.6 %. In Gainesville, the stock decreased by 9 % and spare parts availability increased from 97.0 to 98.4 %. These inventory reductions are largely realized within one year by consuming the overstock.

For the spare parts assortments of Boxmeer and Gainesville, 13.4 % of the spare parts are in common, as shown in Fig. 6. These spare parts represent half of the annual turnover.

By pooling these parts, two stock-lowering effects are identified:

- Slow Moving parts initially stocked at both locations (often 1 piece per location), are only stocked at one of them (resulting in 1 stocked part for both locations).
- Safety stock in a pooling situation is smaller than in the original situation, because the uncertainty of demand (expressed as a normalized standard deviation, i.e. via the coefficient of variation) is relatively smaller.

The business solution
In this experiment the following services were delivered:

- Tactical planning
- Monthly KPI-reporting and bad-actor analyses
- Pool management

In terms of pricing Marel preferred a fixed-fee model. Given the pilot setting, a fixed-fee model suited best. That way the possible risks of the pilot would not have to be carried by Marel.

Fig. 6 Commonality in the assortments of Marel Stork

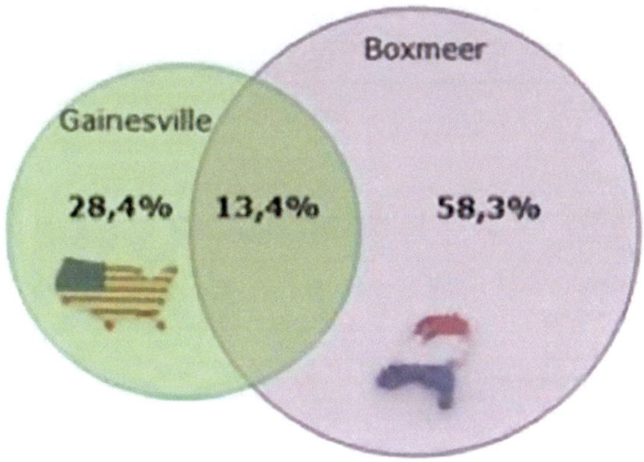

The control tower solution

Processes

Following the mentioned services, the processes in scope are Forecasting and Inventory Planning. The demarcation is as follows. Forecasting and Inventory planning is completely done in the control tower. Deployment is primarily done by the Marel operational planners.

Controls

Pro-active and reactive controls are in place, just as defined in Sect. 4.2.

Organization

The delivery of planning services follows a monthly rhythm. This is done by planners with a MSc degree in the control tower. On a biweekly basis a Planning Services account manager confers with his counterpart at Marel. This meeting was primarily about the controls and how to act upon them.

ICT

The ICT solution is a prototype version of the one presented in Sect. 4.4. However in terms of functionality, we extended the solution for pooling during this experiment. During the experiment, the control tower is physically located at the Marel premises, Boxmeer, The Netherlands.

Results

The stock pooling effects ultimately resulted in an additional stock reduction of 22 % compared to the situation where the stock locations Boxmeer and Gainesville were optimized separately.

The extra priority shipment costs, as a result of additional backorders on one of the sites, are largely determined by the weight of the spare part (due to high priority air transport). The additional annual costs for urgent shipments are only a fraction (12 %) of the reduction in inventory costs by pooling stock.

Table 3 Summary of case study solutions

	NedTrain Haarlem	Naval maintenance company	Marel stork
Services			
Tactical planning	✓		✓
Operational planning			
Monthly KPI reporting	✓	✓	✓
Top 100 management	✓	✓	
Feeding collaborative supply chains	✓	✓	
Logistics maintenance preparation	✓		
Pooling			✓
Control tower			
Location	Remote	Local	Remote
Processes	Forecasting, inventory planning and deployment (partially)	Assortment management, forecasting, inventory planning and deployment	Forecasting, inventory planning
Control	Reactive and pro-active	Reactive and pro-active	Reactive and pro-active
Organization	Account manager and planner	Account manager and planner	Account manager and planner
ICT	Prototype solution	State of the art solution	Prototype solution
Pricing model	Effort based	Effort based	Fixed fee
Commercial contract awarded	✓	✓	

Unfortunately, this project did not result in a support contract. The exact reasons were not revealed, but we expect the pure business case of pooling was not sufficiently convincing. Only for that specific part, Marel explored additional support.

6.4 *Overview of the Case Study Solutions*

For the reader's convenience, we provide a brief summary of the experiments in the Table 3.

7 Conclusions and Directions for Further Research

7.1 Conclusions

In this paper we presented Planning Services as a new and promising concept and business model for managing spare parts using a control tower concept.

The case studies showed that the objectives imposed on Planning Services, as mentioned in Sect. 2, were realized. On the first objective mentioned in Sect. 2, all case studies reported a better balance between spare parts availability and working capital. On the second objective, it appeared difficult to really prove that Planning Services led to lower operational costs. In all cases we proved to reduce the number of stock replenishment orders, being an important driver for operational costs. However, other operational costs were hard to determine and/or companies were hesitant to provide detailed costs figures of personnel and current ICT costs. The third objective was realized: now there is a complete and proven Spare Parts Planning Services solution that can be deployed on a world wide scale.

Van Laarhoven (2007) warned about the low likelihood of outsourcing inventory management. At least from these case studies a more balanced picture arises, possibly after some initial aversion. After consultation with the customers, the following critical success factors made them take the step towards Planning Services:

- Demonstrating significant and continuous improvement in terms of spare parts availability and working capital
- Mutual trust as well as the existence of a friendly and informal relation
- A very clear explanation of the concept Planning Services

Apart from these objectives, the case studies proved the value of well-tested mathematical forecasting and inventory management models. We have been able to relate a substantial part of the inventory reductions to the use of better models as well as a better acceptance of the inventory parameters by the planners.

7.2 Directions for Further Research and Exploration

We aim to explore at least two more areas. The first one is the psychology of (operational) planning. In cases we provide tactical planning services, the customer generates and evaluates purchase requisitions. Currently, the reasons to deviate from the generated advice, are arbitrary (Geertjes 2014). Hence, with Planning Services we want to get grip on the planner's behavior. To realize that we launched a new innovation project, see also Rustenburg (2013).

Further, following demands from the Aerospace and Process industry, we aim to develop pooling concepts in multi-company settings. We encounter very practical obstacles using current models, such as: *suppose there is one part left in the pool, and 2 customers need that particular part, what to do?* We believe that enriching current pooling models with gaming theories combined with additional contractual agreements is a viable direction. To prove this, we intend to start a new international demonstration project in the process industry.

Acknowledgments The authors want to express their deep gratitude to Dinalog, in particular to Wim Bens (former Managing Director) and Henk Zijm (Scientific Director) for stimulating us to start up and continue this ambitious innovation process. We also thank the companies involved—NedTrain, The Naval Maintenance Company and Marel Stork—for allowing us to use their operation as laboratory for Planning Services. Last but not least, the author is grateful to Jürgen Donders, Maarten Driessen, Friso Kuipers and Rick van Urk—all consultants at Gordian—for their critical and constructive comments.

References

Basten RJI, Van Houtum GJ (2014) System-oriented inventory models for spare parts. Surv Oper Res Manage Sci 19:34–55

Cohen MA, Zheng YS, Agrawal V (1996) Service parts logistics: a benchmark analysis. IIE Trans 29:627–639

Cohen MA, Agrawal N, Agrawal V (2006) Winning in the aftermarket. Harvard Bus Rev 192–138

Deloitte (2006) The service revolution in global manufacturing industries. Deloitte Res

De Rooy R, Rustenburg JW (2010) Demonstration project planning services: plan of approach. (Available on request)

Driessen MA, Arts JJ, Van Houtum GJ, Rustenburg WD, Huisman B (2014) Maintenance spare parts planning and control: a hierarchical framework and agenda for research. Prod Plann Control Manage Oper. doi:http://dx.doi.org/10.1080/09537287.2014.907586 (in press)

Geertjes F (2014) A first step towards fully automated spare parts planning systems—an empirical study on ordering behavior. Master thesis

Guide Jr VDR, Srivastava R (1997) Repairable inventory theory: models and applications. Eur J Oper Res 102:1–20

Kennedy WJ, Patterson JD, Fredendall LD (2002) An overview of recent literature on spare parts inventories. Int J Prod Econ 76:201–215

Kim S-H, Cohen MA, Netessine S (2007) Performance contracting in after-sales service supply chains. Manage Sci 53(12):1843–1858

Muckstadt JA (2005) Analysis and algorithms for service parts supply chains. Springer, Berlin

Rustenburg WD (2010) Innovatie in service logistiek: Achtergronden en analyses voor een 'factor 10' service logistiek onderzoeksprogramma binnen Dinalog, SLF research project, Januari 2010. In Dutch. (Available on request)

Rustenburg WD (2013) Ultimate spare parts planning: an intervention project. 2013. (Available on request)

Sherbrooke CC (2004) Optimal inventory modeling of systems: multi-echelon techniques. Wiley, New York

Soalheira J, Timbrell G (2014) What is shared services? In: Bondarouk T (ed) Shared services as a new organizational form, pp 67–84

Strikwerda J (2014) Shared service centers: from cost savings to new ways of value creation and business administration. In: Bondarouk T (ed) Shared services as a new organizational form, pp 1–16

Van Houtum GJ (2010) Maintenance of capital goods, inaugural lecture. Eindhoven University of Technology

Van Laarhoven PJM (2007) Innovatie in beweging, de logistieke toekomst van Nederland. Official report of the committee Van Laarhoven (topsector logistics), in Dutch

Van Wingerden E, Basten RJI, Dekker R, Rustenburg WD (2014) More grip on inventory control through improved forecasting: a comparative study at three companies. Int J Prod Econ doi: http://dx.doi.org/10.1016/j.ijpe.2014.08.018

Impediments to the Adoption of Reverse Factoring for Logistics Service Providers

Christiaan A.J. de Goeij, Alexander T.C. Onstein and Michiel A. Steeman

Abstract In this chapter we discuss the main impediments to the adoption of reverse factoring (RF) by suppliers in the logistics services business. Although the usage of RF is ascending, empirical evidence on RF and its implementation, especially from the point of view of suppliers, is scarce. The project reported in this chapter provides in-depth insights from seven case studies based on 20 interviews. Employees representing different departments within seven logistics service providers (LSPs) were interviewed. Interviews were also conducted with shippers, financial service providers, industry associations and university experts. The LSPs included in our research are mainly SMEs. Our results show that the main impediments to adopting RF for LSPs result from a lack of knowledge of RF, obstacles related to the collaboration with buyers and inefficiencies in the payment process. This article clarifies the practical implications of adoption of RF for LSPs and suppliers in general. Our results show that LSPs need to improve their knowledge of RF and the quality and efficiency of their invoicing processes if they are to benefit from RF. Furthermore, this research shows a need to find solutions to lower the cost to buyers of onboarding suppliers, so that they can open up RF to more suppliers—not just those that are most significant strategically.

Keywords Reverse factoring · Supply chain finance · Working capital · Logistics service providers · Case studies

C.A.J. de Goeij (✉) · M.A. Steeman
Windesheim University of Applied Sciences, Zwolle, The Netherlands
e-mail: caj.de.goeij@windesheim.nl

M.A. Steeman
e-mail: ma.steeman@windesheim.nl

A.T.C. Onstein
Amsterdam University of Applied Sciences, Amsterdam, The Netherlands
e-mail: a.t.c.onstein@hva.nl

H. Zijm et al. (eds.), *Logistics and Supply Chain Innovation*,
Lecture Notes in Logistics, DOI 10.1007/978-3-319-22288-2_15

1 Introduction

Since the credit crisis, financial institutions have been required to increase their liquid assets. This, in turn, has led to a tightening of lending terms, and financing costs for companies have grown. With liquidity scarcity as a result (Ellingsen and Vlachos 2009), companies have found it difficult to invest in inventories and raw materials. This has affected the performance of individual companies and the entire supply chains they are part of More and Basu (2013).

The credit crisis has encouraged some companies to devise more aggressive strategies for cash management as a way to increase liquidity. Extending supplier payment terms is one example of this practice. Suppliers now have to wait longer to get paid, which results in a demand for capital to maintain adequate cash flow (Hofmann 2009). These new strategies have also affected relationships between buyers and suppliers adversely (More and Basu 2013). Cash tied up in slow-moving receivables creates working capital problems for suppliers (PricewaterhouseCoopers 2009). Furthermore, extended payment terms can negatively affect buyers by increasing their supply risk (Aeppel 2010).

In this environment, supply chain finance (SCF) solutions can provide suppliers with the liquidity they require. This prevents disruption to production lines and optimises total costs within the supply chain, thereby stabilising relationships between buyer and supplier (Hofmann 2013). This chapter focuses on reverse factoring (RF)—the most widely used SCF solution. The research is part of the Dinalog project 'Expedited Payment', which aims to increase participation in RF among Dutch logistics service providers (LSPs) as a solution to cash problems. LSPs in the Netherlands are mostly small to medium-sized enterprises (SMEs) (Panteia 2010). In the Netherlands, LSPs have been hit hard by the credit crisis, especially those that are concerned mainly with transportation (Graydon 2013). Late payments and working capital problems have resulted in more bankruptcies in the sector. We focus on LSPs as suppliers to shippers. We refer throughout this chapter to shippers as buyers of logistic services. RF is not yet widely used by LSPs in the Netherlands. As a starting point for broadening RF among Dutch LSPs, we analyse current impediments to adoption. Our main research question is:

What are the impediments to the adoption of reverse factoring for logistics service providers?

This research question and the answers to this question contribute to filling multiple existing research gaps described in the literature review in Sect. 2. The research question will be examined by means of an exploratory multiple case study. We explain the methodology in Sect. 3. Section 4 presents the impediments that have had an impact on the adoption of RF for LSPs, resulting from our research. Section 5 ends with suggestions for further investigations based on the results and the limitations of this case study.

2 Literature Review

Academic papers on supply chain management (SCM) mostly discuss the flow of goods, materials and information (D'Avanzo et al. 2003; Fellenz et al. 2009; Pfohl and Gomm 2009). Financial flows in supply chains are often neglected (Ceccarello et al. 2002; Fairchild 2005; Fellenz et al. 2009). It is only recently that financial flows have been given proper consideration in academic papers. These papers observe a lack of coordination between financial and physical supply chains (Mentzer et al. 2001; Pfohl and Gomm 2009; Hofmann and Kotzab 2010; Gupta and Dutta 2011). This disparity causes cash to be trapped in the supply chain. Financial institutions and supply chain service providers have introduced many solutions to release this cash, leading to the emergence of the field of SCF (Robinson 2007; Demica 2007; More and Basu 2013). Unlike SCM, finance, or logistics, SCF is a relatively young discipline (Hofmann 2005); this explains why so far it has little empirical foundation (Wuttke et al. 2013a) and lacks clear definitions (Hofmann 2013). Defining the true nature of SCF appears to be challenging, because experts disagree on whether SCF is a discipline, technique, product or program (Templar et al. 2012). Many academic publications mention supply chain finance while actually referring to reverse factoring.

Templar et al. (2012) discern three schools of thought on SCF. In the first, named financial supply chain management, SCF is described broadly, and is considered to cover all activities that can be related to finance in supply chains, from financial processes (transaction processes, data processing, invoice matching etc.) to financing techniques. The second, denoting SCF as supply chain financing, is more narrowly described and concerns the financial instruments that can be used to optimise financial flows in supply chains. Fields that are covered in this school of thought are trade financing, fixed-asset financing, working- capital financing and supplier financing. The third perspective sees SCF as buyer-driven payables solutions, focusing mainly on one SCF instrument: reverse factoring. This instrument forms the basis of the current chapter.

Reverse factoring is an arrangement through which a buyer and its financier offer a supplier credit for the period of the payment term against the credit rating of the buyer (Demica 2007). If the buyer has a high credit rating, the supplier enjoys low short-term financing costs. Often providers of electronic platforms that supply real-time transactional information to all parties concerned partner with financiers (Wuttke et al. 2013a). The buyer sends the approval of the invoice to the platform, which is accessible by all three parties. This approval is a payment guarantee allowing the supplier to receive funds directly from the financier. When the supplier wants to access these receivables, the financier advances a payment with a discount based on the buyer's borrowing rate. This credit allocation uses the buyer's payment guarantee as collateral, leading to a favourable borrowing rate. At the end of the credit period the buyer pays the financier, irrespective of the financial situation of the supplier (Tanrisever et al. 2012). Figure 1 shows the typical steps in RF.

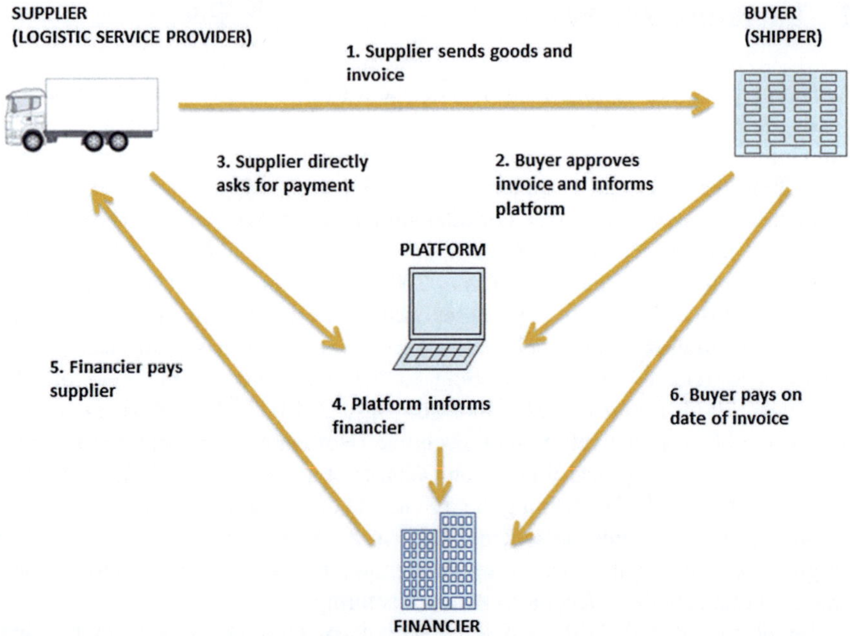

Fig. 1 Explanation of reverse factoring

Suppliers in an RF arrangement gain access to lower financing costs. Suppliers' cash flow also becomes more predictable, because they have more clarity about *when* they can access receivables. By offering financial support to suppliers, buyers can decrease supply risks and stabilise supply chains. For buyers, RF is a way to maintain or extend the payment term, while still leaving net benefits for the supplier. If the buyer extends the payment term, he reduces his need for short-term financing (Tanrisever et al. 2012). Automated transaction platforms can bring down costs for both buyer and supplier. Financiers (usually banks) benefit because they charge a fee amounting to the discount agreed for advance payment. Financiers are also able to create or extend relationships with suppliers without taking on additional risk. They can build a credit history with the supplier that may lead to additional lending (Klapper 2006).

As noted in the introduction, RF is the most commonly used SCF solution, and is common in emerging economies (Klapper 2006). In the last five years, adoption of RF has increased in other economies. Both academic and non-academic publications show a growing number of companies using RF. Philips, Heineken, Unilever, Volvo and Motorola are just a few examples (Seifert and Seifert 2011; Blackman et al. 2013; Steeman 2014).

Recently, a series of case studies on SCF and RF have appeared. Templar et al. (2012) have performed a multiple case study on the adoption of SCF looking at four companies in four different industries. Wuttke et al. (2013a) address adoption of

SCF using a multiple case study on 'the innovation adoption strategy for SCF'. They employ the innovation adoption model proposed by Rogers (2003) that states there is an initiation phase before every implementation phase in innovation adoption. More and Basu (2013) and Blackman et al. (2013) both report single case studies. More and Basu (2013) investigate the challenges of SCF by examining an Indian firm with global operations. Blackman et al. (2013) are not researching SCF instruments as such, but aim to 'explore the concept of financial supply chain strategy in a global business environment' by completing a detailed case study of Motorola. All of these case studies are written from the perspective of the buyer. Many other academic papers on SCF and RF take the perspective of the supply chain as a whole but do not explicitly consider the perspective of the supplier. This paper focuses on the supplier in the adoption of RF, and therefore contributes to closing that research gap. While most academic papers on SCF and RF mention suppliers, they do not specify whether these suppliers are materials or parts suppliers, LSPs or other service providers. Until now, segmentation of supplier types in academic papers has mostly been based on differences in creditworthiness, the purchasing volume they represent to their buyers, and their strategic relevance to buyers (Wuttke et al. 2013a, b).

We focus explicitly on LSPs as suppliers in RF. Few authors consider LSPs in relation to SCF. Hofmann (2005) sees new responsibilities emerging at the intersection of finance and supply chain management, not only for financial service providers, but also for logistics service providers. These concern inventory financing. Hameri and Hintsa (2009) and Hofmann (2009) also conclude that LSPs can increase their service provision in the field of inventory financing. However, there are hitherto no academic papers dealing with LSPs as suppliers in RF arrangements. This is another literature gap we will contribute to closing.

Academic and non-academic literature shows that four types of impediments are most prevalent in the adoption of RF. These impediments are based on papers that do not explicitly focus on LSPs in RF schemes. We use literature on suppliers in general (without specifying whether these are materials or parts suppliers or service providers), supported by insights into the adoption of SCF from the point of view of the buyer. All the categories and subcategories of impediments are depicted in Table 1, and function as propositions in our research. In Sect. 4 we ascertain

Table 1 Categories and subcategories of impediments in the literature

Lack of knowledge	Unfamiliarity with RF
	Lack of skilled personnel and training on RF
Payment process	Poor visibility
	Lack of standardisation in exchange of invoices
	Mistakes in invoices
	Disputes
Inter-firm collaboration	Ease of substitution
	Business volume of supplier
Intra-firm collaboration	Lack of coordination between departments

whether the impediments in Table 1 also apply specifically to LSPs. In the remaining part of the literature review we describe each impediment in further detail.

2.1 Lack of Knowledge

There is a lack of general awareness about SCF programs like reverse factoring, which negatively affects their adoption (Hofmann and Belin 2011). There is a significant lack of knowledge regarding regulations and legal mechanisms (M3 Consultancy and Zanders 2014) and automated digital payment systems (Fellenz et al. 2009). Companies providing external legal or accounting assistance are generally also unfamiliar with RF. Research by M3 Consultancy and Zanders (2014) shows that the underlying reasons for this ignorance are attributable to the fact that the concept is new and to a general lack of existing standards concerning RF.

This lack of knowledge causes fewer companies to consider adopting RF seriously. This is exemplified by the relationship between the supplier and the bank in RF constructions. RF is generally arranged with the bank of the buyer. Wuttke et al. (2013a, p. 158) report on suppliers that anticipate "…a catch to SCF, as they did not trust an external bank they are not currently doing business with". This shows that, for suppliers, it is unclear what the consequences of RF are when buyers enlist a different bank. Lack of skilled personnel and training on SCF tools and techniques within companies adds to the challenges that are faced when considering adoption of SCF (More and Basu 2013).

2.2 Payment Process

For RF, fast approval of invoices by buyers is important, so that suppliers can be financed quickly for the payment guarantee. A lack of standardisation in the exchange of invoices is often one reason why approval of invoices takes a long time. According to Fellenz et al. (2009) major delays occur when buyers and suppliers do not use the same payment systems. Many companies use paper-based manual processes that make administration less visible (Fairchild 2005; Aberdeen 2006). Companies often struggle to accept electronic invoices even though they offer speed, transparency and cost savings (Berez and Sheth 2007). A lack of automation along with poor visibility leads to difficulties for SCF providers in employing third-party financing programs like RF (More and Basu 2013). Mistakes, which can be the result of a lack of standardisation in exchange of invoices, are also a major cause of slow payments. Such errors, along with a lack of standardisation in invoice exchange and poor visibility, increase the chances of disputes between buyer and supplier. And disputes further increase the time it takes to approve invoices (Fellenz et al. 2009).

2.3 *Inter-firm Collaboration*

Hofmann (2013) presents an overview of the literature concerning SCF and states that, in the majority of papers, inter-firm collaboration is an important factor. RF is a way for buyers to decrease supply risk, optimise financing costs, and at the same time support their suppliers financially. However, for buyers, collaboration with certain suppliers is more important than with others. M3 Consultancy and Zanders (2014) state that because of a lack of standardisation, the onboarding costs of RF are still quite high for all parties involved. Onboarding costs are related mainly to changes in the administrative process that have to be implemented. These costs could influence a buyer when choosing suppliers for RF. For financiers, the cost of onboarding is fixed for every supplier. The higher the purchasing volume a supplier represents, the more attractive it is for buyers and financiers to include this supplier in an RF arrangement. Most buyers that already use RF have a strict threshold for the purchasing volume a supplier needs to represent, ranging from one to five million euros. In The Netherlands, this is the major reason why very few SMEs participate in RF (M3 Consultancy and Zanders 2014). Furthermore, to decrease supply risk, a buyer looks at the strategic value of a supplier's products or services. If there are many companies that can easily replace a supplier, supply risk is lower than when the supplier is one of a kind. A unique supplier has a better chance of being included in RF schemes (Steeman et al. 2014).

2.4 *Intra-firm Collaboration*

In academic papers, a lack of intra-firm collaboration is recognised as a clear impediment to the adoption of SCF. Adoption of SCF by buyers requires cross-functional collaboration between finance, procurement and logistics departments within companies (Stefansson and Russell 2008; More and Basu 2013; Wuttke et al. 2013a). Stefansson and Russell (2008) state that cross-functional collaboration is needed because the finance experts are in the finance department, while contact with suppliers is managed by the procurement and logistics departments. Financial managers can make decisions that constrain operational managers and vice versa (Protopappa-Sieke and Seifert 2010), as finance and SCM departments 'do not speak each other's language' (Timme and Williams-Timme 2000). So far, we are not aware of academic papers dealing with the adoption of SCF for suppliers, while, arguably, suppliers should also introduce cross-functional collaboration. However, such a collaboration might concern different departments, as suppliers would involve sales instead of procurement.

3 Research Method

In order to understand the impediments that exist for LSPs in adopting RF, we carried out an exploratory research in the form of a multiple case study to contribute to filling in a number of research gaps. We checked for construct validity, internal validity, external validity and reliability, following (Yin 2009; Baxter and Jack 2008; Gibbert et al. 2008).

3.1 Case Selection

Seven LSPs from the 'Expedited Payment' project were selected as cases for this research. To enhance external validity we ascertained whether the cases were representative for LSPs in the Netherlands. Case demographics can be seen in Appendix 2. Size is an important factor to consider, as research shows size is related to purchase volume and ease of substitution (M3 Consultancy and Zanders (2014). According to research by Panteia (2010), 99.5 % of Dutch LSPs are SMEs. Six out of seven LSPs in our research are SMEs. The remaining case is the only large company, serving as a polar type, as suggested by Eisenhardt (1989). Along with size, the primary activities a company performs are important as our literature review shows that these affect ease of substitution. Five of the seven LSPs offer transportation and storage as their main products; the other two LSPs are forwarders. To further increase external validity, we carried out three interviews with domain experts from an LSP industry association representing more than 6000 members.

3.2 Data Collection

20 semi-structured interviews were conducted to collect data. Interview questions addressed the impediments that the literature describes, but were sufficiently open to elicit new findings. Of the 20 interviews, nine were conducted with seven LSPs. In the case of one of the LSPs, only the director was interviewed. In the case of four LSPs, a number of employees from different departments took part in a combined interview. Employees from different departments were interviewed separately in the other two LSPs. Following the interviews, transcripts and conclusions were checked by the LSPs for truthfulness and completeness. Data triangulation was achieved by carrying out interviews with shippers and financial service providers, which constitute additional relevant parties next to suppliers in RF arrangements. Furthermore, interviews were conducted with industry associations of both LSPs and shippers, and a university expert with experience in adopting RF in a company. This improved the construct validity of our research (Yin 2009). The interviews

were conducted with the help of students from the Amsterdam University of Applied Sciences.

Reverse factoring is multi-disciplinary. That is the reason why a number of people with different functions in a given company were interviewed. For most of the smaller SMEs, one person at director level was interviewed, and one employee dealing with accounts receivable. In the largest of the six SMEs, a CFO and one of their departmental employees were interviewed. The interviews with the large LSP involved the financial director and head of corporate social responsibility. More detailed information about interviews and case demographics can be found in Appendices 1 and 2.

3.3 Coding and Data Analysis

All interviews were fully recorded and transcribed to create a rich database and to improve reliability. Transcripts of the interviews were coded. We used five code categories, which were the four impediments from the literature review and one extra category for 'other impediments'. We completed axial coding by making subcategories within these categories. For internal validity we matched patterns among cases. Our strategy for coding, as suggested by Baxter and Jack (2008), was first to allow several researchers to code the data independently. We then met to come to a consensus on categories and groups. The categories and their subcategories are listed in Table 2.

Analysis of results followed the codes. Initial results were reviewed by the researchers and their peers, and participants in the case studies. These persons were all members of the project 'Expedited Payment', thus reducing the likelihood of

Table 2 Categories and subcategories of impediments in our research

Lack of knowledge	Unfamiliarity with RF
	Lack of skilled personnel and training on RF
Payment process	Complicated pricing structure[a]
	International transportation documentation[a]
	Multiple types of invoices[a]
	Lack of standardisation in exchange of invoices
	Poor visibility
	Mistakes in invoices
	Disputes
Inter-firm collaboration	Ease of substitution
	Business volume of supplier
	Lack of trust[a]
Intra-firm collaboration	Collaboration sales and employees handling invoicing[a]

[a]Impediments which are found in our research, but not in the literature

false reporting (Yin 2009). Alternative explanations were sought and considered before we arrived at our final conclusions. In this paper, confidentiality and anonymity of participants are ensured to prevent dissemination of sensitive financial information. Names of companies and organisations are not mentioned in this paper.

4 Results and Analysis

This section elaborates on all categories and sub-categories of impediments to the adoption of RF found in our research. It concludes with Table 2, which is an appraisal of the impediments from literature, that function as propositions in Table 1.

4.1 Lack of Knowledge

None of the seven LSPs is currently involved in a reverse factoring scheme. Before the interviews, only a few of the respondents knew what RF was. Only one out of the seven LSPs, i.e. the large enterprise, had a customer who had discussed the possibility of establishing an RF scheme. However, both this LSP and its customer were still at the stage of investigating the possibilities of RF. This means that no opportunity for RF training for employees could be said to exist within any of the seven LSPs.

The LSP industry association confirmed that knowledge of RF is rare among its members, especially for personnel below director level. In particular, there is a lack of knowledge in LSPs about the degree and the nature of information that needs to be shared in an RF scheme. In an RF arrangement information that is not relevant to the specific transaction between buyer and supplier need not be shared. However, LSPs are concerned about sharing detailed information on returns, pricing structure and financial situation. They are worried that a shipper might cease trading with them if, via an RF scheme, they could compare LSPs more easily. The interviews also show that LSPs do not have a clear idea of the type and the duration of contracts that need to be signed for RF. Furthermore, LSPs are uncertain as to whether entering into RF agreements with the shipper's bank would have any consequences for the relationship with their own bank.

4.2 Payment Process

Quick approval of invoices is a key element in Reverse Factoring. However, this quick approval is difficult when invoices are complicated. The pricing structure of invoices that LSPs send to shippers can be complex. LSPs must consider many factors when invoicing. There is, for example, always the question of how and by whom the waiting time for truck drivers, fuel and tolls should be paid. When shipping a container that is less-then-container-load, the value of the goods inside can be determined in many different ways, e.g., by weight or by cubic meter. These are all examples of matters that make invoicing more complicated, thereby increasing the chance of mistakes. When such factors are not elucidated in contracts, which is often the case, invoices are open to dispute.

The payment process becomes more complicated when goods are shipped across borders. A CMR waybill is needed in this case, which is a proof-of-delivery document that must be signed directly after shipment of goods. A shipper (being the buyer) will generally receive a copy of the waybill. European law prevents this waybill being signed digitally. When payment takes place after 30, 60 or 90 days, shippers sometimes delay payment even further on the pretext that they need the original waybill before they can proceed with the payment.

The interviews revealed that LSPs use many different types of invoices. Some customers demand an invoice for every single order. However, LSPs frequently complete many ‘small jobs’ comprising different orders from one specific customer. Sometimes customers demand LSPs to send an invoice every 7, 14, or even 30 days, with all orders being charged at once. If work is done at the beginning of any 7, 14, or 30-day period, the LSP needs to wait almost the full period before even starting the billing process, let alone to be paid. Additionally, such a range of invoices introduces yet more complexity to business, thus increasing the likelihood of mistakes and disputes in the payment process.

Interviews with both LSPs and shippers make it clear that a lack of standardisation in the exchange of invoices is a major reason for more days of sales outstanding. Many invoices are still sent by post. If invoicing is done digitally, LSPs and shippers often struggle with incompatible payment systems. LSPs usually send a PDF or Excel file via e-mail. When the shipper receives the invoice, they must convert the PDF or Excel file into another format, which can be time consuming. E-invoicing, where a shipper and an LSP share a payment system and receive real time information about transactions, is not in use among the companies interviewed.

The majority of LSPs have several employees, in different departments, each responsible for one specific part of the invoicing process. This can result in a lack of visibility, making it difficult to determine what the status of an invoice is, and causing delays in completing the process.

Mistakes in invoices are a reason for more days of sales outstanding. The interviews showed that a complicated pricing structure, international transportation documents, different types of invoices, lack of standardisation in exchange of invoices and lack of visibility in the payment process can increase the chance of making mistakes.

These mistakes increase the likelihood of the LSP and the shipper entering into disputes. The majority of the LSPs interviewed did not have any method for resolving disputes, which often causes further delays in payment.

4.3 Inter-firm Collaboration

The vast majority of the companies interviewed said that the shipper (i.e. the addressee) was the dominant company in the supply chain. As mentioned earlier, only one of the seven LSPs had a customer that discussed the possibility of joining an RF program with them. This was the largest of the seven LSPs in terms of the amount of purchasing volume it represented to shippers. RF programs with purchasing volume limits above one million euros excluded the vast majority of LSPs.

LSPs state that they operate in a very competitive market with low margins, where shippers are dominant. Shippers and industry associations of both shippers and LSPs agree on this observation, and add that LSPs are easy to substitute, especially where a company's services are limited to transportation and storage. Because of the limited supply risk for shippers, these LSPs are not the first companies that come to mind for inclusion in an RF program.

LSPs reveal a lack of trust towards dominant shippers that control purchasing conditions on which LSPs have little influence. According to these LSPs, such shippers frequently use their power to manipulate the payment process. Since LSPs show low levels of trust towards shippers, they are less disposed towards buyer-driven programs like RF. They generally believe the shipper has a better negotiation position for RF, and therefore that shippers benefit more in a business arrangement by using RF to extend payment terms. As mentioned earlier, LSPs are unaware of the information that must be shared for RF. They are worried shippers will use this information in a way that would be disadvantageous to them. Some LSPs fear that dependency on a specific shipper might increase if they were involved in an RF program.

4.4 Intra-firm Collaboration

The LSPs that were interviewed do not report a lack of collaboration between departments. The literature about intra-firm collaboration as an impediment to RF refers to the internal structures of big buyers. Our research concerns mostly SME suppliers, where finance departments typically consist of up to five people that are located in the same room along with other departments. Collaboration problems between operational and finance departments therefore differ from those described in the literature review. LSPs in our research frequently mention a lack of cooperation between sales staff and invoicing departments as one of their greatest operational impediments. Sales staff at LSPs are concerned with closing deals. They can discuss the conditions under which the transaction takes place but are not responsible for the invoicing process that follows. When there are disputes with customers concerning invoices, the finance, administration or planning departments are responsible. According to many of our respondents, sales staff should play a greater role here, as they are more aware of the history of a client relationship and the specific agreements that are made. In this way invoices can be approved earlier by buyers.

This research considered a series of impediments to the adoption of reverse factoring described in the literature. The four main categories in Table 2 are the same as those in Table 1. Since these categories were already broad there was no need to add a fifth category in Table 2. As mentioned earlier, the series of impediments described in the literature is not based specifically on the supplier/LSP point of view. However, our results show they provide a good basis for determining the impediments to RF in general. Additions from our research are based mainly on sector-specific problems. When it comes to the payment process, our research shows complicated pricing structures, international transportation documentation and multiple types of invoices are particularly relevant for LSPs . A lack of trust in collaboration is found in supply chains with an unequal power distribution between companies. An unequal power distribution is prevalent in the transport sector, where companies are easy to replace. The category of intra-firm collaboration did not appear to be a main impediment in our research. This can be explained by the fact that we consider SMEs predominantly, where different departments are small and often located in the same room. The existing literature deals with larger companies, where the dynamics between departments differ.

5 Conclusions

This paper contributes to academic research carried out on reverse factoring. It takes the perspective of LSPs as suppliers. While most research on RF, and SCF in general, is rather conceptual in nature, we provide empirical data that give in-depth insights. By building on recent studies like those of Wuttke et al. (2013a) and More and Basu (2013), we were able to offer further findings on the adoption of SCF solutions. These are relevant not only to academia, but also to practitioners.

In Sect. 4 we explained that the new impediments found in our research relate mainly to LSPs. However, we have uncovered other impediments that are applicable to other types of suppliers, like a general lack of trust and failure of sales persons and employees handling invoicing to collaborate. In any buyer-supplier relationship where the power is in the hands of the buyer, and where the buyer seems to take advantage of this, suppliers may not trust buyer-led initiatives like RF. Better communication between sales persons and employees handling invoices could be relevant to any supplier. With greater efficiency in the payment process, invoice approval happens more quickly; this is a key element in RF.

Currently, the most significant impediments to LSPs and suppliers in general occur as a result of dysfunctional inter-firm collaboration. Purchasing volume thresholds exclude most LSPs from participating in RF schemes. This is the reality faced by many suppliers that only represent a low purchasing volume to a buyer, especially if they can be easily substituted by a competitor. To make it profitable for buyers and financiers to include suppliers that represent a low purchasing volume, solutions must be found that lower the costs of onboarding. According to M3 Consultancy and Zanders (2014), costs are high for buyers and financiers because there is currently no standardisation in the onboarding process. The management of legal affairs and contracts is one element of RF onboarding that is a victim of this problem. More research is needed to devise methods to standardise this onboarding process, thereby reducing costs and making suppliers that represent low purchasing volumes a more profitable proposition for buyers and financiers. Onboarding costs for buyers and financiers show a close relationship to a supplier's knowledge of RF. The more knowledge the supplier has, the less time it takes the buyer and financier to complete onboarding. Education and training programs for suppliers therefore help to bring down onboarding costs. Furthermore, such initiatives might help suppliers to overcome trust issues by offering access to detailed information on the benefits to all parties involved in RF.

Our research shows that LSPs can make substantial improvements to payment processes. Quick approval of invoices by buyers is important for RF, but this is often impeded by the inefficiency of the invoicing process of LSPs. A higher degree of automation in the payment process might increase visibility and decrease mistakes in invoices. Popa (2013) suggests clarifying internal responsibilities by determining who sets and who monitors payments. This is highly recommended for LSPs as a way to make the collection of receivables less time consuming, and facilitate ease of tracing mistakes. This will enable quick resolution of mistakes and identify commonly recurring errors with a view to eliminating them. Furthermore, we recommend clear dispute resolution procedures to save time.

An important limiting factor in our research is that only one of the seven LSPs interviewed had been approached about the use of RF. None of the seven LSPs is already in the implementation phase of adopting RF. Our research results provide useful insights from the supplier's point of view, but not all of the insights apply directly to all suppliers. Most of the companies we researched offer transportation and storage as their core products, which are services that are easy to substitute. Moreover, most companies we interviewed were SMEs representing low purchasing volumes to their buyers. Future research on the adoption of RF for suppliers would benefit from case studies regarding companies in different commercial environments and addressing suppliers that represent bigger purchasing volumes to their buyers.

Appendix 1: Interviews

Type of company/organization	Number of companies/organizations	Number of interviews
Logistics service providers (LSPs)	7	9
Shippers	2	2
Financial service providers	3	4
Industry association LSPs	1	3
Industry association shippers	1	1
Universities	1	1
Total	15	20

Appendix 2: Case Demographics

Cases	A	B	C	D
Main activities	Transportation and storage	Transportation and storage	Transportation and storage	Forwarding
Size	SME—small	SME—small	SME—small	SME—small
Respondent #1	Director/owner	Director/owner	Director/owner	Owner business developer
Respondent #2	Financial assistant	Financial assistant	–	Financial assistant
Respondent #3	–	–	–	–

Cases	E	F	G
Main activities	Forwarding	Transportation and storage	Transportation and storage
Size	SME—medium	SME—medium	Large company
Respondent #1	Managing director	CFO	Financial director
Respondent #2	Head of credit control	Business controller	Head of corporate social responsibility
Respondent #3	Business controller	–	–

References

Aberdeen Group (2006) Global Supply Chain Benchmark Report. Industry Priorities for Visibility, B2B Collaboration, Trade Compliance and Risk Management. Aberdeen Group, Boston

Aeppel T (2010) "Bullwhip" hits firms as growth snaps back. Wall Street J. Available via http://online.wsj.com/articles/SB10001424052748704509704575019392199662672. Accessed 10 Aug 2014

Baxter P, Jack S (2008) Qualitative case study methodology. Study design and implementation for novice researchers. Qual Rep 13(4):544–559

Berez S, Sheth A (2007) Break the paper jam in B2B payments. Harvard Bus Rev. Available via HBR. http://hbr.org/2007/11/break-the-paper-jam-in-b2b-payments/ar/1. Accessed 10 Aug 2014

Blackman ID, Holland CP, Westcott T (2013) Motorola's global financial supply chain strategy. Supply Chain Manag Int J 18(2):132–147

Ceccarello C, Neser G, Pestre C et al (2002) Financial indicators and supply chain integration: a European study. Supply Chain Forum 3(1):44–52

M3 Consultancy and Zanders (2014) Zonder interventie zal het MKB op korte termijn niet meeprofiteren van de groeiende ketenfinancieringsmogelijkheden in Nederland. Available via Tweede Kamer. http://www.tweedekamer.nl/kamerstukken/detail.jsp?id=2014D13314&did=2014D13314. Accessed 21 Apr 2014

D'Avanzo R, Von Lewinski H, Van Wassenhove LN (2003) The link between supply chain and financial performance. Supply Chain Manag Rev 7(6):40–47

Demica (2007) The growing role of supply chain finance in a changing world. Available via Demica. http://www.demica.com/images/PDFs/supply_chain_finance_steady_supply.pdf. Accessed 2 May 2014

Eisenhardt KM (1989) Building theories from case study research. Acad Manag Rev 14(4):532–550

Ellingsen T, Vlachos J (2009) Trade finance in a liquidity crisis. The World Bank policy research working paper series 5136

Fairchild A (2005) Intelligent matching: integrating efficiencies in the financial supply chain. Supply Chain Manag Int J 10(4):244–248

Fellenz MR, Augustenborg C, Brady M et al (2009) Requirements for an evolving model of supply chain finance: a technology and service providers perspective. Commun IBIMA 10:227–235

Gibbert M, Ruigrok W, Wicki B (2008) What passes as a rigorous case study? Strateg Manag J 29(13):1465–1474

Graydon (2013) De transport and logistiek sector. Graydon Whitepaper. Available via Graydon. https://www.graydon.nl/uploads/files/NL/Whitepaper_Transport_%26_Logistiek_%202013.pdf. Accessed 12 May 2014

Gupta S, Dutta K (2011) Modeling financial supply chain. Eur J Oper Res 211(1):47–56

Hameri A-P, Hintsa J (2009) Assessing the drivers of change for cross-border supply chains. Int J Phys Distrib Logistics Manag 39(9):741–761

Hofmann E (2005) Supply chain finance: some conceptual insights. In: Läsch R, Jancker CG (eds) Logistik management—innovative Logistikkonzepte. Deutscher Universitätsverslag Wiesbaden: 203–214

Hofmann E (2009) Inventory financing in supply chains: A logistics service provider-approach. Int J Phys Distrib Logistics Manag 39(9):716–740

Hofmann E (2013) Research on supply chain finance. A review, a framework and suggestions for the future. In: 25th Annual NOFOMA conference 2013:1–16

Hofmann E, Belin O (2011) Supply Chain Finance Solutions. Springer briefs in business. Springer, Berlin

Hofmann E, Kotzab H (2010) A supply chain-oriented approach of working capital management. J Bus Logistics 31(2):305–330

Klapper L (2006) The role of factoring for financing small and medium enterprises. J Bank Finance 30(11):3111–3130
Mentzer JT, De Wit W, Keebler JS et al (2001) Defining supply chain management. J Bus Logistics 22(2):1–25
More D, Basu P (2013) Challenges of supply chain finance: a detailed study and a hierarchical model based on the experiences of an Indian firm. Bus Process Manag J 19(4):624–647
Panteia (2010) Omvang en prestaties van het MKB in de topsectoren. Available via Topsectoren Onderzoek. http://www.topsectorenonderzoek.nl/wp-content/uploads/omvang_prestaties_MKBinTopsectoren_Fris.pdf. Accessed 13 June 2014
Pfohl H-C, Gomm M (2009) Supply chain finance: optimizing financial flows in supply chains. Logistics Res 1(3–4):149–161
Popa V (2013) The financial supply chain management: a new solution for supply chain resilience. Amfiteatru Econ J 15(33):140–153
PricewaterhouseCoopers (2009) Demystifying supply chain finance—insights into the what, why, how, where and who. Available via PwC. www.pwc.com/en_US/us/issues/surviving-the-financial-downturn/assets/supply_chain_finance.pdf. Accessed 17 July 2014
Protopappa-Sieke M, Seifert RW (2010) Interrelating operational and financial performance measurements in inventory control. Eur J Oper Res 204(3):439–448
Robinson P (2007) The 2007 guide to Financial supply-chain management. Available via Euromoney. http://www.euromoney.com/images/502/51463/financial_supply_chain_management.pdf. Accessed 13 June 2014
Rogers EM (2003) Diffusion of innovations, 5th edn. Free Press, New York
Seifert RW, Seifert D (2011) Financing the chain. Int Commer Rev 10(1):32–44
Steeman M (2014) SCF 2.0: supply chain finance beyond traditional ways. In: The Paypers. E-invoicing, supply chain finance & e-billing market guide 2014: 50–51
Steeman M, De Goeij C, Klapwijk P (2014) An exploratory study into supply chain finance: the relevance of supplier segmentation. Int J Phys Distrib Logistics Manag (submitted)
Stefansson G, Russell DM (2008) Supply chain interfaces: defining attributes and attribute values for collaborative logistics management. J Bus Logistics 29(1):347–359
Tanrisever F, Cetinay H and Reindorp M et al (2012) Value of reverse factoring in multi-stage supply chains. Available via SSRN. http://ssrn.com/abstract=2183991. Accessed 9 Feb 2015
Templar S, Cosse M, Camerinelli E et al (2012) An investigation into current supply chain finance practices in business: a case study approach. Unpublished paper
Timme SG, Williams-Timme C (2000) The financial-SCM connection. Supply Chain Manag Rev 4(2):33–39
Wuttke DA, Blome C, Foerstl K et al (2013a) Managing the Innovation adoption of supply chain finance. Empirical evidence from six european case studies. J Bus Logistics 34(2):148–166
Wuttke DA, Blome C, Henke M (2013b) Focusing the financial flow of supply chains: an empirical investigation of financial supply chain management. Int J Prod Econ 145(2):773–789
Yin RK (2009) Case study research: design and methods, 4th edn. Sage Publications Inc., Thousand Oaks

Part IV
IT-Based Innovation

Towards an Approach for Long Term AIS-Based Prediction of Vessel Arrival Times

Alexander Dobrkovic, Maria-Eugenia Iacob, Jos van Hillegersberg, Martin R.K. Mes and Maurice Glandrup

Abstract The goal of this paper is to conduct a review of existing solutions and related algorithms on maritime route prediction using Automatic Information System (AIS) data, determine to what extent they can be applied to solve the prediction problem, and identify areas that have to be improved in order to get an industry-acceptable solution to enhance various logistics planning processes. The contributions of this paper are: (i) to present the available solutions for trajectory prediction of a vessel; (ii) to identify components that can be used for finding a solution for the identified problem as well as showing the strengths and weaknesses of each available option; and (iii) to propose a new concept for arrival time estimation based on trajectory prediction and the use of algorithms from the included literature review.

Keywords Automatic Identification System · Trajectory analysis · Route prediction · Maritime logistics · Transportation planning

1 Introduction

For any system that requires (periodic) decision making, better policies require solid information on its present and future states. In the case of logistic service providers and their planning process, being able to increase efficiency and customer service relies on the availability of a platform capable of delivering and processing relevant information. To design and develop such a platform, a research project called Synchromodal-IT has been initiated and funded by Dinalog. This project

A. Dobrkovic · M.-E. Iacob (✉) · J. van Hillegersberg · M.R.K. Mes · M. Glandrup
Faculty of Behavioural, Management and Social Sciences, University of Twente,
P.O. Box 217, 7500 AE Enschede, The Netherlands
e-mail: m.e.iacob@utwente.nl

A. Dobrkovic
e-mail: a.dobrkovic@utwente.nl

H. Zijm et al. (eds.), *Logistics and Supply Chain Innovation*,
Lecture Notes in Logistics, DOI 10.1007/978-3-319-22288-2_16

aims to develop IT services for synchromodal transport, which is a form of intermodal transport where the best possible combination of transport modes is selected dynamically based on the available data (Dutch Institute for Advanced Logistics 2013).

The added value expected from such a platform is the ability to provide planners with essential information for process optimization that they either cannot acquire on their own or the expenses of doing so are beyond a single budget. Nevertheless, the potential benefits can be large; for example, using a data mining technique on external data describing events that have impact on the logistic process provides a necessary foundation for optimization algorithms seeking to reduce overall logistics expenses through better planning.

One particular problem has been identified in discussions with logistics providers participating in the SynchromodalIT project: the uncertainty of deep sea vessel arrival in ports. This problem involves three actors: logistic service providers, deep sea vessels and port terminals. The focus is often on adding value to the last actor by providing additional information that can improve planning decisions. In this research we focus on the logistic service provider (LSP). Specifically, we focus on LSPs that rely on barges for the long haul and use truck transport primarily for the first and last mile. In this long haul hinterland transportation, the related planning involves setting the appointment windows at various terminals within the port, during which the goods are loaded from deep sea vessels or trucks into barges and vice versa. Depending on the terminals used, appointments need to be made 24 till 48 h in advance. However, since terminals give priority to deep sea vessels, their unexpected arrival times cause serious disruptions in the planning of barges, which propagate further throughout the supply chain, often forcing LSPs to use trucks for the entire length of a route and cover barge expenses if they have been hired in advance.

LSPs do not have information about deep sea vessels and their arrangements with port terminals. Also, since the timing of these vessels can be restricted by other factors, such as water levels in the port, their arrival is generally subject to change, but will nevertheless have priority over barges (basically because they represent large volumes and generally high-value contracts between terminal operators and shipowners). Even though arrival of deep sea vessels is essential for LSPs, they often are not timely informed and hence are forced to make their planning without it. The planners currently check deep sea vessel position data delivered through various websites that display maritime traffic, but due to an information quality issue, it is not possible to improve the planning, relying on this information in its present state.

From the perspective of LSPs, the arrival of deep sea vessels often results in a disruption of their plan. If there is a method that could identify possible disruptions in advance, it would allow planners to alter plans and avoid duplicating costs that are incurred due to hiring both barges and trucks for the delivery of the same cargo. Since LSPs do not have access to deep sea vessel routes and corresponding times, the solution must be based on publicly available data. By creating prediction services that enable LSPs to mine this available data, it is possible to provide planners with advance warnings of disruptions.

In order to acquire GPS position data of ships, we focus on using Automatic Information System (AIS) data. AIS is a self-reporting system designed to be used for collision avoidance allowing for a higher level of situational awareness (Vespe et al. 2012). According to the SOLAS convention (International Maritime Organization 2002), AIS is required for ships of gross tonnage of 300 and higher in international voyages, while in non-international voyages a gross tonnage of 500 is the lower bound for cargo and passenger ships. The European Union is moving towards making AIS mandatory for fishing vessels sailing in waters under the jurisdiction of the Member States (Commission of the European Communities 2008). Being a cheap alternative to radar, AIS can be used to acquire positions of all vessels in the areas covered by AIS receivers.

The downside of relying on AIS is that, as a self-reporting system, the trustworthiness of information depends on data reported by the vessel, and, as such, is prone to spoofing or intentional incorrect information reporting (Katsilieris et al. 2013). Logistics providers participating in this project claim that spoofing is not a problem, but they also state that not all ships enter valid destination information, and even that in most cases the destination remains a blank field. For the problem formulation we therefore assume that there is no tampering of AIS data, but we also take into account that there are areas not covered by AIS receivers making detection of vessels in those areas impossible. For the purpose of predicting trajectories only, we use AIS packets containing position, speed, and direction.

The goals of this paper are to conduct a review of existing solutions and related algorithms on maritime route prediction using AIS data, determine to what extent they can be applied to solve the prediction problem, and identify areas that have to be improved in order to get an industry-acceptable solution to enhance planning.

For each terminal in the Netherlands, we aim at providing an accurate prediction of arrivals of deep sea vessels in the given vicinity. These predictions have to be updated frequently as vessel positions change continuously. If prediction values get higher than a user given threshold, the system must be able to raise an alarm and inform the logistics planner about it. To achieve that, for every vessel in consideration, its trajectory has to be analyzed and based on the result, probabilities of destination terminals have to be computed together with an accurate arrival time prediction at these terminals.

The methodology followed in this paper is a systematic literature review under the guidelines set by Webster and Watson (2002). The main contributions of this research are threefold: (i) to present the available solutions for trajectory prediction of a vessel; (ii) to identify components that can be used for finding solutions to the identified problem as well as to show the strengths and weaknesses of each available option; and (iii) to propose a new concept for arrival time estimation based on trajectory prediction and using the algorithms mentioned above.

The remainder of this paper is organized as follows. In Sect. 2, we explain the methodology used for the systematic literature review and the selection of relevant literature sources. In Sect. 3, we classify the approaches we identified in the literature review, according to the methodology used to handle the prediction problem. In Sect. 4, we explain what improvements need to be done in order to find a solution to the prediction problem and eventually improve the efficiency of logistics planning. In Sect. 5, we summarize the work done, present our major findings, and discuss future work and impact.

2 Methodology

A systematic literature review according to the methodology set by Webster and Watson (2002) will be used to determine the material to review. The main sources of information are peer-reviewed scientific papers. The search will be conducted using SciVerse Scopus to identify the list of papers to be used. Inclusion criteria are documents that discuss trajectory prediction of maritime vessels and usage of AIS data to determine position and destination. The exclusion criteria will be used to filter documents not being written in English as well as documents being older than 8 years. This is followed by a forward and backward literature search on the given list. Based on the problem definition, we will formulate questions that are used to assess the papers' relevance for the problem under study, and classify them accordingly. Each group will be then thoroughly reviewed and finally a new conceptual model will be discussed.

The scientific knowledge about trajectory prediction using AIS data is retrieved from scientific papers. A quick search on SciVerse Scopus using keyword "*vessel trajectory*" returned 63 documents while using keyword "*ship trajectory*" returned 127 documents. Changing the keyword to "*motion pattern*" gave the list of 52,528 documents that satisfy the criteria. Adding the term "*AIS*" to the query returned the following results: two papers were retrieved using the criteria "*vessel trajectory*" and "*AIS*", "*ship trajectory*" and "*AIS*" returns only one document, while the keyword "*motion pattern*" and "*AIS*" gave four papers as a result.

After combining and reorganizing keywords, the following search query has been entered in Scopus.com: "*TITLE-ABS-KEY(("motion pattern" or "trajectory") AND ("ship" or "vessel") AND "AIS") AND SUBJAREA(COMP OR ENGI OR MATH OR PHYS) AND PUBYEAR > 2006*" returning a total of 17 papers. No language filtering was performed since all papers were written in English. These 17 papers form the basis for the remainder of this paper.

Based on the problem definition, we formulate the following question. Searching for answers in the selected papers will help us determine to what extent the presented solutions can be used to solve the identified problem.

1. What is the object of prediction (trajectory, arrival time, route deviation)?
2. What method is being used to estimate the future state of a system?
3. How far in the future can the method predict?
4. What is the prediction accuracy level?
5. Is data availability/quality taken into consideration and how?

Table 1 provides an overview that shows to what extent selected papers answer these questions.

Table 1 AIS data based prediction arrival literature overview

	Authors	Prediction objective	Prediction method	Prediction timeframe	Prediction accuracy	Data quality/availability
1	Liu and Chen (2014)	p. 84	pp. 92–94	pp. 93 and 96–97		pp. 84 and 92–94
2	Wijaya and Nakamura (2013)	p. 222	p. 223	p. 226	pp. 225–226	
3	Ma et al. (2013)					
4	Wang et al. (2013)	p. 191	pp. 191–192			
5	Pallotta et al. (2013)	pp. 2219–2220	pp. 2221–2222, 2223–2231 and 2234–2240		pp. 2238–2239	pp. 2229–2230
6	Talavera et al. (2013)	p. 95	p. 97		pp. 97–98	
7	Hornauer and Hahn (2013)	p. 103	p. 104			
8	Vespe et al. (2012)	p. 1	pp. 2–4			p. 5
9	Oliveira and Gusovsky (2012)					
10	Lei et al. (2011)	p. 34	pp. 35 and 36–38		pp. 40–41	p. 37
11	Lampe et al. (2010)		pp. 2 and 3			
12	Demsar and Virrantaus (2010)					
13	Laxhammar et al. (2009)	p. 756	pp. 757–758 and 759			p. 759
14	Ristic et al. (2008)	p. 40	pp. 41 and 42–44			p. 43
15	Redoutey et al. (2008)	p. 141	pp. 143–145	p. 148		
16	Uiboupin et al. (2008)					
17	Harati-Mokhtari et al. (2007)					

Since six of the selected articles do not address or answer any of the research questions (or only a single one), we reduce our list to a total of 11 papers.

3 Taxonomy

Ideally, the complete solution should allow prediction of deep sea vessel arrival times 48 h in advance. However, in the reviewed group of articles, the prediction interval is considerably shorter. Depending on the objective, trajectory predictions are done in order to avoid collisions or to detect anomalous behavior. For collision avoidance, predictions are done in a time window that corresponds to immediate future states, in time only minutes away from the current time. Anomalous detection deals with trajectories over a longer time span, and although it could include longer forecast intervals, no attempt has been made yet to predict trajectories several days in advance. Taking into account that prediction methods vary depending on their future coverage, reviewed articles are divided into two groups: short and long term prediction. We use short term prediction to classify all solutions that give the future position or trajectory of a vessel at most 1 h ahead of the current time, while for long term prediction we refer to those being able to predict considerably more than 1 h in the future.

The second element used for classifying solutions is the prediction objective. Looking at the answer to the first question in Table 1, two distinctive concepts are identified. A future position of a vessel is estimated either using its moving pattern for extrapolation or by means of extraction of sea routes used by ships and assigning a probability that a particular vessel's route is one of those. Some articles include both approaches and are classified as a separate group for clarity. Table 2 shows the taxonomy and the associated classification.

Table 2 Taxonomy and classification

Objective		Timeframe	
		Short (less than 1 h)	Long (more than 1 h)
A	Vessel motion patterns	Wijaya and Nakamura (2013), Wang et al. (2013), Hornauer and Hahn (2013), Laxhammar et al. (2009), Ristic et al. (2008), Redoutey et al. (2008)	
B	Route extraction		Vespe et al. (2012), Talavera et al. (2013), Lei et al. (2011)
A + B	Route extraction and motion patterns		Liu and Chen (2014), Pallotta et al. (2013)

Referring to Table 2, a clear correlation is observed between prediction objective and prediction timeframe. Solutions used to give short term predictions exclusively deal with vessel motion patterns while for long term predictions it is necessary to start with route extraction, which can be followed by a motion pattern analysis. We subsequently discuss the short term prediction through vessel motion patterns (Sect. 3.1), the long term prediction through route extraction (Sect. 3.2), and the long term prediction through both route extraction and motion patterns (Sect. 3.3).

3.1 Short Term Prediction Through Analysis of Vessel Motion Pattern

Articles in this group primarily focus on collision avoidance, hence the emphasis is on estimating ship positions in the near future to support steering decisions while the actual destination is irrelevant (Hornauer and Hahn 2013; Wang et al. 2013). Wang et al. (2013) analyze two vessel collision circumstances in inland waters. For that purpose, a two-step process is used to fit a trajectory (spline) between two positions coming from AIS data and then to find a point on the curve equivalent to the distance traversed. Hornauer and Hahn (2013) distinguish between "active" and "passive" vessels while planning trajectories for collision avoidance. "Passive" vessels are those that do not contribute (do not take an active role) to the joint collision avoidance planning procedure, therefore their trajectories have to be estimated. This is accomplished through the sequence of steering decisions controlling the physical model, based on Bayesian model learning applied to historical AIS data.

Ristic et al. (2008) assume that motion patterns are already extracted and focus on identifying anomalous trajectories, as well as estimating the future position of a vessel using a Kernel density estimator. The authors formulate a null hypothesis that a vessel has a normal motion behavior. If it holds, extraction of motion patterns from the training set is performed, where each pattern corresponds to the cluster of trajectories with the same origin.

Redoutey et al. (2008) focus on minimizing AIS data needed to guarantee an accurate short term prediction. To that end, the authors compare three different prediction methods: point-based, vector-based and linear prediction. They show that, when taking into account the speed of the ship, time since the last update and availability of course-over-ground (COG) data, one of the three prediction methods will give better results than the other two, which leads them to propose a combined context-aware algorithm. Wijaya and Nakamura (2013) deal with Big Data, storing static and dynamic AIS information on an experimental Hadoop cluster. Under the assumption that vessels of the same type follow the same route in the vicinity of a port, e.g. cargo ships prefer one terminal, seven factors are used for a given ship to gather its k-nearest neighbor (kNN) ships and obtain the central positions—estimated position with minimum sum distance to all of it's neighbours from the interval of T_0–T, with T_0 being the current, and T future time. These central positions will be used to estimate the trajectory. Laxhammar et al. (2009) compare two probabilistic models—Gaussian mixture model and an Adaptive kernel density estimator with a novel contribution being the measure of anomaly detection performance.

3.2 Long Term Prediction Using Route Extraction

Estimating trajectories or positions for an interval longer than 1 h requires isolation of sea lanes, comparing them with the known trajectories of vessels to determine the closest match, or an anomaly.

Talavera et al. (2013) analyze the distribution of traffic on the waterways and subsequently use it as input for a collision estimation model. The authors' goal is to quantify, and propagate uncertainty from the variability of trajectories by means of the Dempster-Shafer theory, which is used to estimate collision candidates. The method they propose is to group latitude and longitude information for each ship and ordering this information in chronological order, to define trajectories. Routes are represented through geographical clustering of similar individual trajectories. Finally, the distribution is established by dividing the width of a route in intervals of equal length, quantifying the number of ships that have passed through each of those intervals.

Vespe et al. (2012) present an unsupervised learning approach to incrementally learn motion patterns without any specific a priori contextual description. Extraction of waypoints (WP) is the first step followed by defining sea lanes and routes connecting them. The algorithm proposed uses AIS data to detect changes in COG for the vicinity of the observed position; if they are above a given threshold, a new turning WP is added to the list. Similarly, locations where observed vessels have their speed-over-ground (SOG) equal to zero for the defined time interval are identified as stationary WPs. Regions with the first/last sightings of vessels are identified as entry/exit WPs. With all WPs populating the list, sea lanes are added, connecting pairs of WPs. Using historic AIS data, for each lane, a set of attributes such as speed and course are discovered and assigned as stable properties. This concept is improved by Pallotta et al. (2013), stating that the distance-based approach is not always able to distinguish WPs close to each other.

Lei et al. (2011) use a grid-based clustering (DBSCAN) approach to discover "hot regions" in order to constrain the granularity problem. After discovering the "hot regions", modification of the probabilistic suffix tree (PST) is used to convert large sequential data into a variable length tree and extract probability distributions for discrete events occurring in the sequence. Referring to it as the TMP algorithm, the objective of this research is to discover the object's movement behavior.

3.3 Long Term Prediction Using Combined Approach

The remaining two articles by Liu and Chen (2014) and Pallotta et al. (2013) also deal with long term prediction, but they combine both elements: the route extraction and the motion pattern analysis.

The focus of Liu and Chen (2014) is on recovery of missing AIS data using temporal link prediction based on tensor CP (CANDECOM/PARAFAC) decomposition. Although intended for recovery of missing data, this process can be used for the future as well as for the past, and the authors present the results of an experiment with one and three days of data missing. The experiment on AIS data, where three shipping routes are already known, shows that recovered data has tendency to fin into those route patterns, however no accuracy guarantees can be given.

Pallotta et al. (2013) use a "vectoral" approach to represent sea traffic with the intention to be used in decision support systems, evaluation of effectiveness of tracking performance, or anomaly detection. They improve the method presented by Vespe et al. (2012) using the DBSCAN algorithm to overcome the difficulty of distinguishing between close waypoints. The method does not require a number of clusters to be given a priori, and it introduces a way to classify noise points. Similarly to video surveillance techniques, the point-based incremental algorithm is applied while anomalies are detected using a sliding time window. Pallotta et al. (2013) propose the TREAD functional architecture manager, which allows learning based on relevant events related to temporal and spatial characterization of vessel behavior. Clustering those events leads to WP discovery followed by route classification, prediction, and anomaly detection. The accuracy of prediction changes as the ship gets closer to its destination.

4 Research Agenda

Reflecting on the reviewed literature, we may conclude that the current research interest mainly focusses on anomaly detection and collision avoidance. Considering that logistic service providers require knowledge about deep sea vessel arrival time and not its trajectory, we can only use solutions aimed at providing long term predictions. As can be observed from the classification presented in Table 2, all long term predictions rely on sea lane extraction before attempting to estimate a ship's destination. Therefore, using an efficient algorithm that performs this sea lane extraction is an essential starting point. A new solution should incorporate methods such as the vectoral approach described by Pallotta et al. (2013), or the concept of using "hot regions" as proposed by Lei et al. (2011).

In order to provide a powerful decision support platform that meets the industry needs, improvements are required in several areas:

- improving quality handling of received AIS messages,
- fusion of AIS and meteorological data,

- waypoint/route extraction of Big Data sets, and
- improving precision through discovery and inclusion of behavioral patterns.

AIS data is not completely reliable and as stated in research conducted by Harati-Mokhtari et al. (2007), errors range from 1 % for MMSI (ship's unique identifier) to 49 % for destination and estimated arrival time. Considering the possibility that receivers may not receive some or even all AIS messages or that errors occur in the decoding process, it is vital that an assessment of the quality of data precedes any other action. The majority of the papers do not address this issue, while others (e.g., Lei et al. 2011; Pallotta et al. 2013; Vespe et al. 2012) treat errors as noise. We believe that a separate algorithm has to be introduced for independent handling of data quality issues, and possibly preprocessing of data. Such an algorithm should consist of a 3-way check for anomalies. The first one is to immediately discard received data that is out of bounds, such as latitude below $-90°$ and above 90° and longitude below $-180°$ and above 180°. The second one should remove data that is within bounds, but indicate unfeasible trajectories. For example if several consecutive readings indicate that a vessel is somewhere in the Atlantic, only to be followed by a position in the Pacific, such an AIS message is also to be rejected. The last check is to assign reliability to the data source and decide whether to store or reject data if the reliability is above or below the required threshold. As noted by Wijaya and Nakamura (2013), the reliability of destination parameters in an AIS message is correlated with the country of origin of that vessel. Since the volume of the data does not allow assigning reliabilities for each vessel independently, we aim to tackle this problem by classifying vessels into groups according to vessel type and country of origin. This is to be followed by a test on whether each vessel from the group ends in the destination as designated by the AIS destination block. If the test returns a positive result, the reliability factor for that group might increase, while in the case of negative results, it might decrease.

The sheer volume of current and historical AIS data is the second challenge. Anomaly detection and collision avoidance have the luxury of focusing on a specific region. For the problem under study, this is not possible. Only Wijaya and Nakamura (2013) consider putting AIS in the context of Big Data. This is important to consider since DBSCAN is the preferred algorithm for the waypoints/"hot regions" discovery. With the data volume exceeding memory capacity, an alternative has to be considered in particular in relation with distributed data mining and clustering.

A third improvement might be to look at a vessel's historical data to discover behavioral patterns, and the impact they have on the likelihood of deviating from the usual sea lane. Closely related to the last step of the data quality check, the goal is to identify behavioral patterns related to specific regions and their sub-regions. Each pattern is a set of sailing attributes such as average speed, rate of turn and average trip duration in that region. By associating a vessel to one of the discovered patterns, a new predicted route can be constructed and the final position from that route can be used for the arrival time prediction algorithm.

Lampe et al. (2010) show that sea lanes are influenced by weather conditions as ships opt for routes closer to the shore in case of bad weather conditions. To

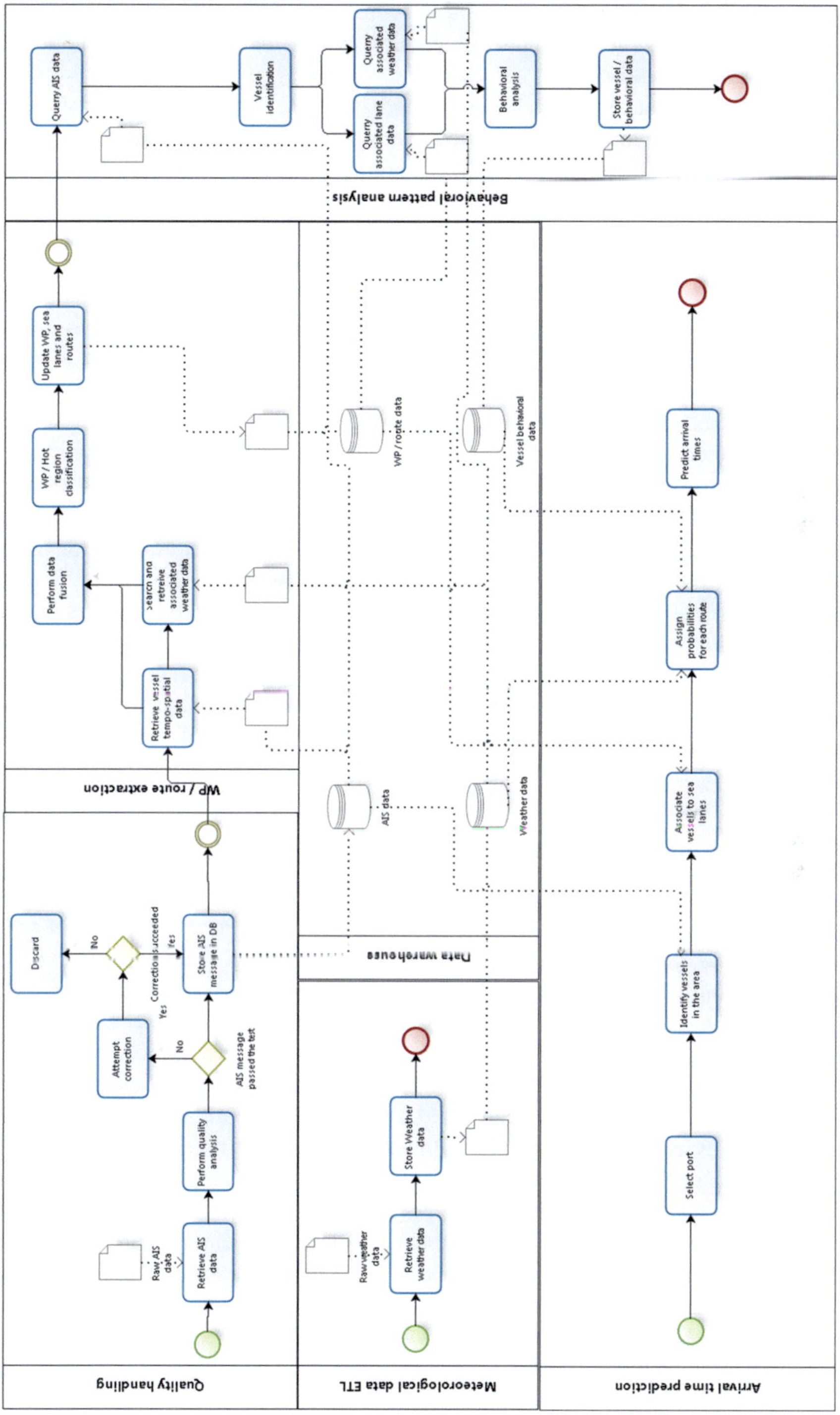

Fig. 1 Research agenda, how to improve the prediction model

increase the accuracy of the prediction, it is necessary to extract weather data and fuse it with AIS data for an improved learning of sea routes that correspond to the actual weather conditions.

These improvements and their interdependencies are shown in Fig. 1. It shows a research agenda following the improvements and relations between the various steps as discussed above, including data handling in data warehouse. At the bottom part, we find the arrival time prediction service, using previously processed data to predict vessel arrival time. We begin the figure with quality handling as the process responsible for retrieving raw AIS data and performing a quality analysis. If no error is found, a retrieved AIS message will be stored in the AIS database. Otherwise, an error correction should be attempted. Parallel to this process, meteorological data is acquired from external sources (either publicly available weather data, or data coming from meteorological services) and stored in a separate database. As a sequel to these two processes we extract the WPs and routes/lanes connecting them. This process retrieves and fuses AIS and weather data preparing them for the unsupervised learning of maritime routes. Since the volume of data required classifies the problem as a Big Data problem, distributed data mining algorithms must be applied to successfully complete the learning process. To additionally improve the prediction, a behavioral pattern analysis has to be performed on existing data sets aiming to discover preferences that each vessel makes when choosing which route to use. Finally the solution will utilize all those data sets to identify all vessels in the vicinity of the desired port, associate them to discovered sea lanes, and assign the probability of following each of them, by also using behavioral data. The final result will be displayed to the user of the system.

5 Conclusion

In this paper, we have presented the problem of estimating the arrival times of deep sea vessels at port terminals, in view of their impact on accurate estimate for hinterland transportation. In order to increase the efficiency of freight operations, planners at logistic service providers need insight into the prediction of unscheduled arrivals, so that barge operations can be re-planned in advance. To accomplish that, we have conducted a literature review on academic papers relevant to this topic and formulated five questions that were used to determine the relevance of each document to this problem. This was followed by a classification of the different approaches into three groups depending on the objective to be predicted and on the timeframe (1 h versus longer horizon). We provided a brief review for each of those groups. Finally, we identified the area that needs improvement and formulated four different objectives that need to be tackled: data quality issues, data volume and distributed data mining, discovery and inclusion of behavioral patterns, and fusion of weather data. It is our expectation that systematically addressing these problems and integrating solutions allows for a more efficient planning.

The final result of this research is to create prediction services. From a research perspective, our objective is to deliver a framework and algorithms for predicting the future state of a system. For the problem on hand, we specifically focused on the prediction of deep sea vessel arrival times, to improve the reliability and efficiency of hinterland transportation plans. It is important to note that other prediction based problems in synchromodal transport networks include similar challenges, such as storing and processing Big Data, and dealing with poor data quality and missing values. Therefore, we aim to transfer the knowledge gained in solving the current problem to other related areas within the Synchromodal-IT project, including road congestion prediction, train and truck arrival time estimation, as well as supply/demand forecasting. Answering these challenges will demonstrate the benefits of the Synchromodal-IT approach to logistics. Implementation of the developed prediction services will be carried out by our project partners that provide consultancy and software development services.

Acknowledgment This work is part of the R&D project Synchromodal-IT that is partly funded by the Dutch Institute of Advanced Logistics (DINALOG).

References

Commission of the European Communities (2008). Document COM(2008) 310 final—2005/0239 COD, 11-06-2008

Demšar U, Virrantaus K (2010) Space–time density of trajectories: exploring spatio-temporal patterns in movement data. Int J Geogr Inf Sci 24(10):1527–1542

Dutch Institute for Advanced Logistics (2013) SynchromodalIT. Retrieved 11 Aug 2014, from http://www.dinalog.nl/en/projects/r_d_projects/synchromodalit/

Harati-Mokhtari A, Wall A, Brooks P, Wang J (2007) Automatic identification system (AIS): data reliability and human error implications. J Navig 60(03):373–389

Hornauer S, Hahn A (2013) Towards marine collision avoidance based on automatic route exchange. Paper presented at the control applications in marine systems

International Maritime Organization (2002) Safety of Life at Sea (SOLAS) convention Chapter V. Regulation 19

Katsilieris F, Braca P, Coraluppi S (2013) Detection of malicious AIS position spoofing by exploiting radar information. Paper presented at the Information fusion (FUSION), 2013 16th international conference on

Lampe OD, Kehrer J, Hauser H (2010) Visual analysis of multivariate movement data using interactive difference views. Paper presented at the VMV

Laxhammar R, Falkman G, Sviestins E (2009) Anomaly detection in sea traffic-a comparison of the gaussian mixture model and the kernel density estimator. Paper presented at the information fusion, 2009. FUSION'09, 12th international conference on

Lei P-R, Su J, Peng W-C, Han W-Y, Chang C-P (2011) A framework of moving behavior modeling in the maritime surveillance. J Chung Cheng Inst Technol 40(2):33–42

Liu C, Chen X (2014) Vessel track recovery With incomplete AIS data using tensor CANDECOM/PARAFAC decomposition. J Navig 67(01):83–99

Ma SX, Sun J, Guan YQ (2013) Detection probability of airborne AIS. Appl Mech Mater 401:1204–1207

Oliveira LF, Gusovsky D (2012) Risk assessment of dropped and dragged anchors to offshore pipelines. Paper presented at the advances in safety, reliability and risk management—proceedings of the European safety and Reliability conference, ESREL 2011

Pallotta G, Vespe M, Bryan K (2013) Vessel pattern knowledge discovery from ais data: a framework for anomaly detection and route prediction. Entropy 15(6):2218–2245

Redoutey M, Scotti E, Jensen C, Ray C, Claramunt C (2008) Efficient vessel tracking with accuracy guarantees web and wireless geographical information systems. Springer, Berlin, pp 140–151

Ristic B, La Scala B, Morelande M, Gordon N (2008) Statistical analysis of motion patterns in AIS data: anomaly detection and motion prediction. Paper presented at the information fusion, 2008 11th international conference on

Talavera A, Aguasca R, Galván B, Cacereño A (2013) Application of Dempster-Shafer theory for the quantification and propagation of the uncertainty caused by the use of AIS data. Reliab Eng Syst Saf 111:95–105

Uiboupin R, Raudsepp U, Sipelgas L (2008) Detection of oil spills on SAR images, identification of polluters and forecast of the slicks trajectory. Paper presented at the US/EU-baltic international symposium, 2008 IEEE/OES

Vespe M, Visentini I, Bryan K, Braca P (2012) Unsupervised learning of maritime traffic patterns for anomaly detection. Paper presented at the data fusion and target tracking conference (DF&TT 2012): algorithms and applications, 9th IET

Wang Y, Zhang J, Chen X, Chu X, Yan X (2013) A spatial–temporal forensic analysis for inland–water ship collisions using AIS data. Saf Sci 57:187–202

Webster J, Watson RT (2002) Analyzing the past to prepare for the future: writing a literature review. Manage Inf Syst Quart 26(2):3

Wijaya WM, Nakamura Y (2013) Predicting ship behavior navigating through heavily trafficked fairways by analyzing AIS data on apache HBase. Paper presented at the computing and networking (CANDAR), 2013 first international symposium on

Modular IT-Support for Integrated Supply Chain Design

Matthias Parlings, Tobias Hegmanns, Philipp Sprenger and Daniel Kossmann

Abstract Supply chain design deals with long-term and strategic decisions on the structural design and configuration of supply chains. Available supply chain design tools offer particular planning functionalities. They provide only little support for linking all activities within the workflow of the user. Bridging this gap is the core innovation of the new approach presented in this article. The developments come along with the idea of a Domain Specific Modelling Language for simulation use in supply chain design tasks. This language translates the modeling elements of the application (application language) into the technical modelling elements of the tool (simulation tool language) and vice versa. As a result, the complexity of configuring a simulation model is kept at a minimum in the background. All this is embedded in an service-based tool architecture that enables work flow support of the user and collaborative engineering. Finally, the application of the tool is presented in an practical use case of an automotive first tier supplier.

Keywords Supply chain design · Strategic network planning · Logistics-as-a-Service · Network simulation

This research is a result of the project "Supply Chain Design" (Förderkennzeichen 01IC12L03) funded by the Federal Ministry of Education and Research in the Effizienzcluster LogistikRuhr.

M. Parlings (✉)
Department of Supply Chain Engineering, Fraunhofer IML, Joseph-von-Fraunhofer-Str. 2-4, 44227 Dortmund, Germany
e-mail: matthias.parlings@iml.fraunhofer.de

T. Hegmanns
Division Enterprise Logistics, Fraunhofer IML, Dortmund, Germany

P. Sprenger
Faculty of Mechanical Engineering, Graduate School of Logistics, TU Dortmund, Leonhard-Euler-Str. 5, 44227 Dortmund, Germany

D. Kossmann
Delphi Deutschland GmbH, Delphiplatz 1, 42119 Wuppertal, Germany

H. Zijm et al. (eds.), *Logistics and Supply Chain Innovation*,
Lecture Notes in Logistics, DOI 10.1007/978-3-319-22288-2_17

1 Introduction

Change drivers such as shorter product life cycles, higher frequencies of new product launches as well as the developments in emerging markets lead to increased dynamics and uncertainty in supply chains. In order to secure environmental and economic efficiency, the necessary redesign of supply chain structures and processes has to be realized within a shorter term horizon. Due to their scope and complexity in today's companies, supply chain design (SCD) tasks are often separated and solved in various "planning islands", e.g. different functional or organizational units. That is, distributed planning teams work on dependent sub-tasks, make use of isolated tools and various databases. The individual planning procedures are time consuming, sometimes poorly coordinated and based on local, partially incomplete information and assumptions.

The research project Supply Chain Design embedded in the EfficiencyCluster LogistikRuhr intends to resolve the isolated use of applications in supply chain design procedures. The objective is the development and implementation of a comprehensive and holistic supply chain design approach based on a novel technological basis of a service-based tool landscape. The technical idea amounts to service-based IT support based on adequate tools and a seamless informational backbone.

In this contribution the intermediate results of the project are presented. This includes an approach for the structuring of SCD tasks as a basis for developing a modular IT solution. Furthermore, it is linked to a generic supply chain design workflow as the basis for a process-oriented tool design. The developments come along with the idea of a Domain Specific Modelling language for simulation use in SCD tasks. This language translates the modeling elements of the application (applicational language) into the technical modelling elements of the tool (simulation tool language) and vice versa. As a result, the complexity of configuring a simulation model is kept at a minimum in the background.

2 State of the Art Review

The following sections describe the framework for the developed tool approach. First, typical supply chain design tasks have been analyzed. Then, existing planning approaches and tool support for these task are examined. Finally, based on that, a generic planning approach is presented and an appropriate software solution is sketched in the form of a tool landscape with an underlying design workflow in Sect. 3.

2.1 A Framework for Structuring SCD Tasks

Supply chain design as the subfield of supply chain management (SCM) that integrates long-term and strategic decisions on the structural design and configuration of supply chains is only imprecisely defined in the academic literature. Many different tasks are occasionally subsumed under the term 'supply chain design'. An explicit definition and structuring of these SCD tasks could not be identified, neither in the scientific literature nor in practice. However, for developing a holistic tool approach it is essential to gain a common understanding of SCD and the underlying tasks.

Kuhn and Hellingrath (2002) have developed a task model for structuring SCM tasks and assigning appropriate software solutions that focuses on short-term ('supply chain execution') and medium-term ('supply chain planning') tasks. Supply chain design is roughly described, the underlying tasks remain a 'black box'. As one contribution to this research project, Parlings et al. (2013) have further expanded the model to defining SCM tasks on the design level: the resulting 'Supply Chain Design task model' is displayed in Fig. 1 and briefly described in the following.

Generally, SCD tasks can be separated into three main groups: Superordinate SCD tasks, supply chain structure design tasks and supply chain process design tasks. The

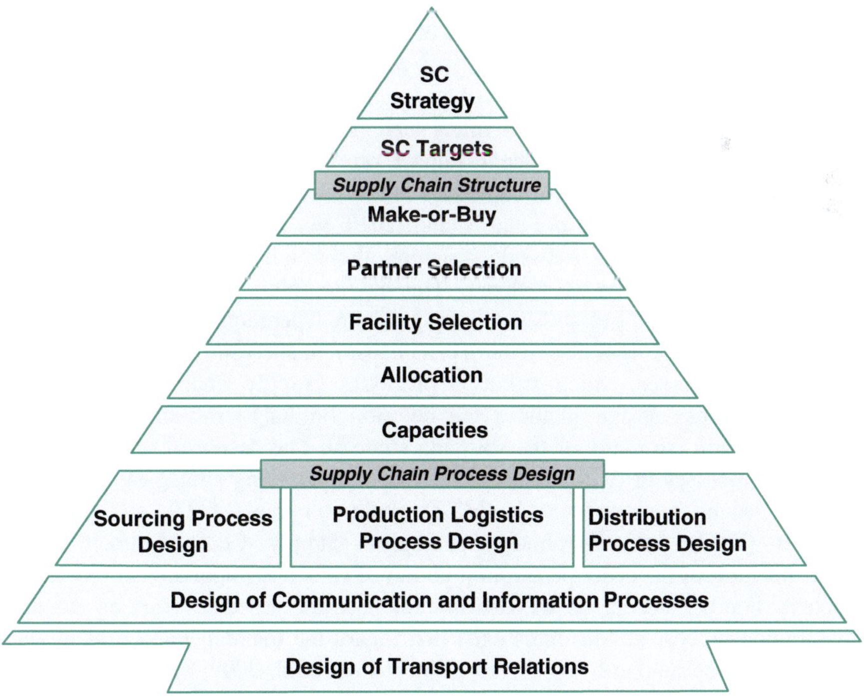

Fig. 1 The supply chain design task model (Parlings et al. 2013)

superordinate tasks include the decision on and continuous adaption of the supply chain strategy and the definition and continuous adaption of the target system for SCM. Determining the right SC strategy is a corporate task originating from the corporate strategy process and being a relevant input factor for the subsequent SCD tasks. The overall goal is to design and align the supply chain according to the supply chain strategy (Chopra and Meindl 2010, pp. 37–58). To achieve a satisfactory alignment of the supply chain to the strategy, an accurate measuring of the level of attainment of the overall objectives is necessary. Thus, appropriate indicators need to be defined and arranged in a performance measurement system (Parlings et al. 2013).

The second group of tasks is related to the design of the supply chain structure. After the supply chain strategy has been defined and a target system is installed, it is possible to design the network according to these characteristics. These tasks can be divided into five key sections: the make-or-buy decision, supply chain partner selection, facility selection, product and customer allocation, and dimensioning of capacities.

The first task in designing the supply chain structure is to carry out the make-or-buy decision (Schönsleben 2011, pp. 70–75). It needs to be decided whether value is created by the company itself or should be sourced from a different party (Cuber et al. 2009, p. 70). The selection of supply chain partners is closely related to make-or-buy decisions and is mutually dependent on it. Decisions affecting facility selection include the determination of the role, number and location of facilities (Chopra and Meindl 2010, p. 125).

This decision leads to the allocation of products to locations for production or warehousing purposes. The allocation of products to factories deals with determining which facilities are responsible for the specific value creation process. Besides the role and location of facilities such as warehouses and production sites, their capacities have a significant impact on the supply chain performance. Therefore, the dimensioning of capacities is the third major task in designing the supply chain structure (Chopra and Meindl 2010, pp. 125–126).

Based on the created network structure, the rough definition of intra- and inter-company processes is realized. This is the third section of SCD tasks. According to the well-established Supply Chain Operations Reference (SCOR) Model the related process design tasks can be distinguished into sourcing processes, production processes, and distribution processes (Supply Chain Council 2012). Sourcing process design includes the decision on long-term inbound logistics concepts within the range of the sourcing strategy. The design of production and logistics processes in SCM involves the decision sourcing strategies, production concepts such as make-to-stock (MTS), make-to-order (MTO), and engineer-to-order (ETO), and distribution strategies (Supply Chain Council 2012). Determination of the order penetration point (OPP) is one important activity in this context. For the design of the distribution network processes various standard distribution process architectures exist that regard the interdependence of product strategy and the market to be served (Simchi-Levi et al. 2009, pp. 209–241).

The design of information and communication processes as well as of transport relations consists of cross-sectional tasks within the process design section. They affect the supply chain's source, make, and deliver processes. The design of information and communication processes is particularly important for SCM, as "information deeply affects every part of the supply chain and impacts every other driver" (Chopra and Meindl 2010, p. 69). The design of transport relations is closely linked to the design of the procurement, production, and distribution processes. Transport processes connect network locations and therefore need to fit the appropriate processes mentioned above. The determination of the suitable means of transport, which generally follows from certain relations within the logistic system, is an SCD task (Simchi-Levi et al. 2009, pp. 85–87; Chopra and Meindl 2010, p. 25).

2.2 *Planning Approaches and Tool Support for SCD*

In the literature multiple approaches to supply chain design problems are presented. The aim of the next section is to deduce the key steps and activities of supply chain design procedures. These serve as a basis for the development of a generalized SCD planning workflow.

An analysis of the planning approaches for supply chain design reveals that nearly all approaches show a similar kind of problem-solving process structure. Based on the supply chain strategy and targets the SCD-planning problem is defined and the planning potentials are analyzed in the first place (Wolff and Nieters 2002; Chopra and Meindl 2010; Kuhn et al. 2010; Straube et al. 2011). The next step is the collection and generation of planning data to define static and dynamic parameters and the required level of abstraction (Wolff and Nieters 2002; Seidel 2009). The level of detail varies significantly throughout the approaches. Reiner and Schodl's (2003) approach for example represents a very high level of abstraction and displays a more or less generic framework. In contrast, Chopra and Meindl (2010) or Freiwald (2005) present more specific models. Several of the reviewed approaches consider quantitative methods like optimization or simulation for the analysis of different scenarios in the next step (Goetschalckx 2000; Wolff and Nieters 2002; Seidel 2009). The last process step can be described as the evaluation of the optimization or simulation results and the suitable preparation of the results for subsequent decision support. Specific advices on how to design a report are not in the scope of the reviewed approaches.

We have undertaken an in-depth review of planning approaches which shows that most of the planning approaches for Supply Chain Design follow a process-oriented structure, with the main steps being strategy and target definition, problem definition, data collection, model building, scenario analysis, evaluation and reporting.

The design approaches do in most cases not address general tool support. Authors use specific stand-alone models to proceed from task to task. However, the analysis methods used can generally be classified within the two general approaches of analytical methods (material flow calculation and mathematical optimization) and simulation (Seidel 2009, p. 97).

In order to gain a comprehensive overview of the state of the art with regard to software tools for supply chain design, an internal and external review has been carried out. An extract of the results is presented in Table 1. The first group represents the tools that are currently used by the project partners for SCD purposes. The second group contains SCD tools that are offered by external software producers and research institutes. The tools are classified by the primary analysis method the evaluation is based on and the supply chain design tasks the tools are addressing.

The research results show that only a few tools are sufficiently flexible to cover all supply chain design tasks defined above. Supply chain design tools offer particular planning functionalities. They provide only little support for linking all activities within the workflow of the user. Bridging this gap is the core innovation of the new approach presented in the next section. The aim is to create tool support covering all phases of the supply chain design process from strategy definition through data collection and modelling to evaluation and reporting. The tool handling should become more process oriented following the workflow of the design procedures. Also, it should link the spatially distributed planning teams of partial designs in a collaborative mode and with a commonly used information backbone. Specific tools do not have to be replaced by this approach, but the separate tools used in the planning process should be linked by means of a service-oriented platform.

3 Planning Approach for Integrated Supply Chain Design

In this section a generalized planning workflow for supply chain design is presented and the integration of the SCD planning workflow in a service-based tool environment is outlined.

3.1 A Generalized Planning Workflow for Supply Chain Design

The state of the art analysis reveals that various data (sources), evaluation methods and key competences need to be integrated when preparing a design decision for supply chains. As has been indicated earlier, the core idea of this research project is to develop a generalized planning workflow that supports the integration of different tools and methods in order to individually solve the respective design task. Based on the review of planning approaches in research as well as on an internal

Table 1 Overview SCD tool landscape

	General information			Applicability to SCD tasks					
	Tool	Software producer	Primary analysis method	Facility selection	Allocation	Capacities	Sourcing process design	Distr. process design	Design of transport relations
Project internal supply chain Design Software	Network analyzer	Daimler AG	Math. optimization	X	X	X	–	–	–
	Electronic supply chain calculator	Daimler AG	Material flow calculation	–	–	–	X	X	X
	Total cost Application	Delphi	Material flow calculation	X	–	–	–	–	(X)
	DISMOD	Fraunhofer IML	Math. optimization	X	X	(X)	(X)	X	X
	OTD-NET	Fraunhofer IML	Simulation	X	X	X	X	X	X
Project external supply chain design software	4flow-vista	4flow AG	Math. optimization	X	X	X	X	X	X
	Ncdis	Fraunhofer FCM	Simulation	X	X	–	–	X	X
	Supply chain Guru	LLamasoft	Simulation and optimization	X	X	X	X	X	X
	Xcargo	Locom	Math. optimization	X	X	X	–	–	X
	OMP plus	OMP	Math. optimization	X	X	X	X	X	X

(continued)

Table 1 (continued)

	General information			Applicability to SCD tasks					
	Tool	Software producer	Primary analysis method	Facility selection	Allocation	Capacities	Sourcing process design	Distr. process design	Design of transport relations
	Oracle SCM R12	Oracle	Math. optimization	X	X	(X)	X	–	–
	Quintiq supply chain design	Quintiq	Math. optimization	X	X	X	X	X	X
	PSIglobal	PSILogistics	Math. optimization	X	–	X	X	X	X
	SimChain	SimPlan	Simulation	X	X	X	X	X	X
	Viewlocity control tower platform	viewlocity	Math. optimization	X	X	X	X	X	X

best practice review among the project partners, the planning workflow displayed in Fig. 2 has been developed.

An external event, a defined planning interval or an indication from supply chain performance monitoring can be potential triggers for initiating the (re-)design of the supply chain. After such a planning impulse has been detected, the SCD process starts with the target definition phase. This phase mainly covers the superior SCD tasks described in the SCD task model. However, it is not considered mandatory for all SCD projects as changes in the supply chain strategy and the targets are not necessary as frequent as structural and process design tasks might have to be executed. This phase is typically not supported by IT systems and has therefore not been focused on in the research project.

In the next phase, the project scope is defined. First, the SCD tasks to be carried out need to be identified and specified. The SCD task model can serve as a useful tool for clarifying those tasks. This is followed by a rough pre-investigation of possible solution alternatives. General parameters and limiting factors such as supplier black lists or global economic developments are considered for narrowing

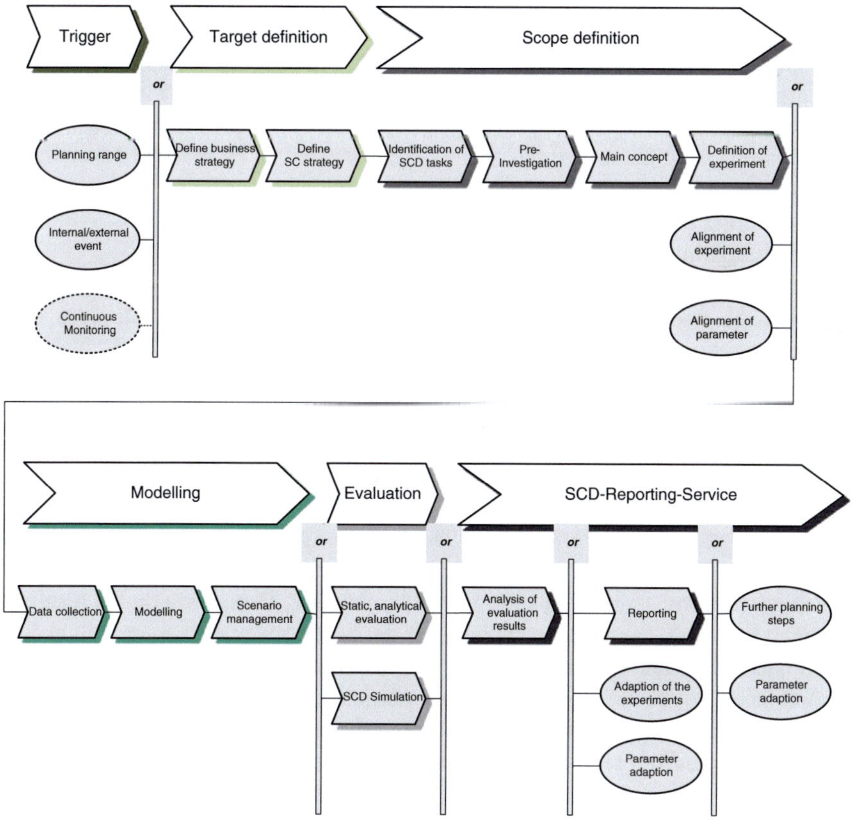

Fig. 2 Generalized SCD planning workflow

down the solution space to generally acceptable scenarios. This step typically requires high manual efforts and is supported by company-specific check lists. After the main constraints have been identified, a rough basic concept for the evaluation is defined. Usually this is done by specifying different supply chain design scenarios for comparative analysis. Based on this concept, the main experiments need to be described. The resulting plan for experiments marks the end of the scope definition phase.

Since experiments on the real system are usually impossible in Supply Chain Design, a representation of the system in a virtual model needs to be built. Therefore, the main data for defining the supply chain need to be collected and stored in an appropriate data model. The collection of all relevant data from various systems is often a very time-consuming process for companies when evaluating different supply chain design scenarios. Hence, the development of a standardized storage for supply chain data should be an integral part of the supply chain design tool framework.

Building a model of the supply chain generally is a complex, time-consuming task as individual models need to be built based on the technology to be used in the evaluation phase. The modelling phase has been identified as being a major limiting factor for shortening the time required for decision making in supply chain design problems. Therefore, one of the major objectives of the SCD research project is to develop a generic modelling service that uses standardized modules for modelling a supply chain. This concept is introduced as the 'domain specific language' in the next section. As different supply chain scenarios need to be evaluated in order to identify the most favourable solution, an intelligent scenario management is required in the SCD process. Scenario management offers functions for conducting series of experiments and comparing alternatives.

Once the supply chain has been modelled and the evaluation scenarios have been prepared the actual evaluation of possible solutions is carried out. Two main alternatives are intended in the workflow. Static evaluation methods based on material flow calculation are mainly be used for the cost assessment of the different alternatives. For complex investigations under consideration of dynamic effects, a simulation service is being developed. It is based on the simulation software OTD-NET that has been developed by Fraunhofer IML.

After the simulation and/or static evaluation of the different design scenarios the results are validated and verified first. This contains standardized validation checks that are included in the simulation service as well as the individual verification of the simulation and static evaluation results by the user. Depending on the results of the analysis phase, the model parameters or the experiments themselves might need to be adapted. If the evaluation results are validated and meaningful, task-specific reports need to be compiled. For this purpose, a reporting service has been developed which supports the preparation of reports based on evaluation results. It contains standardized report categories that can be individualized depending on the design task and the requirements of the decision makers.

The introduced planning workflow is characterized by its independence from specific methods. In combination with the corresponding SCD services the user can reduce the time and resource expenditures when carrying out structure and process design tasks. The general approach for integrating the planning workflow and the tool support is summarized in Fig. 3. Within the research project four main SCD services are developed for supporting SCD decision finding: The 'Scenario Service' is used for defining and managing the experiments and their corresponding scenarios. For modelling the underlying supply chain structure and processes, the 'Modelling Service' has been developed. This service is further described in the next section. The 'Simulation Service' carries out the simulation itself. Supply chain structure and process data as well as the simulation results are stored in a generic database for supply chain simulation, the so-called "SCDB" (supply chain database) that has been developed at Fraunhofer IML. For the evaluation of the simulation results a 'Reporting Service' is developed that allows for an SCD task and target group specific configuration of reports. The services are connected via a configuration database, the so-called "LAS-Bus". This database administers the experiments and scenarios, the report configuration and interfaces to various data sources across the company.

3.2 A Domain Specific Language for Modelling Supply Chains

Experience from previous efforts reveal that model-based approaches like simulation or optimization are often time-consuming. This drawback can now be overcome by new available IT technologies. First, processor power and cloud

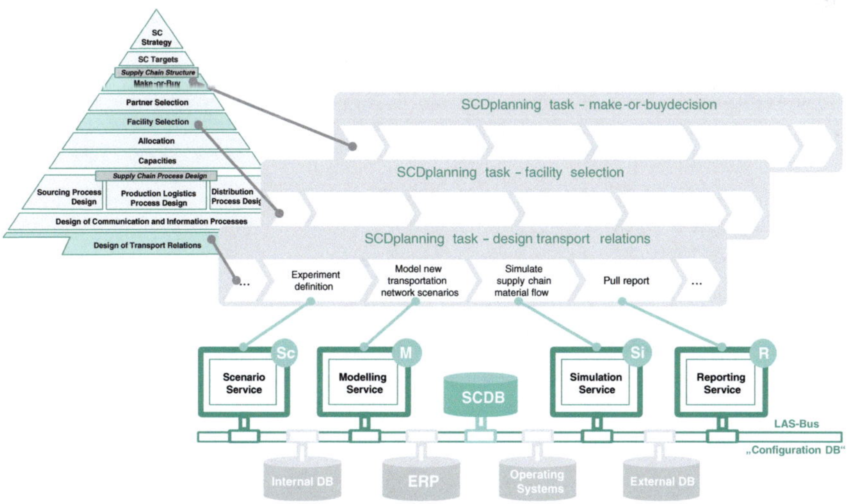

Fig. 3 General approach: planning workflow and tool integration

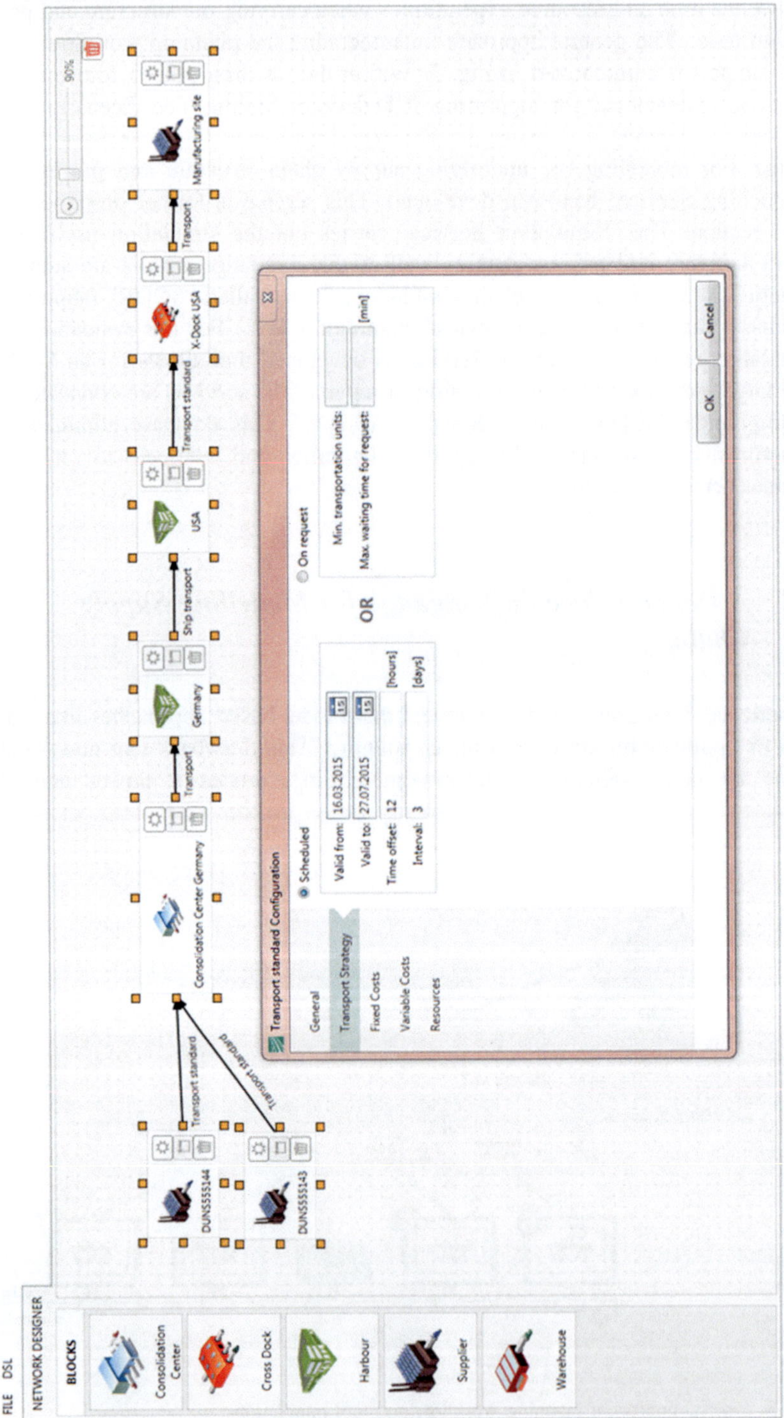

Fig. 4 Exemplary network design within the SCD modelling service

computing help to reduce computational time. Second, intelligent modelling approaches may help to reduce time for preparing, adapting or maintaining models. Below, we present the concept of a Domain-Specific Modelling language as a means to speed up modelling processes for simulation. The Domain-Specific Modelling language provides building blocks for creating the supply chain model. These building blocks incorporate knowledge of the real world processes on an aggregated level. They capture relevant parameters describing the structure, internal control logic and the interactions of essential real-world supply chain elements, e.g. plants, transportation, warehouses, cross-dock facilities.

This means that the Domain-Specific Modelling language translates the real-world application into the technical elements of a simulation model. Transportation channels, for example, are characterized by different capacities, means of transportation, schedules and lead times. When using this modelling element the user is guided through a dialogue helping him to specify all required model parameters to design different transportation channels without deeper technical understanding of the simulation tool itself. The complexity of configuring the simulation is therefore kept in the background and requires minimum effort from the user. To this end, the typical modelling elements have to be anticipated and translated into simple user interfaces, guided menus and user dialogues for pre-defined supply chain building blocks.

Within this research project, the major building blocks 'Consolidation Center', 'Cross Dock', 'Harbor', 'Supplier' (resp. production), 'Warehouse' and 'Transport' have been identified as being sufficient for modelling supply chains against the background of supply chain design. Figure 4 displays the model of an exemplary supply chain in which parts are consolidated in a German consolidation center and shipped via ports in Germany and the USA to a cross dock in the USA. From there the manufacturing site in the USA is supplied. Each node of the supply chain can be parameterized in a menu that is displayed in the center of the figure. After having designed and parameterized the network, the model can be exported as an xml-file (DSL-export).

The application of this innovative modelling approach as well as the other intermediate results presented within a use case of the research project is demonstrated in the following chapter.

4 Demonstration of Intermediate Results in a Use Case

In the following sections the application of the developed solution approach is presented in an practical use case at an first tier automotive supplier.

4.1 Actual Situation and Problem Description

The developed solution approach is demonstrated in the use case of the research project partner Delphi Automotive PLC. Delphi Automotive PLC is a leading global supplier of technologies for the automotive and commercial vehicle markets. In the case on hand, the supply chain design problem is making a robust footprint decision for the division of Electrical and Electronic Architectures in the region of Europe, Middle East and Africa. The core product in this case is a wiring harness, linking all electrical and electronic connections in a car environment.

Delphi as a 1st Tier supplier is affected by a high level of OEM diversity, in line with increasing product complexity and enhanced customer requirements like just-in-sequence supply of the vendor parts, shorter product lifecycles and an increasing number of models and platforms. On the one hand, the production process of wiring harnesses contains high manual labor content. This leads to driving the manufacturing footprint to regions of low labor rates usually located in Eastern Europe or Northern Africa. On the other hand, requirements like just-in-sequence supply limit the distance between Delphi's manufacturing sites and the OEM plant. Thus, a trade-off between cost efficiency and supply chain performance requirements needs to be achieved.

In relation to the SCD task model introduced in Sect. 2.1, the decision problem considers the supply chain structure (facility selection) and process design (transport relations) in a multi-criteria evaluation of different manufacturing footprint scenarios. The decision is to be made under project and market-related uncertainties for a longer planning horizon (Sprenger et al. 2014).

4.2 Adapting the SCD Planning Workflow

Delphi's footprint planning process is triggered by new business opportunities and an internal quotation process. The first planning step is part of the pre-investigation and should reduce the number of potential manufacturing locations to a manageable level (Sprenger et al. 2014). The next planning step is the definition of experimental scenarios by using the 'Scenario Service'. This service enables planners at Delphi to take structural premises such as footprint changes of OEMs or suppliers and/or parameter variations like changing exchange rates or variations of project volumes into account.

Based on a set of scenarios the Delphi real-world supply chain will be transferred into a simulation model by using the 'Modelling Service'. The transfer is realized by means of building blocks of the Domain-Specific-Language. The 'Modelling Service' enables planners to consider typical transport strategies for the raw material supply like direct transports, milkruns or consolidated transports and direct and sequences finished goods transports on the downstream side.

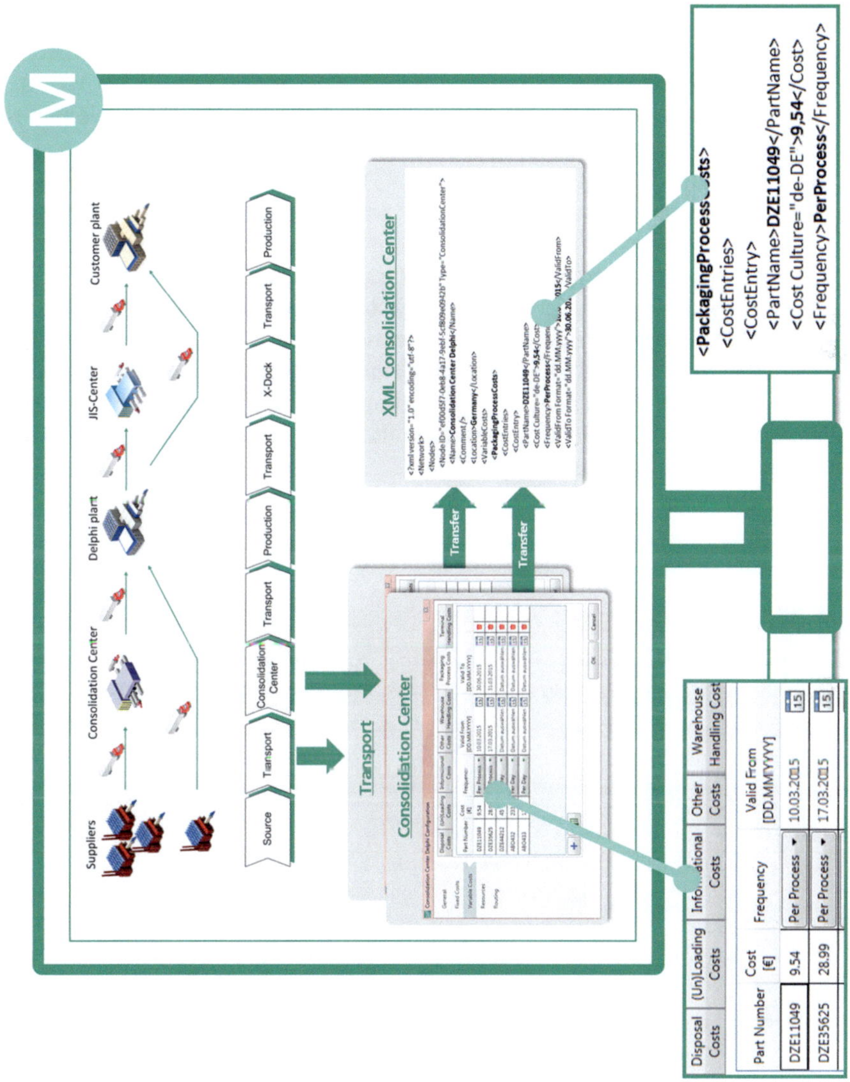

Fig. 5 Delphi supply chain model transfer with DSL building blocks

The network displayed in Fig. 5 can be represented by the elements 'Source', 'Transport', 'Consolidation Center', 'Cross Dock' and 'Production' which are provided as pre-defined building blocks with configurable parameters. For the example of the consolidation center, there is a pre-defined set of structural and process-based data to be entered in order to parameterize this network node. The structural data comprises routing information such as the locations of the consolidation facility and the subsequent transport routes. Typical process-based data such as loading, unloading, warehousing or repacking costs, as well as the underlying process time can be inserted manually or via a data import, pulled up from databases, which are backboned connected with the 'Modelling Service' (see Fig. 6). Examples for databases at Delphi are SAP, the corporate 'Logistics Cost Application Tool', which provides several value stream data (e.g. costs and lead-times), and the 'Sales Planning System' with the latest information of the order situation. After the modelling, this service combines all populated parameters of the specified model data and provides an exact translation in an XML-file.

The next planning step is supported by the 'Simulation Service' which runs different Delphi supply chain network scenarios using the OTD-NET simulation engine. The final step of the footprint planning process is the composition of analysis reports. With the help of a "Reporting Service" the simulation results for all simulated scenarios will be pulled from the SCDB and aggregated to management reports to assist decision makers.

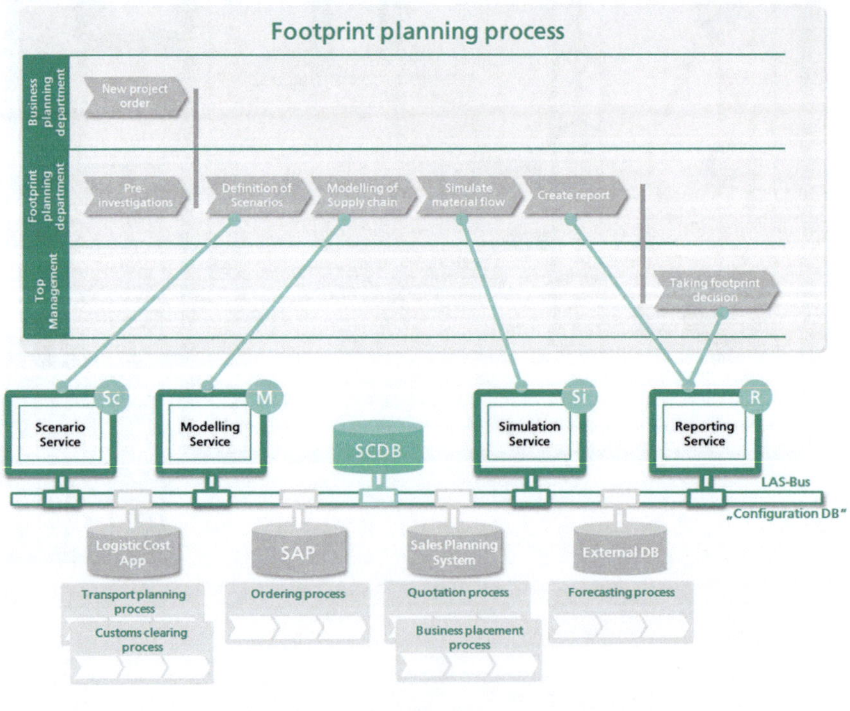

Fig. 6 Integration of Delphi's planning workflows in modular IT-infrastructure

This modular IT-support allows planners at Delphi with limited simulation skills to use simulation as an evaluation method even in a very dynamic planning environment. With the help of the developed "Modelling Service" parameters supply chain structures can be adjusted in a quick and efficient way. Next to the support of Delphi's planning process for footprint decisions, other planning processes as for example the quotation process or transport planning process are integrated.

5 Conclusion and Outlook

For solving long-term and strategic supply chain planning problems an overall SCD-planning approach as well as supporting methods and IT tools are required to get rid of planning islands and enable an adaptable and permanent planning. The initial investigation showed that supply chain design tools offer particular planning functionalities but do not cover the complete workflow for supply chain design problems.

To bridge this gap a generalized planning approach has been developed, which supports the integration of different tools and methods for the respective supply chain design decision. The representation of complex systems like SC networks in virtual models is one of the central steps within the generalized planning approach, but the most complicated and time-consuming one as well. To enable decision makers to use model-based approaches like simulation a rapid modelling support is required. This is achieved by means of a so-called domain specific modelling language which offers building blocks for simulation modelling on an aggregated level.

The vision fueling this research is a service-based network of tools, databases and data sources required to solve real-world supply chain design problems. Practice shows that various tools and data sources are involved in the problem solving process of typical supply chain designs. Therefore an interoperable and linked tool landscape is considered an important step forward, speeding up problem solving steps. Correspondingly, when cutting down time requirements for executing planning tasks, this gain can be invested in useful considerations of design alternatives, sensitivity and scenario analysis. By this, the planning quality is improved as well.

Within this contribution the application of the generalized planning workflow and modelling with the help of the domain specific language has been demonstrated using the example of footprint planning in automotive supply chain networks. The benefits for the tool have been proven in the ease of modelling and convenient scenario benchmarking. Future development will lead to even more collaborative supply chain design engineering. Objectives for the further development of the tool approach address multi-user functionalities, integration of data sources and support of the design-life cycle. The latter means to support the design process throughout phases from first supply chain design drafts to detailed models for start-of-production and during operations.

References

Chopra S, Meindl P (2010) Supply chain management: strategy, planning, and operation, 4th edn. Person Education, Upper Saddle River

Cuber S, Helmig J, Quick J, Wienholdt H (2009) Supply chain design: Methoden zur Gestaltung und Optimierung von Wertschöpfungsketten. UdZ - Unternehmen der Zukunft 9(1):69–71

Freiwald S (2005) Supply chain design—Robuste Planung mit differenzierter Auswahl der Zulieferer. Peter Lang, Frankfurt am Main

Goetschalckx M (2000) Strategic network planning. In: Supply chain management and advanced planning—concepts, models, software and case studies. Springer, Berlin, pp 79–96

Kuhn A, Hellingrath B (2002) Supply chain management: Optimierte Zusammenarbeit in der Wertschöpfungskette. Springer, Berlin

Kuhn A, Wagenitz A, Klingebiel K (2010) Praxis Materialflusssimulation – Antworten, zu oft zu spät? In: Jahrbuch der Logistik 2010. Korschenbroich: Free Beratung, Germany, pp 206–211

Parlings M, Cirullies J, Klingebiel K (2013) A literature-based state of the art review on identification and classification of supply chain design tasks. In: Proceedings of the 17th cambridge international manufacturing symposium

Reiner G, Schodl R (2003) A model for the support and evaluation of strategic supply chain design. In: Strategy and organization in supply chains, Physical. Heidelberg, pp 305–320

Schönsleben P (ed) (2011) Integrales Logistikmanagement: operations und supply chain management innerhalb des Unternehmens und unternehmensübergreifend, 6th edn. Springer, Berlin

Seidel T (2009) Ein Vorgehensmodell des softwaregestützten supply chain design. Unternehmenslogistik, Dissertation, University of Dortmund

Simchi-Levi D, Kaminsky P, Simchi-Levi E (2009) Designing and managing the supply chain: concepts, strategies and case studies, 3rd edn. McGraw-Hill, Boston

Sprenger P, Parlings M, Hegmanns T (2014) Planning approach for robust manufacturing footprint decisions. In: Kersten W et al. (eds) Next generation supply chain—trend and opportunities. epubli, Berlin

Straube F, Doch S, Nagel A, Ouyeder O, Wuttke S (2011) Bewertung öko-effizienter Logistikstrukturen in global agierenden Wertschöpfungsketten. Flexibel - sicher - nachhaltig. In: 28. Deutscher Logistik-Kongress. Bundesvereinigung Logistik. Hamburg: Dt. Verkehrs-Verl., 19–21 Oktober 2011. Berlin, pp 201–227

Supply Chain Council (2012) SCOR 11.0 model reference

Wolff S, Nieters C (2002) Supply chain design—Gestaltung und Planung von Logistiknetzwerken. Praxishandbuch Logistik, Sect. 3.5

Intelligent Control of Freight Services on the Basis of Autonomous Multi-agent Transport Coordination

Frank Arendt, Oliver Klein and Kai Barwig

Abstract In a highly competitive market, freight forwarders face a fierce pressure to reduce costs by optimizing their dispatch and planning processes. An increasing share of smaller shipments, dynamic markets, traffic problems and a growing variety of special equipment and vehicle types for general cargo render the manual planning of transport logistics a prohibitively complex optimization challenge. The autonomous coordination of transport services and planning processes can help to cope with the dynamics and distributed nature of logistics networks. In this paper, we introduce a multi-agent based approach that enables an autonomous dispatch process in a realistic transport scenario. The presented approach has been used and validated as an appropriate way to solve resource allocation problems when new transport orders can appear at any time. Simulation experiments with real data from the logistics partner STUTE show that the procedure outperforms the previous distributed manual dispatching process significantly in terms of flexibility and speed, leads to a reduction of empty mileage and increases capacity utilisation of trucks. Additionally, the system is designed to serve as a decision-support system (DSS) which provides proposals for allocations of transport orders to trucks to support the decision process of a human dispatch manager.

Keywords Vehicle routing · Multi-agent systems · Decentralised systems · Logistics dynamics

F. Arendt · O. Klein
Institute of Shipping Economics and Logistics (ISL), Universitätsallee 11/13,
28359 Bremen, Germany
e-mail: klein@isl.org

K. Barwig (✉)
BACO - Dr. Kai Barwig Consulting & Interim Management, Hollen 32,
27327 Martfeld, Germany
e-mail: b.a.c.o@t-online.de

H. Zijm et al. (eds.), *Logistics and Supply Chain Innovation*,
Lecture Notes in Logistics, DOI 10.1007/978-3-319-22288-2_18

1 Introduction

Commercial traffic has a significant share of the traffic in the Federal Republic of Germany. Given the limited traffic capacity of the German highway infrastructure and the dynamic growth of commercial freight traffic, a major concern of the Federal Government of Germany is the improvement of commercial freight forwarder dispatch concepts in order to lower the burden on the available infrastructure.

Fulfilling a given amount of orders with fewer kilometres to drive comes along with a competitive advantage: freight forwarders in a highly competitive environment face a fierce pressure to reduce costs by optimizing their dispatch planning processes—a highly complex task predominantly done manually. In recent years this planning process has become increasingly complicated and difficult to optimize due to the mandatory application of the European Community Social Legislation for drivers working hours, which renders the manual dispatch planning a prohibitively complex optimization challenge. For a detailed algorithmic solution including a quantitative analysis, we refer to Meyer et al. (2011). The main reasons for the increased complexity of the planning process are:

- Smaller shipment sizes and an increasing number of shipments
- Quicker market developments implying fast changes
- Increasing traffic problems like congestion harming determinability
- Growing variety of special equipment and vehicle types for general cargo

In this context the objective of the research project AMATRAK has been to reduce traffic and to achieve more efficient vehicle capacity utilization in the procurement and distribution logistics, based on autonomous multi-agent transport coordination. The transfer partner of the project was STUTE Logistics. As a logistics service provider with multiple branches all over Germany, STUTE served as an ideal partner for the practical examination of real world transport logistics processes.

The initial situation is as follows: The logistics service provider maintains a number of offices in Germany and every branch has its own customers and conducts its own dispatch process. The customer and cargo structure is very heterogeneous. Parts of the orders are less than truckload, others are full truckload shipments and the cargo ranges from bulk goods to steel coils to palletized goods. So far there is little coordination between the different branches. The whole system is subject to high dynamics primarily due to the short planning horizon and the current traffic situation.

By means of autonomous control in a cooperative logistics system, which is based on multi-agent technology, a more efficient dispatch process, higher transport efficiency and better vehicle capacity utilization will be enabled by reduction of redundant traffic and improved grouping of small goods in highly complex networks.

In order to handle growing dynamics and complexity of logistics systems one possible strategy is to shift from central planning to a decentralized, autonomous control strategy. Distributed routing concepts, as proposed by Rekersbrink et al. (2009), have proven their potential to solve routing problems. Using *TabuSearch* as a well-established algorithm to solve combinatorial optimization problems, the resulting planning solutions have shown that an autonomous cooperating system with a distributed logistics routing protocol (DLRP) leads to better results with an increasing number of shipments. The planning results changed in favour of DLRP due to the increasing complexity. These findings illustrate that autonomous cooperating systems like multi-agent systems need a certain degree of complexity to show their strengths. The main advantages are primarily evident for dynamic and close to reality scenarios (Rekersbrink et al. 2009).

A brief introduction into agent-based systems and their relevance in the field of transport logistics is provided in the next section. After that, we present a solution that has been implemented to enable an autonomous dispatch process in a realistic transport scenario. A discussion of the main simulation results is then followed by the conclusions and, finally, next steps of further research are outlined.

2 Agents and Multi-agent Systems

The term agent is used in a broad context and for a variety of applications. In this paper the term refers to agents as software programs. So far there is no agreement on a single definition of the term (Wooldridge 2009). But most definitions contain similar properties or abilities an agent must have to be considered as an agent. An often cited definition of agents is given by Wooldridge and Jennings (1995). This so called weak notion of agency consists of the properties autonomy, social ability, reactivity and pro-activeness.

There are many more definitions that—depending on the point of view of the researcher—stress different aspects of what an agent should be. Depending on the application for which an agent is used, it should have different properties. This aspect is considered in definitions that distinguish between properties that are essential for an agent and properties that are optional. For a detailed overview of these properties see Lockemann (2006) who based the properties on Wooldridge (2002).

The next step is to examine the interaction of many agents in a multi-agent system (MAS). A MAS consists of several agents that can be heterogeneous and are working on the solution of a specific problem, e.g. on a schedule for machine utilization or the negotiation of a price for a certain service. It is characterized by Jennings et al. (1998) as follows:

- Each agent has incomplete information or capabilities for solving the problem thus each agent has a limited viewpoint
- There is no global system control
- Data is decentralized
- Computation is asynchronous

The need for the agents to be able to communicate arises with these characteristics. There are different forms of communication for agents. Which one is best applicable depends on the environment, the coordination form and the problem that has to be solved. Possible communication forms are blackboard communication (Timm et al. 2006), broadcasting, direct communication and communication via a third party—possibly an agent, labelled as facilitator or yellow pages service (Eymann 2003).

3 Advantages of Multi-agent Applications for Transport Logistics

The contract net protocol (Smith 1980) shows one of the application scenarios of MAS: *resource allocation*. Resource allocation problems illustrate the type of scenarios for which MAS can be used and demonstrate how the use of MAS improves the problem solving process. The use of MAS is best suited for scenarios that include distributed information, problems or resources. Such decentralized scenarios are in addition often dynamic. Using a number of agents to address the different aspects of the problem offers a high degree of flexibility. The ability to use more than one problem solving agent also provides a higher level of security through redundancy or an increase of computational speed. Furthermore MAS typically offer high scalability due to their modular structure (Eymann 2003). The modular structure makes the programming of complex problems easier and changes to the program are simpler to introduce (Dangelmaier et al. 2004).

Centralized systems are often more efficient than MAS because distributed problem solving limits the optimization (Chaib-draa and Müller 2006). But MAS are more flexible and can find solutions for problems that are too complex to be handled by one central instance, especially if the problem itself is distributed. According to Van Dyke Parunak (1999) many industrial problems can be formulated in different ways. If the characteristics of applications are *modular*, *decentralized*, *changeable*, *ill-structured*, and *complex*, an agent-based solution can be more robust and adaptable than one supported by other technologies.

The use of MAS in transport logistics is suitable because it can deal with problems that are non-deterministic and distributed. Most transport logistics applications actually fit the above mentioned characteristics. Nevertheless, a decentralized design is often not chosen by actual projects adopting an agent-based approach (Davidsson et al. 2005). Davidsson et al. provide a survey of existing research on agent-based approaches to transportation and traffic management. As shown in their conclusions, only a few field experiments have been performed and very few deployed systems could be found. A more recent work from Schuldt (2011) describes the implementation of *autonomous control* with multi-agent technology. The proposed system delegates process control to objects like shipping containers in order to cope with dynamics. The actual application of autonomous control is examined by multi-agent based simulation and focuses on the cooperation of logistics entities.

A human dispatcher is not able to consider all possible options for transport schedules because of their high number and frequent changes, especially when some kind of disturbance arises. The planning of transportation routes is a dynamic and complex process which requires a certain level of support and justifies the vital role of information technology in the field of transport. The complexity is determined by a variety of factors and restrictions that have to be considered. First of all there has to be an efficient allocation of goods or transport orders to trucks and allocation of trucks to routes. These allocations are limited by restrictions such as the capacity of the trucks, time windows of the customers, time restrictions of the drivers and special requirements of the cargo. The dynamics are based on factors such as new incoming orders, traffic jams or the breakdown of trucks which require a fast rescheduling of the original plan. Furthermore the planning process is carried out under a latent uncertainty and with incomplete information.

4 Amatrak Mas

To support the dispatch manager, a MAS system will be introduced. With the support of the AMATRAK MAS an overall more efficient dispatch and transport process will be enabled.

The present manual planning process is limited in terms of information that can be evaluated during the available time until a dispatcher must come to a decision. Even if electronic information systems are used, timely access to all necessary and useful information during a particular decision process is hardly achievable, especially when each dispatch manager is responsible for a dedicated set of trucks and transport orders are managed at different branches. The AMATRAK MAS should therefore provide an efficient information flow between human dispatchers and software agents acting on behalf of logistics objects like trucks and other organizational units like branch offices. The advantage of the AMATRAK MAS compared to the present dispatch process arises from its ability to take into account all trucks with their current workload, even if they are controlled by another dispatch manager or at another branch office (see Fig. 1).

The system is designed to serve as both a decision-support system (DSS) and a fully automated system. As a DSS it provides proposals for allocations of transport orders to trucks to support the decision process of the dispatch manager. However, the dispatch manager can also enable a fully automated allocation for any selected transport order.

4.1 System Integration

To ensure that the implementation of the MAS can be conducted without major technical difficulties, the existing IT infrastructure of the logistics service provider

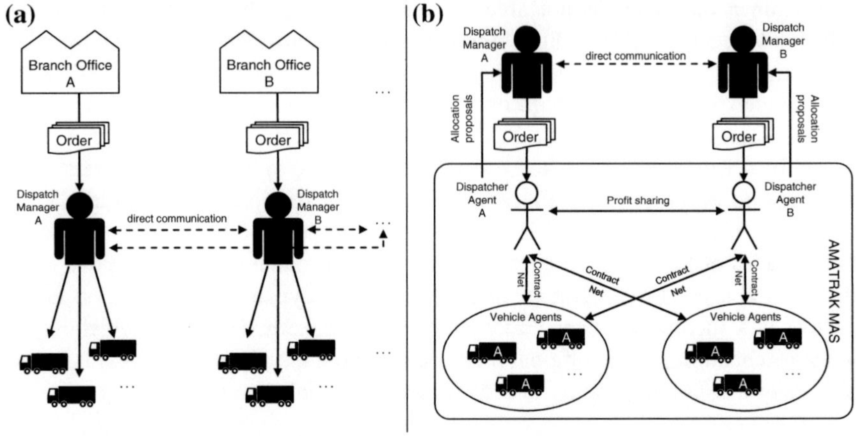

Fig. 1 **a** Present versus **b** MAS-based dispatch process control

has to be considered. The following systems are used by STUTE: active-m-ware: a software that supports the dispatch process; PSV3: a communication system integrated in a PDA (Personal Digital Assistant) that is used for the communication of data between driver and dispatch manager (via active-m-ware); FleetBoard: a system that traces the trucks, controls the driving time and the utilization of the truck. The interaction of the different systems including the MAS is depicted in Fig. 2.

The customer orders being filed in active-m-ware are handed over to the MAS where the orders are processed and the computed tours are reported back to active-m-ware. If the dispatch manager accepts the proposed tours the information is sent to the PSV3 PDA of the driver. Changes on the status of the shipment are reported back by the driver the same way. If problems like a breakdown of a truck occur, the information is handed to the MAS and an alternative schedule will be generated. Besides that, the FleetBoard system monitors the truck. The dispatch manager has a complete and nearly real time overview of the shipments and is still in control of the whole process.

Active-m-ware has a central position in this process. It is the interface for the dispatch manager, responsible for the communication between MAS and PSV3 and in addition it is connected to the database that holds the master data, for example of the customers.

4.2 Multi-agent-Based Design

The MAS is built using the agent framework Jade, a Java based and FIPA (Foundation for Intelligent Physical Agents) compatible platform (Bellifemine et al.

Fig. 2 Interaction of systems

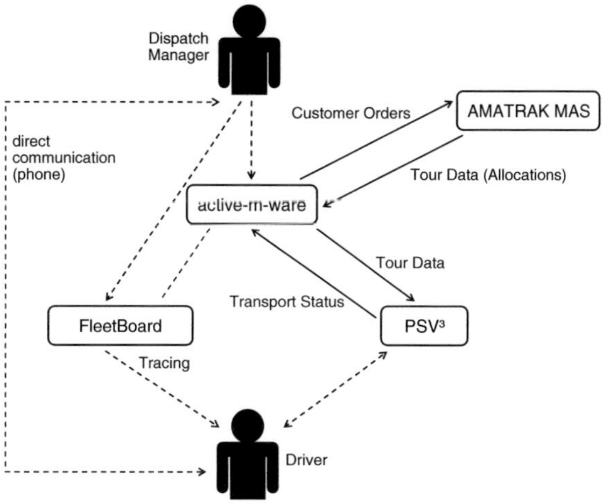

2007). The central communication protocol used by the MAS is based on the contract net protocol which is slightly adapted in order to allow for the direct influence of the dispatch manager. The most important agents of the MAS are the dispatcher agents and the vehicle agents.

4.2.1 Contract Net Based Vehicle Allocation Process

The dispatcher agent is the direct representative of the dispatch manager inside the MAS. Through the dispatcher agent the manager selects the orders to be processed next. Negotiations between the vehicle agents and the dispatcher agent are started by the dispatcher agent. A dispatcher agent can choose the best proposals from vehicle agents and present them to the manager who decides which one will be taken. The manager might agree with the best offer the agent has chosen, but he still may overrule this decision. An autonomous dispatcher agent usually selects the best proposal (Fig. 3).

A vehicle agent will always reply to a call for proposals request (CFP) of a dispatcher agent. This includes the information on whether the agent is able to take the offered order and an individual bid for the order. In addition, vehicle agents also trigger information functions, for example to inform other agents if the tour schedule is changed.

The negotiation process starts with the dispatch manager offering the transport order to all truck agents. The truck agents check if they meet the requirements to fulfil the order. Requirements are typically constraints which are modelled as basic rules e.g. to check if the trailer is suitable for the transportation of the specific cargo, the capacity is sufficient, the truck is allowed to serve the customer, the driver has necessary permissions to handle specific load types, etc. However, some constraints, like preset time windows for pickup and delivery, driving time and working

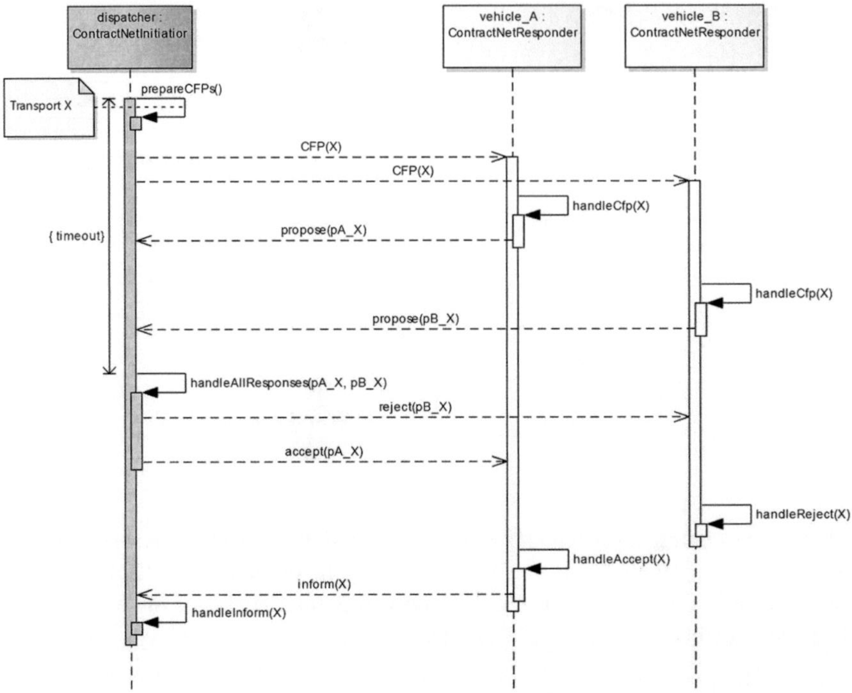

Fig. 3 Contract net interaction between a single dispatcher agent (initiator) and two vehicle agents (responder)

hour limitations of the driver, can only be validated after a concrete potential route is calculated. If the truck is able to fulfil the request, an algorithm is run which determines the "price" of the truck's bid. The result of the algorithm is not a monetary value but an abstract unit that could be regarded as costs. In order to find the best matching route, a cost function to find the optimal route is applied by each truck. This cost function takes into account the actual total distance as well as empty mileage of the route since the trucks are usually not directly located at the pickup location of the transport order. However, in case of additional load alongside the current route, if a trailer is not yet fully loaded, the extra distance can affect the costs of the existing load. In order to get comparable bids from all participating vehicle agents, the cost function calculates the difference of the costs before and after the allocation takes effect, hence the increase of total costs will be minimized while new orders can be allocated incrementally at any time.

For the fully automated allocation method the dispatcher agent chooses the lowest bid to ensure an economical order allocation to the trucks. Alternatively, the system can present the bids as proposals to the user (decision support mode) so that the dispatch manager keeps full control over the actual allocation. The algorithm is designed in such a way that a cost-effective solution for all order allocations is achieved with the decision for the lowest bid, which, additionally, can be calculated within very short time.

However, it has to be considered that all dispatch managers now in general have access to all trucks, or, from the agents' perspective, all vehicle agents potentially take part in negotiations started by any dispatcher agent. This leads to problems if several of these interactions take place concurrently, since transport allocations to the same truck mutually influence each other. In order to cope with concurrent access issues and to enable a robust multi-agent coordination, vehicle agents implement an optimistic *first-come, first-served* approach in their contract net behaviour (Fig. 4). In any case, proposals are submitted independently on the basis of the current route of the truck when the agents receive the CFP message. If the vehicle agent receives the first positive response on one of its proposals, it finally confirms the allocation with an *inform* message as usual. All other acceptances of pending proposals will be answered with a *failure* message so the responsible dispatcher agent can either restart a new CFP for the transport order or select an alternative responder.

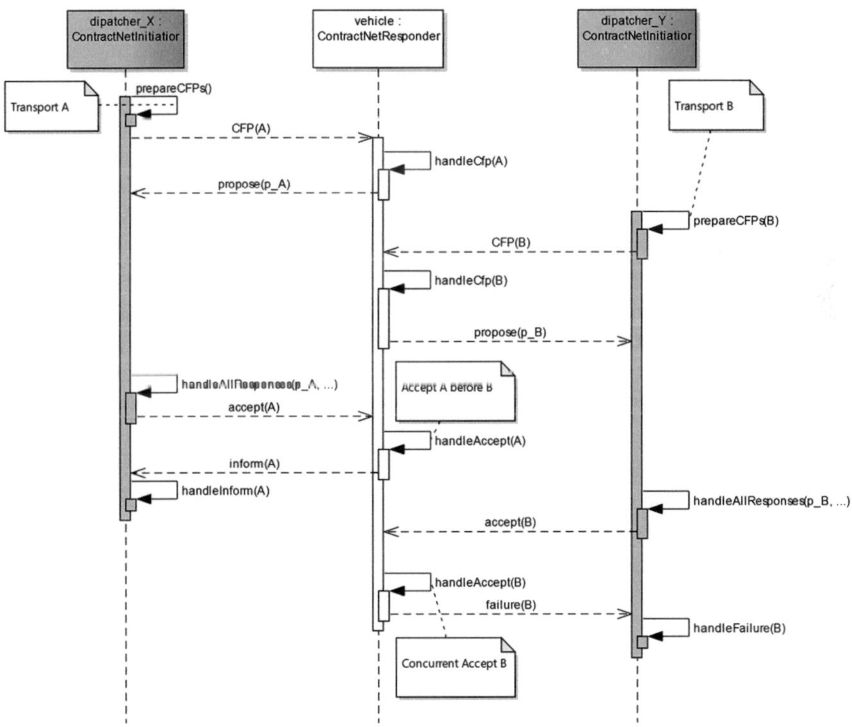

Fig. 4 Concurrent contract net interaction between two dispatcher agents and a single vehicle agent

5 Simulation

To evaluate the MAS, the system has been tested with real data from the transfer partner STUTE. The tests are based on past customer orders. For the simulation two types of test runs are generated, one is a rebuild of the orders in the way the dispatchers allocated them to the trucks, so actual performance figures can be generated. In the other test runs the orders are allocated by the MAS so that a comparison of the different runs can be used as an indicator for the MAS' performance.

The results for one example day with 70 trucks and 160 orders are shown in Table 1.

The total distance travelled by the trucks can be reduced by about 10 %, though the freight mileage remains nearly at the same level. The decrease in the overall mileage comes mainly from the reduction of empty mileage. This reduction is not only important for the logistics service provider that is not paid for these tours but also on a social-ecological level because the empty mileage doesn't produce any value but is a burden for the infrastructure and emission limits. A slight increase in vehicle utilization can also be recognized. The improvements can be generated because the system is able to dispatch all trucks for the whole company and achieve synergies that are lost when the different branches each conduct their own dispatch process. This example shows that the MAS is able to process real data and achieve efficient results. Test runs with data from other days lead to similar performances. A considerable influence on the results comes from the order structure and sequence in which the orders are dispatched. The geographical distribution of the pick-up and delivery locations, the share of less-than-truckloads and full truckloads and the way the tours are dispatched in (the rebuilt) reality have an impact on the level of mileage reduction that can be achieved. Although the simulation is based on real orders, there is a small degree of fuzziness in the results shown in Table 1 due to the fact that some aspects could not be quantified and modelled such as the personal knowledge and experience of the dispatchers. In contrast to a computer system a human being is always able to deal with irregular or special orders.

The sequence in which the orders are processed by the MAS also has an influence on the quality of the results. During the evaluation experiments best results were achieved when full truckload and long distance orders were put first. Nevertheless, the results achieved by the MAS show, that it is an applicable and useful tool for dispatchers. Future work will focus on different strategies to arrange the orders before they are processed by the MAS.

Table 1 Simulation results

	Rebuild	AMATRAK
Total mileage	35.637	31.798
Freight mileage	27.558	27.052
Empty mileage	8.079	4.746

6 Conclusions and Outlook

In this paper an agent based approach to enable the intelligent allocation and control of freight transport services has been presented. Software agents and multi-agent systems have been introduced as an appropriate means to cope with resource allocation problems that involve distributed information and process control in a dynamic environment. These potentials have been aligned with real-world requirements and approved by the logistics service provider STUTE. The AMATRAK research project facilitates this transfer into a practical application. The simulation results of the AMATRAK MAS with its distributed problem solving approach shows that the procedure outperforms the previous distributed manual dispatching approach significantly in terms of flexibility and speed, and results in a severe reduction of empty mileage and a slight increase of capacity utilisation. The MAS is able to deal with a number of dynamic influences due to its good scalability, modular structure and distributed problem solving capability. The case study shows that the AMATRAK MAS is able to process real data and achieve efficient and effective results. Due to the increased visibility enabling a global planning perspective, benefits are particularly demonstrated in a distributed setting with multiple branch offices.

There are still questions and possible adjustments to the proposed approach open for future research. The possibility to arrange orders for their later sequential negotiation has been mentioned briefly but a comprehensive evaluation of possible strategies is a pending task. Another well perceived aspect is the possibility to incorporate real-time data from trucks in order to adjust route schedules and to automatically react on deviations from the current routing plan. Additionally, a future extension of the agent model could also take an inverted allocation control mechanism into account, such that vehicle agents can actively call for additional load if a waste of spare capacity is imminent. The capabilities of this mechanism have been investigated generally with respect to the possible integration of web-based freight exchange services.

A market research carried out by STUTE shows a lack of mature applications with comparable features, particularly with regard to supporting dynamics and concurrent interaction between multiple autonomous branch offices. Currently the project partners ISL and STUTE are developing AMATRAK MAS into iTL|dispo, a new system based on the MAS approach as described above. Mid of 2015, the first module will be released and put into service by STUTE.

Acknowledgments The work presented in this paper was co-funded by the German Federal Ministry of Economics and Technology (BMWi, 19G7028B).

References

Bellifemine F, Caire G, Greenwood D (2007) Developing multi-agent systems with JADE. Wiley, Chichester

Chaib-draa B, Müller JP (eds) (2006) Multiagent-based supply chain management. Springer, Berlin

Dangelmaier W, Pape U, Rüther M (2004). Agentensysteme für das Supply Chain Management. Grundlagen – Konzepte – Anwendungen. Vieweg+Teubner, Wiesbaden

Davidsson P, Henesey L, Ramstedt L, Törnquist J, Wernstedt F (2005) An analysis of agent-based approaches to transport logistics. Transp Res Part C Emerg Technol 13(4):255–271 (Elsevier)

Eymann T (2003) Digitale Geschäftsagenten. Springer, Berlin

Jennings N, Sycara K, Wooldridge M (1998) A roadmap of agent research and development. Int J Auton Agent Multi-Agent Syst 1(1):7–38

Lockemann PC (2006) Agents. In: Kirn S, Herzog O, Lockemann P, Spaniol O (eds) Multiagent engineering. Theory and applications in enterprises. Springer, Berlin, pp 17–34

Meyer CM, Kopfer H, Kok AL, Schutten M (2011) Distributed decision making in combined vehicle routing and break scheduling. In: Kreowski H-J, Scholz-Reiter B, Thoben KD (eds) Dynamics in logistics, second international conference, LDIC 2009, Bremen. Springer Berlin/Heidelberg, pp 125–134

Rekersbrink H, Makuschewitz T, Scholz-Reiter B (2009) A distributed routing concept for vehicle routing problems. Logistics Res 1(1):45–52

Schuldt A (2011) Multiagent coordination enabling autonomous logistics. Springer, Berlin

Smith RG (1980) The contract net protocol: high-level communication and control in a distributed problem solver. IEEE Trans Comput C 29(12):1104–1113

Timm IJ, Scholz T, Herzog O, Krempels K-H, Spaniol O (2006) From agents to multiagent systems. In: Kirn S, Herzog O, Lockemann P, Spaniol O (eds) Multiagent engineering. Theory and applications in enterprises. Springer, Berlin, pp 35–51

Van Dyke Parunak H (1999) Industrial and practical applications of DAI. In: Weiss G (ed) Multiagent systems. MIT Press, Cambridge, pp 377–421

Wooldridge M (2002) An introduction to multiagent systems. Wiley, New York

Wooldridge M (2009) An introduction to multiagent systems. Wiley, Chichester

Wooldridge M, Jennings N (1995) Intelligent agents: theory and practice. Knowl Eng Rev 10 (2):115–152

Smart.NRW—RFID as Enabler for an Intelligent FMCG Supply Chain

Kerem Oflazgil, Christian Hocken, Fabian Schenk, Oliver Teschl, Thorsten Lehr, Mareike de Boer, Christoph Schröder and Rainer Alt

Abstract RFID is a key technology for an increase of transparency in logistic networks and can serve as a solution for unsolved problems in FMCG industry like excessive inventory levels, inefficient capacity utilization and poor service levels. The research project Smart.NRW studies the application of RFID on case level as well as the interorganizational demand planning. The project goal is to integrate high resolution information into the planning and control processes to maximize the on-shelf availability of consumer goods in the wholesale market. The research project includes three major work packages. The first one aims at the definition of the suitable product-specific RFID transponder and its position. The objective of the second work package is to develop a manufacturing process providing cardboard packaging of FMCG articles with RFID transponders. In the third work package real-time data driven methods for adaptive planning and control in the supply chain are developed. To validate the project results two field trials have been carried out,

K. Oflazgil (✉) · C. Hocken
Institute for Industrial Management (FIR) at RWTH Aachen University, 52074 Aachen, Germany
e-mail: Kerem.oflazgil@fir.rwth-aachen.de

C. Hocken
e-mail: Christian.hocken@fir.rwth-aachen.de

F. Schenk
European EPC Competence Center GmbH, 41469 Neuss, Germany

O. Teschl
METRO Systems GmbH, 40235 Düsseldorf, Germany

T. Lehr
Mondi Eschenbach GmbH, 92676 Eschenbach, Germany

M. de Boer
Mars GmbH, 41751 Viersen, Germany

C. Schröder
ESM Ertl Systemlogistik Minden GmbH & Co. KG, 32423 Minden, Germany

R. Alt
METRO Cash & Carry Deutschland GmbH, 40235 Düsseldorf, Germany

H. Zijm et al. (eds.), *Logistics and Supply Chain Innovation*,
Lecture Notes in Logistics, DOI 10.1007/978-3-319-22288-2_19

for which an RFID infrastructure consisting of reader gates and handheld devices has been installed at several points along the supply chain. This contribution reports in detail about the results of Smart.NRW, by describing the process of RFID application in an FMCG supply chain.

Keywords Supply chain management · RFID · Collaborative planning · Fast moving consumer goods

1 Introduction

Companies belonging to the fast moving consumer goods (FMCG) industry face a high pressure to decrease costs and to secure a high service level in a heterogeneous range of products (Geißdoerfer 2009). In addition, a delay in the flow of information between supply chain partners often leads to strongly increasing stock levels, inefficient capacity utilization and decreasing service levels. The information exchange in supply chains is still limited (Bottani et al. 2010).

This paper focuses on the question how supply chain partners can continuously improve the flow of goods and information in order to optimize the planning and control processes in the entire supply chain. Based on this, new methods and tools need to be developed to sustainably improve internal as well as inter-organizational logistics processes.

An approach to answer this question is the implementation of new information technologies. One of these technologies is Radio Frequency Identification (RFID), consisting of a chip with an integrated antenna, also called tag, which can be attached to the items. Since unique number combinations are sent to contact free reading points, the challenges mentioned above can be addressed with the application of RFID (Emde et al. 2012). Thus, based on the higher information transparency supply chain partners can exchange more data and intensify the degree of inter-organizational collaboration. RFID technology makes it possible to capture each item at several supply chain stages in a rapid and automatic way and to increase the transparency considering goods and information flows throughout the supply chain (Bhadrachalam et al. 2009). Consequently, RFID technology seems to be a key enabler to increase information density and to integrate real-time data into logistics networks (Gille 2010).

Nevertheless, there are some technical barriers to introduce RFID in a supply chain, especially on case level. Since the physical parameters of each product and each package vary significantly it is important to implement an efficient and quick process to select the right tag and to define its right position. Considering the broad product range existing in the FMCG industry, this step will ensure an effective reading of RFID on case level.

The problem of finding the optimal tag type and position as well as the automated integration of RFID tags into cardboard packages and the development of

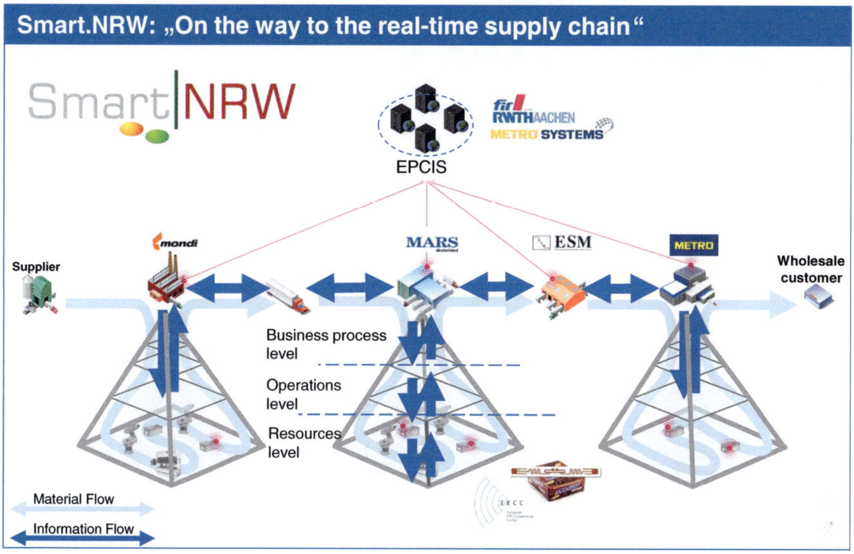

Fig. 1 Smart.NRW supply chain

adaptive planning and control algorithms based on real-time data belong to the main targets of the research project Smart.NRW. The Smart.NRW supply chain consists of Mondi (manufacturer of cardboards), Mars (confectionery manufacturer), ESM (logistics service provider) and METRO Cash & Carry (wholesaler). Besides, the European EPC Competence Center (EECC) develops a procedure for the Optimal Tag Type and Position (OTTP), while METRO Systems provides software solutions for implementing developed algorithms and supports the construction of an RFID infrastructure. The Institute for Industrial Management (FIR) at the RWTH Aachen University coordinates the project and develops methods to optimize the supply chain planning (Fig. 1).

In the next section, the results of the technical work packages will be described. Subsequently, the benefits derived from Smart.NRW are summarized for each supply chain partner. Finally, both field trials and further research needs are discussed.

2 RFID Application in a FMCG Supply Chain

In this section, the development of the research results regarding the OTTP, the implementation of a supply chain platform as well as the development of a manufacturing process will be illustrated.

2.1 Development of the OTTP by the European EPC Competence Center EECC

The FMCG market is characterized by a large variety of packages (Clarke 2008). Since the purchase decision of customers at the point of sale is determined to a large extent by the package (which hence is an important marketing tool), product packaging is undergoing changes continuously. This poses challenges for the UHF-RFID based product identification on case level. The package form and content as well as the RFID transponder position operating in the UHF band strongly influences its performance (Regattieri and Santarelli 2013). After package revisions, the type, position and orientation of the transponder often need to be redefined to ensure a reliable reading.

Due to the large number of FMCG manufacturers, products and product variants and rapidly changing outer packaging, determining the optimal tag position with the conventional method in an anechoic chamber is not feasible. The lack of specialized laboratories that can perform the measurements and the high costs associated with measuring different tags on different positions of an FMCG case creates the need for an innovative and new tag finding procedure. In the future more and more FMCG cases will be equipped with UHF-transponders. Thus, an economic method is needed to find the right transponder and the right transponder position in an automated way.

The optimal tag type position (OTTP) procedure should solve this problem. As part of the Smart.NRW project the EECC together with METRO Systems developed a method to decide upon the right position of a tag without measuring it attached to the product. About 150 UHF labels were measured on eight reference materials with a distinct permittivity. The permittivity as a measure of how an electric field affects, and how it is affected by a dielectric medium, is the main parameter influencing the performance of UHF-RFID tags.

Mathematical Model From these measurements, a mathematical model has been created to predict the performance of each tag on materials with permittivity values different from the ones of the reference materials. With this model on hand, a way had to be found to automatically predict the read range of UHF-RFID tags on different positions of the FMCG package. A permittivity map of the tagged consumer goods was the missing link to realize the OTTP.

The EECC developed a new permittivity probe, designed especially for the needs of using it on FMCGs. It can measure the complex permittivity with both components, the real and imaginary part, and scans an area of size similar to the area which affects the performance of an UHF tag. The real part of the permittivity leads to a "detuning" of the transponder, whereas the imaginary part stands for the absorption of energy by the material, what leads to a direct decrease in read range.

The probe was attached to a positioning system with the ability to automatically scan the surface of different FMCG packages of any size in a pre-defined raster. At each point of the raster the permittivity is measured and connected with the coordinate.

Both the positioner and the vector network analyzer connected to the probe are controlled by a PC running LabVIEW. LabVIEW retrieves the permittivity measures of the tags on the reference materials from the tag database and calculates the corresponding read range for all tags via the above mentioned mathematical model using a MATLAB plugin. The predicted read range for a tag computed from the measured permittivity deviates only slightly from the real read range values of the tags attached on the same coordinate to the product and measured in the anechoic chamber.

As an outcome of the OTTP development in Smart.NRW it is now possible to determine the ideal type of UHF tag and its position for an FMCG product with a reasonable effort and reasonable loss of precision. It allows an economic tag selection for future use of RFID in the FMCG sector. The selection of tags, performance-, size- and pricewise, for FMCGs is now feasible. The OTTP has removed the obstacles that blocked an economic tag finding.

2.2 Implementation of Planning and Control Algorithms in a Supply Chain Platform by METRO Systems

In Smart.NRW, transactional data for all stations and read points of the supply chain have been gathered and stored. In order to use this new information granularity and transparency, a supply chain reporting platform has been developed which enables its user to perfectly control its supply chain processes and pro-actively react on critical situations.

For displaying and analyzing all relevant information, the transactional event data of all RFID read points along the supply chain had to be transformed into their business representation using the so-called EPCIS (Electronic Product Code Information Services) standard. This business event information represents the necessary KPI- and transactional data for the web-based reporting platform. Possible use cases for the platform include order planning based on stock and sales information as well as process throughput time statistics and the reaction on critical situations like out of stock (OOS) or NOSBOS (not on shelf, but on stock). Situations that require an active intervention can be supported by the notification framework. In that way, the involved stakeholders can get information or actions are even automatically taken in case of certain supply chain KPIs are reported to be beyond their limits.

Methodology Electronic Product Code Information Service (EPCIS) is a global standard for exchanging value chain events between business partners in real time (Zheng and Ge 2013). These events typically inform on what, as well as when and why something has happened. The software implementing this standard is called an EPCIS system with the task to administer all relevant value chain events in a central repository.

The reports which were developed for the supply chain platform have been defined in the platform‘s functional specification. Technically, they are fed by the EPCIS events which are generated for all supply chain processes and stored in the EPCIS repository.

Report Application The report application displays all dynamic business intelligence data. It is web-based and enables an authenticated access for all project partners on all internet-ready devices using a web browser. In a user authentication system all legitimate credentials are stored in order to avoid access fraud by third parties.

Based on the authentication credentials, the individually accessible reports for every user are controlled. In that way, all project partners have access to common reports, whereas company-internal process flows are only accessible for the partners themselves. The list of accessible reports is shown to the user after login.

As a report development framework, the BIRT-API (Business Intelligence and Reporting Tools) was used. When a report is selected, all necessary business intelligence data from EPCIS events are retrieved from the database, processed, filtered, sorted and finally displayed in a tabular as well as graphical representation. The user can manipulate the results using report parameters for time, article, location and some additional optional filters. In all cases, the system already proposes a sensible default selection.

The reports provide an extensive list of analytical functions on the supply chain, specifically concerning the following information:

- Carton and product stock information in every supply chain location
- Sales in MCC stores
- Process order throughput times for cartons and products, inside any supply chain partner as well as across all partners
- Order proposals and planning information based on current stock information and historical process data
- Track & Trace search functions (e.g. locations, time periods, orders)
- Track & Trace for every product instance
- Order process quality (time, quantity, quality)
- On-shelf availability in the stores

Some reports also make use of best before date (BBD) information for each product instance. In that way also product quality (“sellable stock”) can be classified.

Notification Framework In order to optimize the operational process quality, an automatic alert and notification system has been developed so that e.g. minimal stock or BBD situations are actively communicated almost in real-time. This enables direct interventions at the first moment possible.

Technically, a combination of database structures, trigger mechanisms and constant script-routines (Cron-Jobs) has been used. From a business data perspective, alert criteria are calculated and continuously updated. In addition, message

Table 1 Key performance indicators

Current details	Event details like timestamp, type × number for articles, order reference
Completeness	Current and historic ratio of order completeness (store-based and global)
Fulfilment	Current and historic order fulfilment KPIs (store-based and global)
Write-off	Number or articles in stock of which the BBD (best before date) residual time is less than 119 days
Reordering number	Number or articles to be ordered as well as the timing of the order (create new order proposal) for target store
Replenishment	Number and location of articles to be transferred from high rack to sales floor
OOS (Out-of-stock)/OSA (On-shelf availability)	OOS event for article type, current and historic OSA figures for article type(s)
NOSBOS (Not on shelf but on stock)	NOSBOS event for article type, current and historic NOSBOS figures for article type(s)
Overstock	Overstock event for article type, overstock ratio and time period (cumulative)
Sales peak	Sales peak event for article type, peak ratio and time period (cumulative)

types and recipients are associated with each notification type. Using Cron-Jobs, every notification is transmitted to the defined recipients in the event of an alert. The Key Performance Indicators (KPI) have been included in the notification framework (Table 1).

2.3 Development of a Manufacturing Process for RFID Case Level Tagging by Mondi

The production process of corrugated cardboards offers different possibilities to apply an RFID tag. In Smart.NRW, the entire procedure to find the optimal technical and economical solution has been analyzed. The following assumptions hold:

- No speed restrictions for the production machines
- Applying without mechanical contact to secure product quality
- Writing of transponders "inline"
- Automatic outplacement of defect transponders
- Separated setup of labeling machine
- Full flexibility considering the positioning

Starting with the corrugator many possible places to apply the tag were found. Applying the tag during manufacturing of the raw corrugated board brings several advantages, especially in terms of flexibility. Both, high pressure (up to 80 bar) and high temperature (up to 160° steam temperature) at this step cause major problems with the tag. Even after having solved pressure and temperature problems several challenges regarding production numbers and costs as well as resulting waste still exist. For the conversion of the raw board there are three standard ways that are described in the sequel.

Inline Converting An inline machine includes every standard working step into one single line. Since this represents an all in one machine there are some limitations in comparison to separated single steps. Die cutting will never be as perfect as it is with a flatbed die cutter. The biggest challenge for inline machines is the high speed. Depending on the box size modern machines are able to convert up to 24,000 boxes per hour. Synchronization and accurate applying lead to high costs.

Die Cutting Die cutting is separated into the two areas flatbed and rotary die cutting. The first one is a very precise process. The advantages of this method include high precision, high accuracy in scoring and high flexibility considering possible geometric forms. The tool in the rotary die cutting is fixed at an axis and has an opposite axis with die cut rubber. The sheet is moved through and cut by the rotated tool. This process is not as precise as the first one. However, higher production speed and rates as well as bigger sheet sizes belong to the main advantages.

Glueing Glueing and folding are the final steps to transform the open sheet to the final cardboard box. The RFID tag is applied in this step. Therefore, a standard applicator needs to be severely modified. The applying head is equipped with an RFID antenna. Regarding the limited space the entire dispenser with the raw material needs to be positioned much higher so that a longer apply tongue is required. The most critical point is the electrical field. To avoid reflections and writing of wrong transponders in sequence, metal parts have to be reduced to a minimum. As technical highlight, the most loaded parts are made of carbon fiber.

3 Supply Chain Benefits of RFID Case Level Tagging

In this chapter, the benefits to be realized by RFID case level tagging in the FMCG industry are pointed out for each supply chain actor.

3.1 Benefits for the Confectionery Manufacturer Mars

Today, the knowledge of what happens at the point of sale is rather limited. This leaves us with different issues regarding forecast quality, unused opportunities in managing marketing activities and inefficient handling of quality issues.

Forecast Quality The goal is to serve all customer orders perfectly. To achieve this, the supply chain will be managed in advance by forecasting the potential demand based on historical data. Unfortunately, this approach still provides an inaccurate forecast of demand, as the real time data of stock in the customers' warehouses are not incorporated. This results in high inventory levels to respond to unexpected demand or in out of stock situations, as it is no longer possible to respond to large demand fluctuations. With a real-time tracking of products via RFID the transparency and real-time data on sales in the customer stores and on remaining stock levels in their warehouses are given for the first time. Consequently, the forecast quality can be improved by using real-time data that reflects potential future customer demand more accurately. This knowledge facilitates managing the supply in advance and respond to unexpected demand fluctuations faster.

Marketing 2.0 Confectionery is highly driven by impulse business. Most shoppers buy chocolate spontaneously while they pass through the shop instead of having noted it on their shopping list. Thus, placing chocolate products not only in the regular shelves but also on secondary placements is critical to success in this sector. Nonetheless, there is no accurate way to measure the impact of second placements to date. Due to reading points on secondary placements and on shelf in the Smart.NRW pilot stores, plus the RFID case level tagging of the products, we are able to trace if a sold product was taken from shelf or from secondary placement. One step further this also enables to compare the efficacy of different second placements, different promotional displays, or to track the on-time delivery and optimal execution of promotions in store. RFID case level tagging provides a new transparency on what is happening in the shop that leads to a better, more data-driven marketing.

Recalls on Single Item Basis When recalling goods, RFID case level tagging offers two benefits. First, it is possible to capture where and when exactly a certain product went through which production process and might have been affected by a certain quality issue (e.g. misprinted packaging, temperature fluctuations). This enables to tackle a quality issue more efficiently and more effectively by narrowing down the recall to only affected items. The second benefit comes from using real-time data on the location of items. Thereby, locating items affected by a recall is incredibly fast and makes it possible to react immediately.

On-shelf-availability (OSA) FMCG companies in most cases do not sell highly anticipated products for which people make an urgent effort to buy them, wait for delivery for a long time or search shops to find the specific product. When it comes to FMCG products buyers behave differently. If they might not come across the specific candy bar they wanted to buy, buyers most likely just buy a candy bar from a different brand. Therefore, the non-availability on shelf leads to a loss of customers to the competition. If products are out of shelf for a longer time period customers will shift their buying intention to other brands, which is precisely what leads to sustained customer losses. These out of shelf situations can be triggered by low delivery performance or by NOSBOS incidents, where products that are actually on stock have not reached the shelf. To ensure the OSA of certain products, NOSBOS incidents should be avoided by supporting the trade. With RFID reading events on the sales floor and RFID case level tagging, alert functions to detect NOSBOS incidents can be implemented (Sundene and Merete 2014).

In all cases, RFID case level tagging provides real-time data to make a step forward in supply chain transparency. Thereby, it helps to realize a shopping experience in which consumers find their desired product in great quality available in store enriched by a marketing experience tailored to their preferences.

3.2 Benefits for the Logistics Service Provider ESM

The main benefits of the results reached in the course of the research project Smart. NRW for the logistics service provider are summarized to four basic processes in terms of product handling.

Goods Receipt The first focus has been on the goods receipt inspection. Often every pallet is controlled manually. RFID case level tagging improves this essential part. Faster inspections lead to faster handling and availability in the warehouse. As a result of this, a second benefit on storing and taking out of store is noted. That means that the holding area can be entered automatically after the goods receipt inspection.

Commissioning and Shipping Another important part is the commissioning and shipping. It is possible to optimize the system with a better fault prevention.

Service The last benefit which can be realize with the help of RFID case level tagging is to improve the service quality. The handling of returns and reclamations has a great potential. Thus, there is a higher visibility regarding the transport of single goods, which helps to optimize the entire transport and customer support.

3.3 Benefits for the Wholesaler METRO Cash & Carry

Four METRO Cash & Carry stores in Germany participated in the research project Smart.NRW. For them, the value drivers for having RFID technology on case level are identifiable in the areas goods receipt, stock management and write offs, as explained in more detail below.

Goods Receipt "Did we get the orders we requested in the right quality, quantity and within the right time-frame?"

This is one of the key questions of goods receipt. Without the usage of RFID tagged goods, the answer to the question above needs detailed manual checking. After the validation, the goods coming from the producer have to be checked in into the goods management system. This leads to additional manual effort, even when using optical scan methods. RFID on case level helps to reduce the effort spent on checking the goods received and afterwards their check-in into the goods management system (Miragliotta et al. 2009). The RFID tag on each case makes this case uniquely identifiable in the in-store process chain.

Stock Management "Do we have sufficient stock?", "By what time do we have to reorder to avoid an out of stock situation?", and "Where is the stock located?"

In order to fulfill customer needs, it is necessary to ensure that enough stock of a specific product is in store. This may lead to conflicts between the availability of sufficient products on the shelf and the capital lockup they represent.

RFID tagging enables to find the right balance between availability and inventory capital because the information on stock levels and sales is precisely and in real time available (Vlachos 2013). This helps to have an accurate reorder prediction and range calculation, respectively, which helps to reduce the order amount together with avoiding both out of stock and over stock situations.

Another advantage is to know exactly where stock of a specific product is located (Rhensius 2010). RFID based real time information considering the location of a specific product enables to reduce replenishment efforts and to avoid NOSBOS situations.

Write-Off Write-offs are generally unwanted because they have a negative impact on revenues. For instance, in the area of fresh products or ultra-fresh products with a small durability time lag, the risk of write-offs is often very high. With RFID case level tagging it is possible to identify such products quickly and to place them on the shelf in ascending order of their best before date. This reduces the risk of these kinds of write-offs and especially the costs that are caused by write-offs.

4 Conclusion

Before summarizing the main contents of this contribution will be summarized, both field trials conducted in Smart.NRW are briefly described.

4.1 Field Trials

The first field trial aimed at the validation of the OTTP results and the technical feasibility of the concept. Six reader gates, handheld devices and label printers have been installed at several points along the supply chain and connected to several EPCIS instances, one for each firm to avoid the central storage of data. Furthermore, new item codes were developed for targeted orders of test products. The packaging with new codes and RFID tags were qualified and approved by the producer.

During the test, the RFID tags were attached to the cartons and also encoded. ESM used the tagged cartons to repack the test items. At this stage, the RFID tags were read for the first time. Other reading points were located at the outgoing goods area of the logistics service provider's and at the incoming goods area of the wholesaler.

The developed production methods were successfully validated during the first field trial. The chosen infrastructure proved to be suitable for the project objectives.

Significant results were reached in the area of (bulk) reading of the packaging which can be used for the OTTP refinement and the hardware configuration improvement.

The second field trial included additional reader gates, readers integrated into high rank shelves and a larger number of test items. Its aim was the implementation, refinement and validation of the achieved results considering the planning and control algorithms. The second field trial also served as a basis for a possible roll out.

The field trial data was collected in the Smart.NRW platform. It was used by the firms to quantify the RFID based measures for raising the supply chain efficiency. Moreover, individual questionnaires were prepared for each firm for quantitative and qualitative analysis and evaluation of the major aspects of the whole research project.

4.2 Outlook

In Smart.NRW the RFID application on case level in a real FMCG supply chain was analyzed. As a key information technology, RFID is used to solve still existing logistic challenges like uncertainty and lack of transparency. A higher information density in the entire supply chain enables the actors to react faster to fluctuating customer demands as well as to decrease stock levels while improving service levels at the same time (Wang et al. 2011). Additionally, a higher degree of collaboration is simplified by improved transparency. In summary, RFID case level tagging leads to more information quality improving sustainably the performance and competitiveness of FMCG supply chains.

However, research activities in this area need to be continued. Smart.NRW focused on one specific product category. The discussions in the project consortium about a 100 % implementation of RFID case level tagging shows that further logistic potentials can be realized. An important step to future FMCG supply chains has been done by Smart.NRW.

Acknowledgments The research project Smart.NRW was supported for North-Rhine Westphalia (Germany) in the context of the operational program "regional competitiveness and employment" 2007–2013. This program is co-financed by the European Regional Development Fund (ERDF) "investment in our future".

References

Bhadrachalam L, Chalasani S, Boppana R (2009) Impact of RFID technology on economic order quantity models. In: 3rd annual IEEE international systems conference, Vancouver, Canada

Bottani E, Montanari R, Volpi A (2010) The impact of RFID and EPC network on the bullwhip effect in the Italian FMCG supply chain. Int J Prod Econ 124(2):426–432

Clarke R (2008) Smart packaging technologies for fast moving consumer goods. Wiley, New York

Emde B, Höfinghoff J-F, Overmeyer L (2012) Kennzahlenbasierte Bewertung von RFIDSystemen. Logistics J LJ:1–14

Geißdoerfer K (2009) Total cost of ownership (TCO) und life cycle costing (LCC). Einsatz und Modelle: Ein Vergleich zwischen Deutschland und USA, 1. Aufl. LIT, Münster, Westf

Gille D (2010) Wirtschaftlichkeit von RFID-Systemen in der Logistik, 1st edn. Breisgau, Gabler, Wiesbaden, Freiburg i

Miragliotta G, Perego A, Tumino A (2009) A quantitative model for the introduction of RFID in the fast moving consumer goods supply chain: are there any profits? Int J Oper Prod Manag 29 (10):1049–1082

Regattieri A, Santarelli G (2013) Manufacturing logistics and packaging management using RFID. In: Dr. Reaz MIB (ed) Radio frequency identification from systems to applications. InTech

Rhensius T, Dünnebacke D (2010) RFID—business case calculation. 3-stufiges Vorgehen zur Planung und Bewertung des RFID-Einsatzes. Vol. 1. G Schuh and V Stich FIR e.V. an der RWTH Aachen (ed.), Aachen

Sundene H, Hole M (2014) Cost-benefit assessment of item level RFID in apparel retailing: a case study of an RFID pilot project in moods of Norway, Institutt for industriell økonomi og teknologiledelse

Vlachos IP (2013) A hierarchical model of the impact of RFID practices on retail supply chain performance. Expert Syst Appl 41(1):5–15

Wang S-J, Wang W-L, Huang C-T, Chen S-C (2011) Improving inventory effectiveness in RFID-enabled global supply chain with grey forecasting model. Department of Industrial Engineering and Management, National Chin-Yi University of Technology, Taiping City, Taichung County 411, Taiwan, ROC

Zheng S, Ge Y (2013) The information discovery service of electronic product code network. In: Emerging technologies for information systems, computing, and management, chapter 13. Lecture Notes in Electrical Engineering

Developing Support Tools for Compliance in Supply Chains

Melissa Robles, Juan Diego Serrano, Maria Laura Maragunic and Bernd Noche

Abstract Decisions about the planning, organization, execution and control of supply chains are to a vast extent based on economic, efficiency and time factors. Multiple methods and tools have been developed to support supply chain managers in this task. With new expectations and demands from society and governments regarding security and trade compliance, the complexity of managing value chains, especially international ones, has increased, and responsible operations are needed at all company levels. This imposes, among others, new challenges when designing networks and selecting partners or serving customers. To cope with these new demands, tools should be developed to support supply chain managers to ensure more secure and compliant operations. In this chapter, a prototype tool for meeting compliance requirements and increasing security in supply chains is presented.

Keywords Compliance · Supply chain security · Sanction lists · Levenshtein

1 Introduction

In the last decade, supply chains worldwide have suffered from more frequent and deeper disruptions in their channels. For example, the major earthquake and tsunami Tohoku in 2011 caused breaches to the auto manufacture industry when suppliers and ports in Japan were shut down in the aftermath. Hurricanes like the super storm Sandy in 2012 caused widespread damage to logistics and transportation networks in the region, forcing carriers to reroute shipments and confronting them with delays. On the other hand, not only natural disasters can cause logistics problems; hazards of another nature such as accidents (fire), political (sanctions), social (strikes) or technological (cyber-attacks) events can shut down supply chains

M. Robles (✉) · J.D. Serrano · M.L. Maragunic · B. Noche
Department of Transport Systems and Logistics, University of Duisburg-Essen,
Keetmanstraße 3-9, 47058 Duisburg, Germany
e-mail: melissa.robles-aguirre@uni-due.de

H. Zijm et al. (eds.), *Logistics and Supply Chain Innovation*,
Lecture Notes in Logistics, DOI 10.1007/978-3-319-22288-2_20

and interrupt the flows of materials, resources, persons, information and money. Note that even if a region or company is not directly affected by one of these events, the resulting government's response or collateral consequences can affect firms because of closed borders, air traffic shutdowns and inspections. Therefore, the global risk landscape has changed throughout the years and has become increasingly complex.

With globalization, international flows of goods and persons increase in speed and volume while the recent development in Information and Communications Technologies (ICT) allows for the administration of far more data of these movements. This phenomenon has changed the way distribution networks are operated nowadays. Supply chains are progressively more dependent on each other. Practices like Just-in-Time and Lean Production, the last offshore supplier's upward trend and global operating companies stress the vulnerability of supply chains.

Logistic flows, modes of transportation and infrastructure are vulnerable to misuse or attacks. This is part of the nature of any supply chain which transports goods and persons across regional and international boundaries through network facilities and transshipment points. The same flows might be damaged or used to transport resources intended for illegal activities. Threats to supply chains can be seen as deterministics (natural disaster, product contamination, etc.) or rational (social-political attacks, theft, etc.) (Ekwall 2012). Companies should be prepared for both types of threats.

A study of the insurance company Allianz yields the three top business risks for the year 2014 as: operation and distribution network disruptions, natural disasters and fire or explosions. The three top risks are the same for the different regions of North and South America, Europe, Middle East, Africa and Asia-Pacific. Especially in the logistics and transport branch were theft, fraud, corruption, mark stagnation or the economy's downswing on the top 5 risks identified (Kidston and Piro 2014).

With the raising number of social-political attacks, new and strict regulations have been implemented to guarantee secure operations and to protect societies. These regulations have a permanent impact on supply chains and shape how they are managed. Guaranteeing secure flows is no longer an exclusive responsibility of governments. Supply chain actors are as well requested to comply with new regulations and to adapt their processes, culture, technologies and relationship with partners and authorities to cope in an effective and efficient way with these requirements. This trend requires the development of new concepts for supply chain management to succeed under these conditions.

As a result of these events, compliance in supply chains is receiving more attention through the years. Compliance is a nebulous concept which only has a meaning with respect to a given function and can be understood as the control and adherence to laws, guidelines and codes as well as internal corporate rules. Compliance in supply chains entails identification, assessment, consistency, accuracy, proactive attitudes, training and awareness which ensure that the company's supply chain is strictly regulatory adherent while at the same time ensuring a timely, safe, and cost-effective logistics.

Compliance reduces risks on supply chains (Cook 2011; Andrukonis 2009; Megede 2010). Thus, the approach of supply chain security in logistics companies should gain a more comprehensive and integrated focus than just protecting assets against theft. According to Ekwall (2012), the approach requires the successful integration of security throughout the entire supply chain and not only at specific stages. This in particular requires the involvement of small companies because they often represent the weakest link in the chain (Closs and McGarrell 2004).

This paper aims to discuss elements of supply chain security needed by companies to comply with regulations. It presents a tool prototype developed to tackle security requirements and suggests directions for future research in this field. The paper is organized as follows. First, supply chain security is analyzed and its relationship to risk and compliance will be explained based on a literature study. Second, a tool prototype will be presented which has been developed after studying the cases of a global industrial goods producer, a carrier association, a packing and transportation service provider, and a logistics service provider. Finally, key findings are summarized and reflections on future related research will be outlined.

2 Literature Review

The literature in Supply chain security (SCS) has flourished recently in the academic community and reveals a need for applied research. A number of contributions have been published dealing with themes such as resilience in supply chains, disaster management, compliance, quality and performance improvement and collaboration. Nevertheless, the literature in SCS is predominately normative with little quantitative research, according to an analysis of 32 papers written between 2001 and 2008 (Williams et al. 2008). As compliance is a part of any security initiative, the study began with a literature review on contributions with a conceptualization of SCS, identification of its components, formulation of solutions and analysis of their effects on supply chain performance. A variety of terms are used in any SCS discussion. At a first stage, we define security, risk, resilience, compliance, and explain their differences as well as their convergences based on the literature.

Given the recent increase on the number of approaches aiming to achieve resilient and compliant supply chains, the terms risk and security have been commonly misunderstood as being synonyms. It could be mistaken that threats to security are risks and thus, security is the same as risk. However, the literature analyzed reveals that both terms entail different concepts. Several definitions have been found in our investigations. In the following we discuss the definitions and related terms which are the basis for this research.

2.1 Security Terminology

Zsidisin (2005) defines S*upply chain risk* as "the extent to which supply chain outcomes are variable or are susceptible to disruption, and thereby lead to detrimental effects on the firm". With this definition, a risk implies the possibility of negative consequences for business. Closs and McGarrell (2004) define *Supply chain security* as "the application of policies, procedures, and technology to protect supply chain assets from theft or damage and to prevent the introduction of unauthorized contraband or people into the supply chain" which implies actions to protect outcomes and resources. Shifting from a theoretical definition, some authors have formulated approaches that describe how risk should be treated and security in supply chains should be implemented. Jüttner (2005) describes companies that identify and manage the risks in their supply chains through a coordinated approach among all supply chain members with the objective to reduce the vulnerability of the entire supply chain and not only of their own operating range (*Supply chain risk management*). Hintsa et al. (2009) point out that companies guaranteeing security in their supply chains also enhance and embed the traditional security measures into a holistic approach that integrates supply chain local and offshore actors, especially when operating on international markets (*Supply chain security management*).

An extended notion brings risk and security under one concept and suggests an approach for *Total security management* (TSM). For Ritter et al. (2007), companies that practice TSM in a comprehensive way implement best practices of risk management and security management along the entire value chain. Hence, they evaluate suppliers, distribution channels as well as internal policies and procedures on their preparation for disruptive events such as political upheaval, natural disasters, and significant accidents. This approach, however, only entails the participation of economical agents such as firms, suppliers and carriers.

Ten years ago, security was seen as the responsibility of governments mainly. Nevertheless, this obligation has been placed on organizations as well, to protect their value chains (Williams et al. 2008). The integration of firms and governments as administrators for security is defended by Autry and Bobbitt (2008) when proposing a *Supply chain security orientation* (SCSO) that foster the propensity of a firm to "partner, plan, adapt, collaborate, and communicate, both internally and with external trading partners and governmental entities, towards the durable goals of strategically preventing and responding to potential security breaches threatening supply chain assets and the minimization of risks that threaten the performance and/or continuity of supply chain operations." With such a comprehensive approach, firms can improve the ability of their supply chains to withstand and recover from an incident, in other words building up their supply chain resilience, and their performance through the measure and comparison of the achieved security levels (Yang and Wei 2013). Parallel, they must be responsible and compliant with regulations. This emphasizes the need to enhance supply chain management which after globalization focuses mostly on efficiency towards resilient and secure supply chains that comply with legal, social and environmental requirements.

2.2 Related Work

Risk and security in supply chains have been studied in several ways, e.g. by exploratory and conceptual research which identifies causes and drives of this orientation, identifies outcomes, state of the art, conceptualized concepts and formulate future research themes. Additionally, one interdisciplinary study mixing criminology and logistics and one empirical research deploying a multiple regression analysis to identify crucial factors on security performance have been identified. It should be pointed out that no contribution was found on the development of tools to support supply chain security. The contributions to the body of knowledge of these authors will be presented in the following paragraphs.

One of the pioneer contributions to security in supply chains is Sheffi (2001). He addresses the industrial sector to better prepare for the future challenges coming as a result of the aftermath of attacks, and to operate under heightened security. Instruments such as core suppliers programs, dual response manufacturing strategies and strategic emergency stocks measures are presented. Finally, it is argued that firms aiming to overcome interrupts on their supply chains will need to find a trade-off between the following factors: repeatability versus unpredictability; lowest bidder versus known supplier; centralization versus dispersion; managing risk versus delivering value; collaboration versus secrecy; redundancy versus efficiency; government cooperation versus direct shareholder value.

Some authors argue that companies should move from a "four walls" security perspective to a broader one (Closs and McGarrell 2004). The report of the IBM Center for The Business of Government adapts the model of Helferich and Cook (2003) and outlines the supply chain security challenge as the effective management of: product velocity, reduction of supply chain variability, provision of customers with the best value for their money, management of supply chain vulnerability and achievement of appropriate supply chain visibility.

Based on these challenges, firms may adopt one of four perspectives (Williams et al. 2008). The first one is an intra-organizational perspective, in which firms adopts SCS between internal organizational functions. The second one concerns an inter-organizational perspective in which other supply chain members, public entities and even competitors stand by for achieving security. The third view is a combination of intra- and inter-organizational perspectives and finally (fourth) the firm may opt to ignore the threats. In this last perspective the firm assumes that the company will not be affected by any threat and in the case there is a threat all companies will be equally affected.

Through interviews with 31 logisticians, supply chain managers and executives working in several countries, Autry and Bobbitt (2008) define a typology of SCSO in 4 dimensions. In the first dimension, firms prepare for and plan security through reactive measures such as risk sharing and insurance. In the second dimension, firms maintain continuous and transparent information exchange channels with suppliers and customers to identify address and jointly combat potential security breaches. In the third dimension, the company adapts its procedures, facilities and systems to increase security. Finally, in the fourth dimension, firms integrate up-to-date technology such as software to guard supply chain processes.

The proper definition of Supply chain security management (SCSM) was sharpened through consecutive papers. In this direction, Hintsa et al. (2009) contribute to its further conceptualization. They elaborate a list of programs, regulations and standards that have emerged since 2001 like the Free and Secure Trade (FAST), the Container Security Initiative (CSI), 24-hour Rule, and the Transported Asset Protection Association (TAPA) standards. According to the authors, these security initiatives can be classified based on the type of originating actor (e.g. border police, transportation authorities, etc.); the transport mode (e.g. sea, air, etc.); the enforceability (e.g. mandatory or voluntary) and the main goal (e.g. reduction of vulnerability, enhancement of customs security control, etc.). Each new measure aims to increase security for cargo, facility, information, human resources and company management systems with possibly overlapping effects between these dimensions.

Although all these regulations demand more responsibility from companies, allocating resources to build more secure supply chains is not easy as the costs are often difficult to justify. It is hard to have a concrete indication that an organization will be target or victim of a threat; it is difficult as well to quantify the return on investment of procedures and technologies for security. Further, a clear link between security and performance is often not obvious (Williams et al. 2008).

To address this last question, two studies analyze how security affects the performance of supply chains. In a case study analysis Hintsa and Hameri (2009) discover that implementing supply chain security initiatives has an immediate effect on the increase of lead times, personal costs and technology investment. On the other hand, these initiatives may motivate the improvement of efficiency, and some companies even believe that these regulations can provide firms with new revenues thanks to new security procedures that may be sold as services. Yang and Wei (2013) perform a multiple regression analysis to examine the effects of security management on security performance. They find that partner relationships have a significant effect on safety performance whereas partner relationship management has a significant positive effect on customs clearance performance.

An interdisciplinary study of Ekwall (2012) uses a routine activity theory from criminology to explain the interaction between the transport network and potential perpetrators and based on this interaction formulates suggestions to prevent crime in supply chains. His contribution focuses on rational risk and the author argues that initiatives to make supply chains more secure should tackle the three elements that are the foundations of crime: motivation of the perpetrator, reward of the crime, and lack of surveillance.

3 Challenges in Supply Chain Security

Based on the literature, the vulnerability on supply chains can be reduced when increasing the visibility throughout the entire chain. This will require the verification of partners as well as frequent communication. The integration of up-to-date technology can guarantee an accurate and efficient information management, an appropriate repeatability of checks while increasing the velocity of processing.

This study aims, through applied research, to develop market-oriented technology which enhances the visibility throughout supply chains and makes them more secure by supporting effective and efficient compliance to regulations. To achieve this purpose, a tool prototype has been developed after studying four cases that were sought to obtain a full picture of the security challenges at different stages of a supply chain. The companies analyzed in these four cases represent different type of actors of the value chain: a producer of industrial goods, a carrier association, and two logistics services providers specialized in packaging, labeling, assembling, storage, and inventory management. The analysis, design and development of this tool have been carried out in the frame of the project *Organisational Innovations with Good Governance in Logistics Networks*.

Among the interviewees were managers for international logistics, executives, the president of a transport association, the head for quality and logistics and managers in logistics projects that work on the planning, organization and control of movements of products and watch over corresponding costs and legal conditions. Interviews were conducted with representatives of these four companies to identify which challenges do they confront to execute supply chains that comply with regulations and to explore which technology development can support SCS in these firms. After a first interview took place, open-question questionnaires were sent to the interviewees. Two of the four companies supplied documentation describing internal processes. The insights gathered through interviews and the questionnaires were validated with the interviewees.

By coding the text of the interviews, problems in the companies' supply chains, named "problematic points", where extracted, analyzed and formulated as challenges. As a result of this exercise, the main challenges in the planning, organization, execution and control of international supply chains were identified and systematically classified into: challenges in compliance, challenges in external processes, challenges in internal processes and challenges in information (Robles et al. 2013). For the purpose of the development of the tool, the challenges addressing compliance issues were selected as the inputs for this study.

Next, a tool for compliance has been developed based on the observations. The development of the tool and performance of the programming results were validated in working sessions and finally a study of success evaluation (market potential) for the tool was conducted by a consultancy firm. In the following there is a brief description of the four case companies that supported the tool design:

- A producer of industrial goods with a global production and distribution network represented in four continents managing standard and customized product solutions. The company operates since more than 25 years and holds several quality and process standard certificates.

- A transport association operating since more than 10 years with more than 60 small carrier firms as members which execute transports at a European level. The association offers advisory services and represents its members on the public and political scene.
- A packing and transportation service provider from small to oversized or hazardous freight with facilities in Germany, Belgium and China. Operations began more than 45 years ago and nowadays the company is certified with several standards.
- A logistics service provider specialized in the automotive components industry; offering assembling, packing, labeling, storage, return handling, disposal, and product quality control delivering goods worldwide.

3.1 Supply Chain Partner Check

With this exercise a number of challenges were collected that have been addressed by these partners in the theme compliance. From the interviews we found that, as companies extend their supply chains to an international arena, they are increasingly demanded to organize and control their processes to guarantee their compliance with laws, regulations and standards. For example, the task of identification, working and collection of custom documents will become more important. This task will be more relevant as missing, incomplete or incorrect custom documents may lead to delays in the transportation of goods. Another example of challenges in compliance is the correct classification of products in an international consignment according to the Harmonized Commodity Description and Coding System (HS). An incorrect HS code assignment may lead to penalties imposed by the custom authorities or the payment of higher taxes and therefore to a disadvantageous cost increment of the cleared consignment. Further, the plethora of laws, regulations and standards across countries for the movement of goods imposes a very hard workload to any company with international supply chains. Besides the challenges mentioned above, the most sensitive one provided to be the check of supply chain partners. In accordance with the strategy of the European Union and the United Nations adopted on 2005 through the document 14469/4/05, new regulations were dictated to avoid that criminal persons or groups use transport networks to sponsor and/or finance illegal actions. Thus, a group of lists, named *sanction lists,* were developed to check the persons involved in any good or financial movement.

Before a company starts the goods transportation or the financial flow, it is important to verify if some action breaks a sanction rule or not. This is the procedure to prove whether the supply chain partner does not belong to criminal groups threatening security, or has financial problems or debts. Every company is responsible for the control of their transactions with other companies, investors, customers or suppliers. When organizing the transport, if the company identifies e.g. that a receiver is "sanctioned", the organization has the obligation to freeze the assets or resources and to inform the authorities.

The difficulty to comply with this partner check is increased by the fact that there are several sanction lists from different banks, organizations and governments that in addition may change even three times per week. Missing a requirement by an inaccurate or not timely check may result in the company losing the freight, paying fines and penalties or even workers being charged with prison sentences.[1]

3.2 Tool Development

In order to provide an application that supports logistics companies to check partners, a software tool has been developed with an advanced fuzzy matching for the mathematical process. The result was named "Sanction-Check List Tool" and represents a first exercise to provide applications that are accessible by small and medium-sized enterprises (SMEs) as well. It is expected that through automation and high performance, such a tool may have an important economic impact on trade and transport facilitation by simplifying procedures and making them more secure for compliant and trusted traders.

The tool should be: (1) attractive for the top management by guaranteeing low investment costs; (2) acceptable by the employees thanks to its user-friendly operation; and (3) accessible for all channel partners by guaranteeing a simple implementation over different operating systems (OS). The first two conditions at the top management and employee's level facilitate the promotion of a security culture within the entire company (intra-organizational). The third condition at the channel partner's level supports the collaboration and wide-spread cover of security measures along the entire supply chain (inter-organizational).

An important input for the design of the tool was the point of view of decision-makers and users in a supply chain named by Autry and Bobbit (2008). At the company's intern level are top management and employees and at the company's extern level are governments (political and legal actors) and supply chain partners (covering degree of security initiatives and sharing risks) representing this decision makers and users.

For the tool, an advanced mathematical process using fuzzy matching was developed which helps to determine the similarities between different data sets when the possible outcome is not only true or false (normal computational logic) or 100 % certain. This should allow for the accurate and fast check of partners along the supply chain. Every time, the data is examined by fuzzy matching, the algorithm returns a probability score which indicates how accurate the data set is.

[1]According to regulation (EG) Nr. 2580/2001.

There are different types of algorithms that can solve this. For the election of the most appropriate one, several were tested such as: Exact Matching, PhonetEx, N-gram or Q-gram-based, Dice's Coefficient, Jaro Distance Measure, Levenshtein Distance Measure, Needleman-Wunsch, Smith-Waterman-Gotoh, MDKeyboard, Longest Common Substring and Frequency.

The best results were achieved with the Levenshtein algorithm in which the similarity of two strings is compared by taking into account the number of character mistakes (incorrect, inserted and deleted characters). It specializes in keyboard typing errors. Thus, this algorithm is useful when possible typing errors are frequent. A confidence level pull-down menu shows different percentages to select how close the compared strings to the reference one should be.

To calculate the Levenshtein distance we assume all operations (insert, delete, and substitute) have a cost or penalty of "1" and a matrix is used to check the input against the lists. The Levenshtein distance is the last number of the last row. The algorithm ends here. However, its result is not easy to interpret because the number alone shows the penalty in letters between the two strings, without taking into account the length of the strings. For this reason it is necessary to divide the distance by the number of letters of the longest string: (maximal length − mistakes)/ maximal length.

The result is the Levenshtein distance as a percentage and with this it is possible to determine the strength of the match between the reference data set and the checked partner. After checking, the user has valuable information to confirm or reject a chain actor as well as a protocol to demonstrate its checks.

3.3 Tool Evaluation

An evaluation of this and other solutions developed in the project *Organisational Innovations with Good Governance in Logistics Networks* was conducted by an independent consultancy firm, see Irowec et al. (2013). For the evaluation, customers of the consultancy firm working on the branches of logistic service providers and producers of metal and electrical parts were considered. These companies did not support the tool design. Irowec et al. (2013) selected these two branches for the evaluation as:

- The logistics service provider branch assures the link between the partners acting in a supply chain. The hard competition in this branch demands differentiation strategies in order to survive on the market. Based on this, it is of utmost importance for logistics service providers to ensure the transparency of logistics processes as well as the establishment of a reliable partner network that complement their own service portfolio, enhancing geographical coverage with optimal costs structures.
- The producers of metal and electrical parts are Germany's biggest industrial branch, highly dependent on the export of their products. Punctual and fast

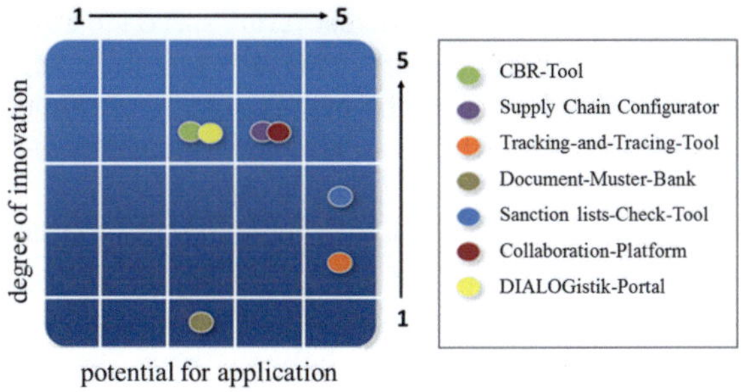

Fig. 1 Graphical representation of evaluation results (adapted from Irowec et al. 2013)

delivery of special materials defines the success chances of companies operating in this branch. This demands efficient, agile and reliable value chains. Thus, the mastery of transparent management of international networks for the procurement and distribution will be a decision factor for the branch success.

The evaluation was based on two factors: the *potential for application* which each potential user expresses from their customer point of view, and the *degree of innovation* which can be deduced from the competitive advantage of the tool. The resulting appreciation of these two factors will determine the *market potential* for each solution. A high appreciation of both *potential for application* and *degree of innovation* can be interpreted as a high *market potential* for the solution.

The evaluation confirms our conclusion from the interviews that the quality of the tool in the context of the operational use depends on the actualization of the sanction lists. The evaluation reveals that "one of the important advantages of the underlying tool is to facilitate its integration in the internal company guidelines". A graphical result of the evaluation is displayed in Fig. 1.

The potential for application of the tool presented in this chapter, i.e., "Sanction lists-Check-Tool" was assessed as "5" and the degree of innovation as "3" where "1" is the lowest grade and "5" the highest. It was noted for both logistics service providers with low and higher structuration of operations, respectively, the toll appears to be highly useful. The company producers of metal and electrical parts assessed the tool as a valuable support for meeting legal requirements when exporting.

4 Conclusions and Future Research

As the interest for more secure supply chains is increasing, companies are confronted with a number of social, ecological and legal requirements which explode the complexity of the management of supply chains, especially international ones. The development of this prototype aims to enhance the attention on applied research to develop methods and tools that support companies by the planning, organization, execution and control of value chains to be secure and in compliance with regulations. Cost, efficiency, and speed are no longer the unique design criteria for supply chain managers, therefore their tools need to encompass a new dimension. Tools like the prototype developed here should be an option for small and medium-sized enterprises to also act responsibly without confronting high costs for these solutions.

Since companies are not willing to invest in expensive technology for security, a web-tool might be licensed as Software as a Service (SaaS). SaaS is a software licensing and delivery model in which the software is accessed by the user using a customer account via a web browser "on-demand".

The functionality of the tool might be enhanced in order to deal with incomplete and uncertain context data. For any evaluation of risks, the lack of information is a major challenge. The incorporation of qualitative information such as management decisions, experiences and reviews might be an additional option provided it can be translated into measurable information with the implementation of fuzzy logic, neural networks, system dynamics, and Bayesian networks. This approach has being used by financial and insurance companies to measure and manage operational risks such as risk capital allocation and insurance fraud risk. As an extension to the work developed in this study Bayesian belief functions linked with fuzzy logic can be used to identify proper mitigation strategies. For this purpose the work of Frez et al. (2014) is a good starting reference. They develop a methodology for using incomplete/uncertain information to answer questions which include uncertainty. By combining fuzziness with a belief function, it is possible to obtain good results with less data. Such an extension might allow the tool to evaluate complex questions such as "What is the probability of finding suspicious consignees in the city area A between periods T1 and T2?"

Similar to this prototype, new tools and procedures should be developed to make processes related to security more efficient, user-friendly and wide-spread covering the complete value chain. This tool helps to quantify the supply chain partner's reliability, but may be extended in future applied research to assess other security risks as well. In this sense, another extension to the work presented in this study may be the conception and development of simulation-based decision making methods for logistics systems that supports the modeling and analysis of uncertain circumstances such as natural disasters, accidents, political measurements (sanctions), social incidents (strikes) or technological failures (e.g. due to cyber-attacks). Simulation can better represent the stochastic nature of supply chains and risk uncertainties due to its modeling flexibility. In this case, a lack of information could be managed with hybrid models that utilize both qualitative and quantitative information with fuzzy logic, decision tree analysis or cluster analysis. Future work may address this matter by applying agent-based modeling and simulation for supply chain risk management.

References

Andrukonis T (2009) Elements of an effective export compliance program. In: Bureau of industry and security U.S. department of commerce. Available via BIS. http://www.bis.doc.gov/index.php/forms-documents/doc_view/244-compliance-pdf. Accessed 15 Jan 2014

Autry CW, Bobbitt LM (2008) Supply chain security orientation: conceptual development and a proposed framework. Int J Logistics Manage 19(1):42–64. doi:10.1108/09574090810872596

Closs DJ, McGarrell EF (2004) Enhancing security throughout the supply chain. Special report to the IBM Center for the Business of Government, Washington, DC

Cook T (2011) Compliance in today's global supply chain. Taylor & Francis Group, Florida

Ekwall D (2012) Supply chain security—threats and solutions. In: Banaitiene N (ed) Risk management—current issues and challenges, InTech, pp 157–175. doi:10.5772/48365

Frez J, Baloian N, Zurita G et al (2014) Dealing with incomplete and uncertain context data in geographic information systems. In: Proceedings of the 2014 IEEE 18th international conference on computer supported cooperative work in design, National Tsing Hua University, Hsinchu, 21–23 May 2014

Helferich OK, Cook RL (2003) Securing the supply chain. Council of Logistics Management, Chicago

Hintsa J, Gutierrez X, Wieser P et al (2009) Supply chain security management: an overview. Int J Logistics Syst Manage 3–4(2009):344–355. doi:10.1504/IJLSM.2009.022501

Hintsa J, Hameri AP (2009) Security programs as part of efficient supply chain management. Supply Chain Forum Int J 10(2):26–37

Irowec M, Rittscher J, Dittrich M et al (2013) Report evaluation OrGoLo. To be published

Jüttner U (2005) Supply chain risk management: understanding the business requirements from a practitioner perspective. Int J Logistics Manage 16(1):120–141. doi:10.1108/09574090510617385

Kidston H, Piro K (2014) Allianz risk barometer: Geschäftsrisiken 2014. In: Allianz risk barometer 2014: Zunehmende Verzahnung von Risiken ist Herausforderung für Unternehmen. Allianz. Available via Allianz. http://www.agcs.allianz.com/assets/PDFs/Reports/Allianz-Risk-Barometer-2014_DE.pdf. Accessed 15 Jan 2014

Megede E (2010) Aufgaben von Compliance – ein Überblick. Presentation at the Deutsch-Nordischen Juristenvereinigung e.V., Augsburg, 1 Oct 2010

Ritter LJ, Barrett M, Wilson R (2007) Securing global transportation networks: a total security management approach. McGraw-Hill, New York

Robles M, Wei F, Noche B (2013) Challenges in the planning, organization, execution and control of international supply chains. In: Clausen U, Ten Hompel M, Klumpp M (eds) Efficiency and logistics—lecture notes in logistics, pp 245–252. doi:10.1007/978-3-642-32838-1_26

Sheffi Y (2001) Supply chain management under the threat of international terrorism. Int J Logistics Manage 12(2):1–11. doi:10.1108/09574090110806262

Williams Z, Lueg JE, LeMay SA (2008) Supply chain security: an overview and research agenda. Int J Logistics Manage 19(2):254–281. doi:10.1108/09574090810895988

Yang CC, Wei HH (2013) The effect of supply chain security management on security performance in container shipping operations. Supply Chain Manage Int J 18(1):74–85. doi:10.1108/13598541311293195

Zsidisin GA, Melnyk SA, Ragatz GL (2005) An institutional theory perspective of business continuity planning for purchasing and supply management. Int J Prod Res 43(16):3401–3420. doi:10.1080/00207540500095613

Cross-Process Production Control by Camera-Based Quality Management Inside a Logistic Assistance System

Markus Zajac and Christian Schwede

Abstract A holistic integration of quality measurement into process control has a huge potential especially for SMEs in the areas of production and logistics. Since poor results of quality measurements can influence the whole order-to-delivery process the quality data should be included as soon as possible in the decision process as part of a superior order control. This paper presents results of the Supply Chain Execution project that developed a low-cost camera-based quality measurement system and integrated it into a Logistic Assistant System that allows for simulation-based process control considering data from the whole supply chain. This lean low-cost approach, which does not depend on sophisticated IT-infrastructures and management systems, yields results that are especially interesting for SMEs.

Keywords Quality management · Production control · Logistic assistance system · Camera-based computer-aided quality assurance

1 Importance of Quality in Production Processes

Globalization and active market dynamics have noticeable impacts especially on small and medium-sized enterprises (SMEs). SMEs form the majority of companies in Europe and thus play a central role in contributing to the overall economic success (Winham 2012). The impacts emerge essentially from shorter product life cycles, the demands for cost-effective products, from a high product variety and increasing competitive pressure. Besides implementing optimal production processes, SMEs have to fulfill associated requirements related to ergonomics and safety at work (Bornewasser and Zülch 2013)—last but not least to comply with

M. Zajac (✉) · C. Schwede
Fraunhofer Institute for Material Flow and Logistics, Joseph-von-Fraunhofer-Str. 2-4, 44227 Dortmund, Germany
e-mail: markus.zajac@iml.fraunhofer.de

H. Zijm et al. (eds.), *Logistics and Supply Chain Innovation*,
Lecture Notes in Logistics, DOI 10.1007/978-3-319-22288-2_21

current norms and laws. A quality management system to ensure the quality of work and product is indispensable. As small companies often develop specialized and customized solutions, the quality of offered products and services determines their competitive position (Mohanty 2008).

A non-existent or insufficient quality management system may lead to various problems and consequently to additional internal and external costs (Dobb 2004). Costs at the production location (internal costs) originate for example as a result of defective goods, rework or product downgrading (Oakland 2014). External costs are customer related and originate for example in returns and recalls, liability claims and special transports (Dobb 2004; Oakland 2014). Hence, a consequent quality management system supports enterprises to fulfill the required product quality as well as legal requirements regarding production processes. The approach presented in this contribution also supports in particular ecological aspects of the supply chain like reducing the number of defective goods, superfluous transports and packaging material. For the establishment of a quality management system in enterprises, IT support is indispensable.

The structure of this contribution is as follows. Section 2 discusses general aspects of quality management systems and computer aided quality assurance. In Sect. 3, we elaborate on Logistic Assistance Systems (LAS) and the utilization of quality management inside LAS. Section 4 describes a developed Logistic Assistance System for production control as part of a project supported by the German Federal Ministry of Education and Research (BMBF). We end with conclusions in Sect. 5.

2 Quality Management and Computer-Aided Quality Assurance

The core task of quality management (QM) consists of the implementation of quality elements or aspects (Benes and Groh 2012). These elements include quality planning, quality control, quality assurance and quality improvement (Brüggemann and Bremer 2012). Table 1 provides a short overview of quality elements.

Since quality control is based on monitored process results and thus plays an essential role in the context of cross-process production control, it will be the focus of this paper. The so-called immediate quality control intervenes in the ongoing production and assembly processes (Benes and Groh 2012). The term production control is closely related to quality control. Production control is integrated in the quality assurance process and can be organized in control loops of varying scope (Rosenberger et al. 2010). Small control loops are used for controlling individual production processes in which current values are compared with nominal values. In case of deviations beyond the tolerance limits, corrective actions can be performed instantaneously. To achieve a cross-process production control (e.g. in a multi-stage production system), quality control has to be applied in all corresponding

Table 1 Quality elements

Quality element	Function
Quality planning	Within quality planning, the quality goals, necessary execution processes as well as corresponding resources are specified
Quality control	Primary tasks are monitoring and if necessary correction of an item to comply with quality requirements. Here the results of quality tests are compared with the goals of the quality planning. In case of deviations, corrective actions are performed
Quality assurance	Quality assurance focuses on providing confidence that quality requirements are fulfilled. Quality planning and quality control are subtasks of quality assurance
Quality improvement	Summarizes all measures to increase effectiveness and efficiency in processes and activities dealing with quality aspects

production processes or activities. In this context, statistical process control (see below) is one method to make the production processes stable by integrating it into an overall quality improvement program of the enterprises (Montgomery 2008).

In case of observed deviations based on a target/actual comparison, appropriate measures have to be taken to ensure immediate corrections by a responsible employee (Benes and Groh 2012). In order to gather actual values, the quality control procedures make use of machine-supported control circuits. Here, samples of the production process are regularly gathered and the measured values evaluated. There are two types of process control (Benes and Groh 2012):

- Statistical process control (SPC): In the course of the production process, samples are taken at certain times. This process control intervenes at an early stage in the production process to counteract negative trends by applying appropriate corrective actions.
- Continuous process control (CPC): With this technique, a 100 % check takes place, resulting in a direct feedback and hence continuous machine control. This type of process control is used, when even the smallest analysis intervals are inappropriate, for example during particle measurement (Dietrich and Wilczek 2002).

To provide automatic acquisition of current quality data within the scope of SPC or CPC for quality tests, Computer-Aided Quality Assurance (CAQ) as part of quality control is applied. CAQ is an IT system for storing and preparing quality relevant data (Benes and Groh 2012). Another important function of CAQ systems consists of test planning as well as providing quality planning and quality test data. Quality data collected along the supply chains can be distinguished in different categories like quality data of suppliers, material testing for incoming goods and quality measurement in production (Brüggemann and Bremer 2012). In production processes, quality relevant machine-data, like rotation, pressure or temperature data can be collected by means of sensors and metrics (Ansari-Ch. et al. 2011). SPC however may become increasingly complex, because of the huge number of sensors and data sources. In such cases, multivariate statistical process control (MSPC) may provide a solution. In MSPC, variables are aggregated in order to master the increasing complexity, as exemplified in the wood pellet industry (Lestander et al. 2012).

Besides collecting machine-data, digital image capturing by camera systems as well as digital image processing are already in use. Camera based systems are replacing the manual control by experts step by step. Megahed and Camelio (2012) show the feasibility of fault detection in a manufacturing environment by using image recognition techniques. Dobrzanski et al. (2007) demonstrate the identification and classification of possible defects occurring in castings of automotive Al–Si–Cu components. In the timber industry, image processing applications for checking the wood quality are widely used (Molder and Martens 2011). To predict the breaking strength of pinewood boards, a real-time camera-based system in conjunction with digital image processing for calculating local grain direction was evaluated positively (Hietaniemi et al. 2014).

3 Quality Management Within Logistic Assistance Systems

To ensure quality requirements within production processes, quality control tasks are performed periodically. In particular, target values of processes are compared against measured actual values (cf. Sect. 2). In case of deviations of actual values from target values, it is necessary to select an appropriate action to ensure immediate correction. Depending on the size of the deviation, a range of possible measures to be selected is available. The selection of production control measures may be quite difficult because of both additional information as well as an impact analysis of the chosen measure, such as adherence to schedules, is needed. For instance, one measure may consist in swapping a material of a current job with a material of another job to complete a customer order in time. For this purpose inventories and scheduled orders have to be taken into account when making the decision. It is therefore necessary, to support the human operator in the choice of a measure by using appropriate IT tools.

To support human planers in decision-making, so-called Logistic Assistance Systems (LAS) have been deployed. These IT systems offer functions such as information processing, information transparency and decision support (Bockholt et al. 2011). LAS integrate all process-relevant data into a logistic process model (transparency) and present this data to decision makers and experts (single-point-of-truth). This data may include inventory levels, customer orders as well as transportation data from the Supply Chain and production programs. As part of information processing, data validation and threshold calculations on integrated data can be performed and subsequently presented. Another major functionality is the support of the human decision process (Kuhn et al. 2008). To this end, effects of decision variants are pre-calculated by using discrete event simulation. Based on the results of the different scenarios the human user can then take the best decision for the current situation. This method combines experience of human decision makers in the selection of basic scenarios with powerful evaluation methods of these scenarios in complex environments. LAS already found their way into the industry in various application fields (Deiseroth et al. 2013; Hegmanns et al. 2014; Müller-Ohe et al. 2014). For example, within collaborative planning in the automotive industry, a LAS is being used to control engine parts and assembled engines in the supply chain (Bockholt et al. 2011).

In contrast to LAS, CAQ-systems are software-based realizations of QM (Brüggemann and Bremer 2012) and provide support in process documentation, in process-integrated quality management, in data-monitoring and in the logging of test results. Thus the focus of CAQ-systems is on carrying out measured value collection as well as data transfer and storage. Yet the required decision support based upon scenario- and process-based measures evaluation is not offered. Hence, a promising approach is the combination of LAS with CAQ systems to support decision makers by offering assessed available options based on collected quality data by a CAQ-system. Since quality data is gathered along the supply chain in large volumes (cf. Sect. 2) and may have an influence on the entire order to delivery process, it is necessary to prepare that data for use in decision making (Brüggemann and Bremer 2012) in LAS. In terms of transparency, quality data have to be transferred into a logistic model first and subsequently included in decision making procedures. For the transfer into a logistic model, LAS can either integrate modules of CAQ systems or communicate with these systems through service interfaces to obtain gathered quality data or the results of material testing. Once the data is transferred, it can be processed to provide scenario-based options to the decision maker. Because quality data is linked to different locations, it can be used to control the material flow in various processes in the supply chain.

4 Supply Chain Execution Project

This contribution describes the results of the research project Supply Chain Execution (SCE). Within this project, that was supported by the German Federal Ministry of Education and Research (BMBF) and is a part of the EffizienzCluster LogistikRuhr research cluster, prototypes of LAS for wood quality measurement based cross-process control have been developed for SME application partners within the furniture industry. After measuring poor quality of certain planks in this industry, a couple of actions are possible depending on the results of the measurement. The focus of the LAS is in supporting the human planner in the selection of an appropriate production control measure.

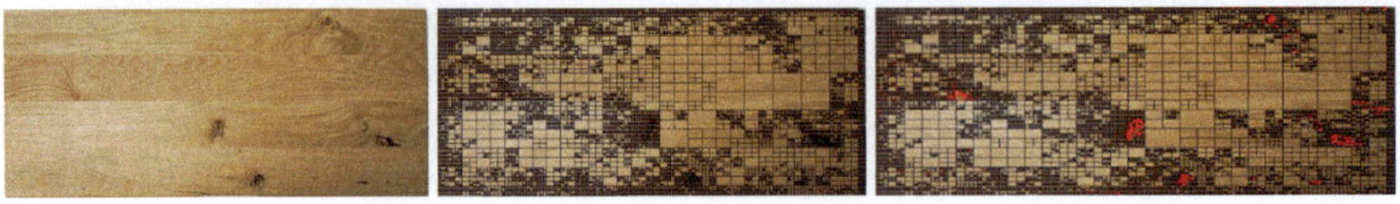

Fig. 1 Image processing in the three steps: unprocessed, segmented and recognized

4.1 Methodical Approach

To enable a cross-process production control, so-called control points are identified by analyzing the supply chains of the furniture manufacturers. At these control points quality control takes place and in case of deviation, appropriate measures are to be taken. The identified control points have the biggest influence on the production control and are typically anchored to production as well as assembly processes (Yüzgülec et al. 2013).

To perform quality measurements to obtain the quality test data at identified control points a camera based CAQ system was developed. In timber production, edge-glued planks have to be tested in various processes of the supply chain. In accordance to SPC (cf. Sect. 2), quality data is collected at pre-determined intervals. To obtain the quality score, a plank is recorded by a camera first. Subsequently, the quality score is calculated by a digital image processing algorithm. Figure 1 shows the three steps of the algorithm.

The first step is to acquire the recorded plank image. By using of grayscale calculation, the image is divided into segments by comparing the surrounding area in the second step. In the third step, the algorithm examines the image on knotholes and calculates the quality score value. In doing so, this CAQ system is able to capture two types of information:

- The quality score of a plank: This score is determined algorithmically, as described above.
- The unique ID of a plank: Besides the quality score, a unique identification number of the plank is also delivered by the system. Therefore, RFID transponders have been integrated in the production of edge-glued planks in order to allow tracing during the entire manufacturing process.

Relative to MSPC (cf. Sect. 2), all captured variables are combined to one unit (cf. the concept of Premium Service in Sect. 4.2). Finally, the gathered information must be further processed to validate the quality score and to provide decision support, if necessary. For this purpose, information is provided to the LAS.

4.2 Logistic Assistance System

The basis of the developed LAS is constituted by logistic service components (further called services), which enable a flexible and customer specific configuration

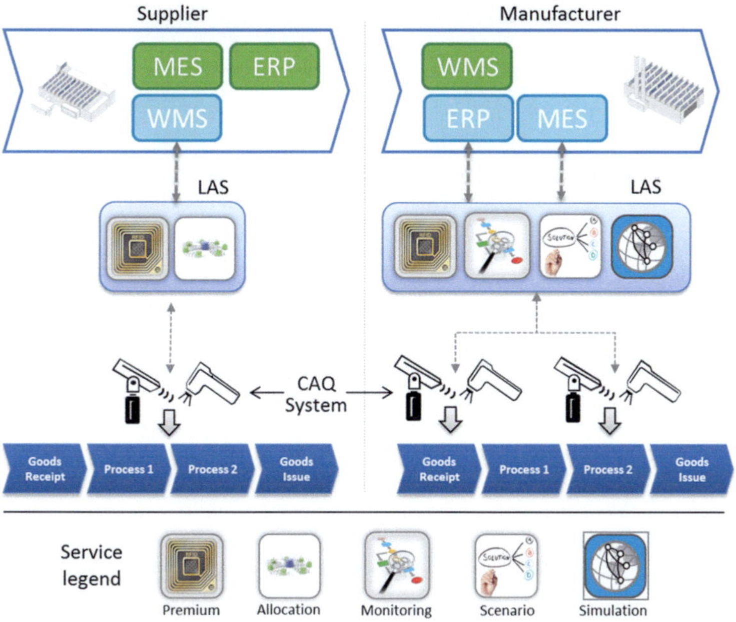

Fig. 2 Schematic figure of LAS integration

of Logistic Assistance Systems. The LAS is integrated in supply chain processes of the furniture manufacturer and imports data from both the CAQ system and various back-end systems (cf. Fig. 2). Subsequently the data are processed by the services of LAS. Besides quality data, back-end systems are requested to receive orders and allocated items (planks). As planks are equipped with an RFID transponder, they will be unambiguously assigned to an order within the order planning. The orders include additionally information about their plank target quality scores, since furniture manufacturers are producing for various customers with different quality needs. Hence, it is possible to use a plank for manufacturing a cupboard door as well as for the rear wall of a cupboard (Yüzgülec et al. 2013).

Table 2 briefly describes selected services which have been customized for the SCE project.

The following section describes the service workflow with the aim to provide production control measures in case of poor quality of a plank. The pressing of planes is one process at a furniture manufacturer which was exemplarily selected as a control point for the demonstrator. At the beginning, quality scores are regularly compared by the Monitoring Service (by requesting the quality data as well as a plank ID from the Premium Service). The goals of the Monitoring Service are compliant to the quality control tasks as described in Sect. 2. For each test, the target quality score is resolved first. For this purpose, the Monitoring Service communicates with back-end systems to resolve the order (and the deposited target quality score) by sending the plank ID. The result of the test is displayed to the

Table 2 Selected services of the LAS in SCE project

Service	Function
Premium service	These services are used to encapsulate data from multiple sources. In the SCE project, one unit describes data of one plank. It consists of captured quality scores and the unique identifier of the plank
Monitoring service	This service performs close to real-time IT monitoring of the quality information from Premium Services. The received quality information is enriched with a predefined quality score of the corresponding order to compare both quality values
Scenario service	In case of a detected quality deviation by the Monitoring Service, the Scenario Service provides a prioritized list with possible control measures
Allocation service	This service detects planks that can be used as alternatives to fulfill the current order. This service supports swapping of planks between orders as well as procurement of stored planks
Simulation service	This service encapsulates an event-discrete simulation tool. By using the proposed control measures as input for this service, possible impacts for the order status as well as other supply chain processes are simulated

decision maker. In case of insufficient quality, the Scenario Service can be called by the decision maker to show appropriate alternatives. The alternatives are presented as a prioritized list. The order of alternatives varies depending on the deviation in quality. In this context, the Allocation Service is a subordinated service, to assist in the allocation of resources (swap or procure material). The following alternatives are offered by the Scenario Service:

- Reworking on/outside the production line: These alternatives are proposed as priorities in case of minor quality deviations.
- Swap with an non-critical order: If reworking is not possible, possible orders for swap are selected by using the Allocation Service. This means that a plank of an non-critical order and of sufficient quality is replacing the disqualified plank of the current order. The dispensed plank should still have the required quality to satisfy the non-critical order.
- Allocation or new production: If a plank cannot be exchanged, or quality deviations are too high, the Allocation Service examines whether quality checked planks are stored in other inventories in the supply chain. If so, the service can attempt to allocate such planks, otherwise a new plank has to be manufactured.

Each alternative represents a production control measure (corrective measure), in case of poor quality of planks. By selecting the first or second measure by a decision maker, the process can by controlled at this control point as planned and the underlying resources are used in an optimal way.

To assess the selected option in an integrated manner, the Simulation Service can be consulted. In the demonstrator, in particular order-based production control is analyzed by consulting the Simulation Service. On the one hand, effects related to customer delivery dates are analyzed, if orders are suspended on short notice. But also future demands of the production process can be resolved taking into account the current order status. This will be the case when orders are rescheduled as the result of selecting an appropriate control measure. By linking the captured quality data of a plank to order and process data, it becomes possible to simulate effects of different control measures in case of poor quality. The effects become transparent and constitute an appropriate basis for the decision maker.

5 Conclusions

In this contribution a Logistic Assistance System for cross-process production control depending on quality data was presented for the timber industry. It integrates a camera-based CAQ-system, which acts as quality data supplier. The integration of the LAS for continuous cross-process quality management enables the consideration of all relevant supply-chain-wide process data to choose the best measure from a holistic point of view in case of poor quality. Thus the presented combination of LAS and CAQ-systems allows for a holistic quality management approach considering effects and resources in the entire supply chain, with possible impacts on transport, stock and delivery date optimization. A coherently and aggregated visualization of large amounts of quality data as well as production control measures within the LAS supports the enterprises to maintain the required product quality as well as to comply with legal requirements.

In the presented scenario the consistent application of the quality management not only leads to the avoidance of defective goods, but also in downgrading of goods to be reused in other processes. In this manner, also additional transports and packaging material can be significantly reduced or even avoided. The presented solution was implemented on low cost hardware and cameras. These cost factors are in particular relevant for SMEs.

Concerning SMEs the next necessary steps might be to connect the LAS not only with low cost quality measuring services, but also with lean quality management systems designed for SMEs, such as the Fraunhofer QUERIS (Vieweg 2014). Such systems allow for a lean administration of quality measurement data, objectives, norms, tasks and measuring devices. This integration would be the last step of a complete SME-focused holistic quality-based process control.

References

Ansari-Ch F, Dienst S, Uhr P, Fathi M (2011) Using data analysis for discovering improvement potentials in production process, Auburn, AL

Benes G, Groh PE (2012) Grundlagen des Qualitätsmanagements, 2, akt. Aufl, Fachbuchverl. Leipzig im Carl Hanser Verl, München

Bockholt F, Raabe W, Toth M (2011) Logistic assistance systems for collaborative supply chain planning. Int J Simul Process Model 6(4):297

Bornewasser M, Zülch G (eds) (2013) Arbeitszeit—Zeitarbeit. Springer, Wiesbaden
Brüggemann H, Bremer P (2012) Grundlagen Qualitätsmanagement. Springer, Wiesbaden
Deiseroth J, Klennert M, Thissen S, Schwede C, Toth M (2013) Virtual experiment fields for logistical problem solving. In: Proceedings of the 17th Cambridge international manufacturing symposium
Dietrich S, Wilczek M (2002) Kontinuierliche Prozessüberwachung—inline-Partikelmessung bei der Tieftemperatur-Mahlung. Verfahrenstechnik 36:48–49
Dobb FP (2004) ISO 9001:2000 quality registration step-by-step, 3rd edn. Elsevier, Amsterdam, Boston
Dobrzanski LA, Krupinski M, Maniara R, Sokolowski JH (2007) Computer aided method for quality control of automotive Al–Si–Cu cast components. J Achievements Mater Manuf Eng 24:151–154
Hegmanns T, Schwede C, Motta M, Henke M (2014) Model-based technologies for tool support in supply chain risk management. In: Delfmann W, Wimmer T (eds) Proceedings of the 7th international scientific symposium on logistics, pp 135–152
Hietaniemi R, López MB, Hannuksela J, Silvén O (2014) A real-time imaging system for lumber strength prediction. Forest Prod J 64(3–4):126–133
Kuhn A, Hellingrath B, Hinrichs J (2008) Logistische Assistenzsysteme. In: Software in der Logistik: Weltweit sichere supply chains. [Anforderungen, Funktionalitäten und Anbieter in den Bereichen WMS, ERP, TMS und SCM], Logistik Praxis, 1. Aufl, Huss-Verl., München, pp 20–26
Lestander TA, Holmberg C, Stenberg L, Lehtonen R (2012) Towards multivariate statistical process control in the wood pellet industry. Biomass Bioenergy 45:152–158
Megahed FM, Camelio JA (2012) Real-time fault detection in manufacturing environments using face recognition techniques. J Intell Manuf 23(3):393–408
Mohanty RP (2008) Quality management practices, 1st edn. Excel Books, New Delhi
Molder A, Martens O (2011) Image processing in the woodworking industry: challenges, solutions and platforms. Electron Electr Eng 113(7):43–46
Montgomery DC (2008) Statistical quality control: A modern introduction, 6th edn. International student version, Wiley, Hoboken
Müller-Ohe H, von der Schwede C, Toth M (2014) A logistics assistant system for customizing market oriented build-to-stock cars for dealers in the automotive industry. In: Delfmann W, Wimmer T (eds) Proceedings of the 7th international scientific symposium on logistics, pp 469–505
Oakland JS (2014) TQM: text with cases, 4th edn
Rosenberger M, Schellhorn M, Linß G, Schumann M, Werner P, Lübbecke S (2010) Quality control for springs during the production process based on image processing technologies. In: Baralt J, (ed) Proceedings of international institute of informatics and systemics, Orlando, Fla
Vieweg O (2014) QUERIS—Das Managementsystem für KMUs. Available at: http://www.queris.de/
Winham I (2012) When SMEs succeed, Europe succeeds. A view from Ian Winham, CIO, Ricoh Europe. Available at: http://www.ricoh-europe.com/about-ricoh/news/2012/When_SMEs_succeed_Europe_succeeds.aspx
Yüzgülec G, Groß S, Ciprina A, Zajac M, Langanke D (2013) Supply chain execution supported by logistics IT services. In: Clausen U, Hompel M, Klumpp M (eds) Efficiency and logistics. Lecture Notes in Logistics, Springer, Berlin, pp 217–227

Logistics Mall—A Cloud Platform for Logistics

Damian Daniluk, Maren Wolf, Oliver Wolf and Michael ten Hompel

Abstract Increasingly, in logistics value chains a shift from core logistics tasks to higher-value logistics services can be observed. Competitive capability is characterized by a rapid design and implementation of these services. The technology of cloud computing enables the provisioning of various services over the internet. Particularly for logistics users a simple and rapid delivery of software components supporting the execution of individual business processes is of high interest. This allows the quick adaption of frequent business process changes along with an accounting on pay-per-use basis. After outlining the current situation of cloud computing in the field of logistics this paper presents the Logistics Mall, an approach for a domain specific cloud platform for the trading and usage of logistics IT services and logistics processes. Finally, results of a study about the acceptance of cloud computing solutions in the logistics domain like the Logistics Mall and an outlook are presented.

Keywords Logistics Mall · Cloud computing · Cloud platform · Business Process as a Service

1 Introduction

For a common understanding of the concept of cloud computing we adopt the working definition of NIST (Mell and Grace 2009) that defines it as "a model for enabling convenient, on-demand network access to a shared pool of configurable computing resources (e.g., networks, servers, storage, applications, and services) that can be rapidly provisioned and released with minimal management effort or service provider interaction. This cloud model promotes availability and is

D. Daniluk (✉) · M. Wolf · O. Wolf · M. ten Hompel
Fraunhofer Institute for Material Flow and Logistics IML,
Joseph-von-Fraunhofer-Str. 2-4, 44227 Dortmund, Germany
e-mail: damian.daniluk@iml.fraunhofer.de

H. Zijm et al. (eds.), *Logistics and Supply Chain Innovation*,
Lecture Notes in Logistics, DOI 10.1007/978-3-319-22288-2_22

composed of five essential characteristics, three service models, and four deployment models." The basic characteristics are on-demand self-service for consumers, broad network access, resource pooling to serve multiple consumers using a multi-tenant model, rapid elasticity of resources and metering capabilities for service provision. The identified service delivery models are software-as-a-service (SaaS) where consumers use a provider's application running on a cloud infrastructure, platform-as-a-service (PaaS) where a consumer can deploy an application on a cloud infrastructure using a provider's tools and platform, and infrastructure-as-a-service (IaaS) where a consumer can run arbitrary software on a provided cloud infrastructure.

Cloud computing is now a reality. After the initial furor and cautious hesitation, the technology was able to hold ground in the market and is now widely in use. Various sectors of the industry have recognized the advantage of distributed computing power and have put it to use in their respective areas of business. The numbers are clear: The rapidly increasing budgets of the past few years give us an indication of the size of the fundamental change in technology that it has brought—cloud computing has shaped and changed the IT market permanently in the same way that a disruptive innovation would. Extremely volatile business areas with strongly fluctuating requirements such as logistics benefit greatly from the flexible distribution of resources in the cloud.

1.1 The New Standard

In many places the use of cloud computing is now viewed as a matter of course and no further discussion is needed. For a few years this topic was still considered to be a future trend and although it caused heated discussions it was almost never viewed as relevant for the policy of a company. This has clearly changed: according to the Federal Association for Information Technology, Telecommunications, and New Media (BITKOM), cloud computing has been selected as the hot topic for the information and telecommunications industry in 2013 for the fourth time (KPMG AG and BITKOM 2013). The market research company Forrester Research predicted that for 2013 almost half of all North American and European companies will have a budget for investing in private clouds and many software development managers will plan to use cloud applications.

1.2 Making the Shift to Cloud Computing

The number of companies making the shift to cloud computing has changed greatly in the past few years: only 28 % of the companies surveyed in 2011 were open minded about cloud computing and interested in it. This number rose to 35 % in 2012. But, surprisingly, the number of sceptics also increased from 38 % (2011) to 44 % (2012).

The number of those who were undecided did go down, which indicates that companies improved their information policies and their uncertainty was dwindling. All discussions about the topic obviously helped decision-makers to make up their minds: the number of those who were undecided shrunk from 33 to 20 %.

The size of a company plays a role in how open they are to adapting the technology: fifty-five percent of companies with more than 2000 employees were open to it while smaller companies were clearly more reticent (KPMG AG and BITKOM 2013).

1.3 Cloud Computing in Action

According to the estimates of the Experton Group in 2013 German companies invest 5 % of their IT expenses in the cloud (Velten and Janata 2013). Compared to 2012 this is an increase of investments of about 52 %. The investments for cloud services, cloud integration and cloud consulting services and cloud technology in the business cloud (B2B) have a sum of 4.6 billion Euros in 2013. For 2014 a growth of 50 % is expected. The cloud investment volume would then be at about 6.9 billion Euros (Bayer 2013).

The strongest increase is expected for investments in cloud services (Mahmood and Hill 2011) like software-as-a-Service (SaaS), Platform-as-a-Service (PaaS) and Infrastructure-as-a-Service (IaaS). The segments cloud technology and integration as well as consulting will also grow at a constant rate, but they will evolve more restrained in comparison to the mentioned cloud services. Until 2017 investments on enterprise cloud services are expected to grow by an average of 50 % (Hackmann 2013).

Some of the reasons for the positive attitudes of many companies towards cloud computing are lower IT administration costs, shorter implementation times for new applications and solutions, rapid scalability of IT services, improved mobility and geographically distributed access to IT resources, lower IT costs, increased organizational flexibility, improved performance and availability, and a higher capacity for innovation. The opportunity to tap into new customer groups and save costs (hardware and personnel) also plays an important role. The main concerns are no longer with data protection but with the loss of data. Other risks that have been identified are a diminishing IT know-how and an unclear legal situation.

2 Logistics Mall

2.1 Vision

Common characteristics of all logistics business processes are individualisation and dynamically changing requirements of the customers' business. A process has to be adapted to every new or changed requirement. Consequently, the underlying IT infrastructure is also subject to these frequently changing requirements.

Today a typical logistics process runs in a heterogeneous IT environment. Such systems are mostly standard software like SAP, Oracle, or more specialized logistics software systems like warehouse management systems (WMS) or production planning systems (PPS). All these systems do not entirely meet the requirements of logistics customers regarding short contract duration, pay-per-use accounting and the provisioning of individual IT services. The majority of the features of standard software is rarely used or not used at all, but has to be paid for, due to a monolithic architecture. Customizing software to adapt to new or changed requirements involves high efforts, costs and risks. Nevertheless the solution's flexibility for upcoming adaptations is not increased.

The introduction of cloud computing is a new opportunity to deliver different IT services over the internet. The idea of this IT paradigm is the abstraction of the underlying software and hardware. Based on this, users do not have to manage and administrate the physical hardware or software they are using. An additional advantage is that software can be acquired on-demand and paid per use (Schuldt et al. 2010).

The concept of providing services over a cloud encompasses three different stakeholders. Basically these are the operator of the cloud computing environment (CCE) being responsible for its administration, providers offering their applications or IT services and customers purchasing and using the services (Kaisler and Money 2011).

Regarding the dynamically changing requirements cloud computing provides the opportunity for a logistics customer to rent IT services only as long as they are needed. Furthermore, using a CCE to provide and use services enables customers and logistic service providers to focus on their key business by outsourcing the IT infrastructure to the cloud. Both customers and logistic service providers do not have to establish and administrate an internal IT infrastructure and only require a connection to the internet for interacting with the cloud. Additionally, pay-per-use accounting is another advantage for the customers. For providers cloud computing is a new opportunity to gain greater market relevance and design new offers by connecting their products with services provided by other independent providers (following the slogan "The whole thing is more than the sum of its components") (Goyal and Mikkilineni 2011; Scholz-Reiter et al. 2011).

2.2 Concept

The main idea of cloud computing—"Anything as a Service"—can be adapted to source complete logistics business processes, designed by connecting single IT services using an appropriate tool that is executed in the browser. Therefore, an adequate process modeling methodology is necessary to offer logistics customers an opportunity to model individual processes themselves. This service delivery model is defined as Business Process as a Services (BPaaS). There are contributions that use BPaaS as abbreviation to indicate models that have the general idea of outsourcing of existing processes into the cloud (Bentounsi et al. 2012). Here, more specifically, the unique feature of this model is that both modeling and process execution are part of the cloud platform. This idea is the basis of the Logistics Mall, a development of the Fraunhofer Innovation Cluster "Cloud Computing for Logistics" (Fraunhofer Innovation Cluster 2013), that focusses on the modeling and execution of processes that are built from several IT services and that can be offered using a cloud platform. Most of the today available approaches are too complex to be used by logistics customers that have minor software engineering skills. On the one hand the vision of the Logistics Mall is to create a methodology that is capable to deliver business processes that can be deployed automatically for execution. On the other hand the logistician may not be overburdened by technical details.

The main idea of the Logistics Mall and the offering of BPaaS is visualized in Figs. 1 and 2. Today's monolithic software solutions are replaced by small, dedicated IT services of different service providers which can be combined to superior services that support the individual business processes of the logistician.

In addition to offering BPaaS solutions the Logistics Mall is also able to provide classic IT services corresponding to the Software as a Service (SaaS) cloud model. The main reason for this is to enable as many software providers as possible to offer IT systems that support a broad range of logistics processes. For an IT service that is offered in conjunction with BPaaS the technical requirements are much higher than for an IT service that is offered only as SaaS in the manner of ASP (application service providing). The reason for this is that BPaaS requires the IT service to use the same business object data model so that in the context of a modeled process IT services of different service providers have a common understanding of the data that is communicated between them. A detailed description of this data model and how it supports logistics business processes and reduces the business-IT gap is provided by Böhmer et al. (2015). Moreover the IT services in the context of BPaaS should

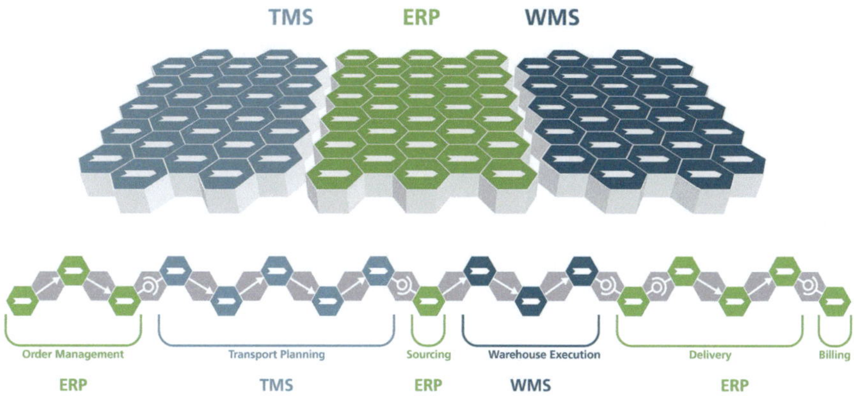

Fig. 1 Today's monolithic software solutions and their interaction within a business process

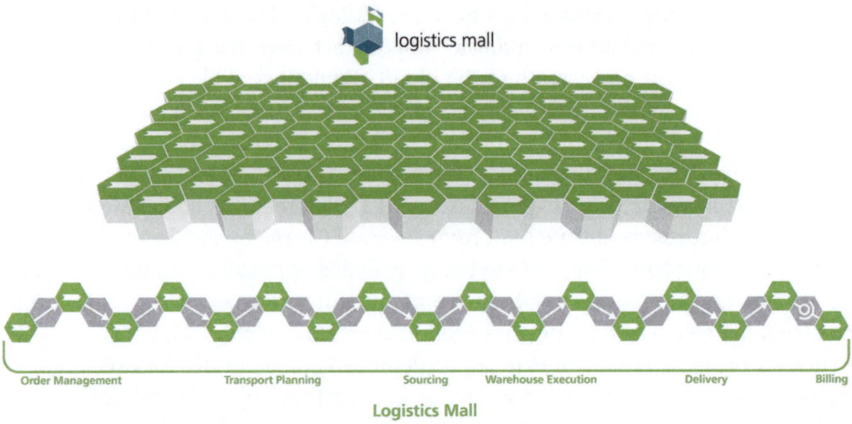

Fig. 2 The Logistics Mall as a cloud platform offers IT-services that can be combined to support individual logistics business processes

have a consistent look and feel of the GUI (Graphical User Interface) when they are used to support one business process. These requirements force service providers to undertake broad modifications on their existing IT services to be offered on the Logistics Mall. In contrast, providers have to only slightly modify their web-based IT services to offer them as SaaS products on the Logistics Mall platform.

Part of the concept of the Logistics Mall (Fig. 3) is that *IT developers* can offer and operate IT services in the cloud. Another target group is companies that have the function of integrators. As *logistics process designers* they combine IT services,

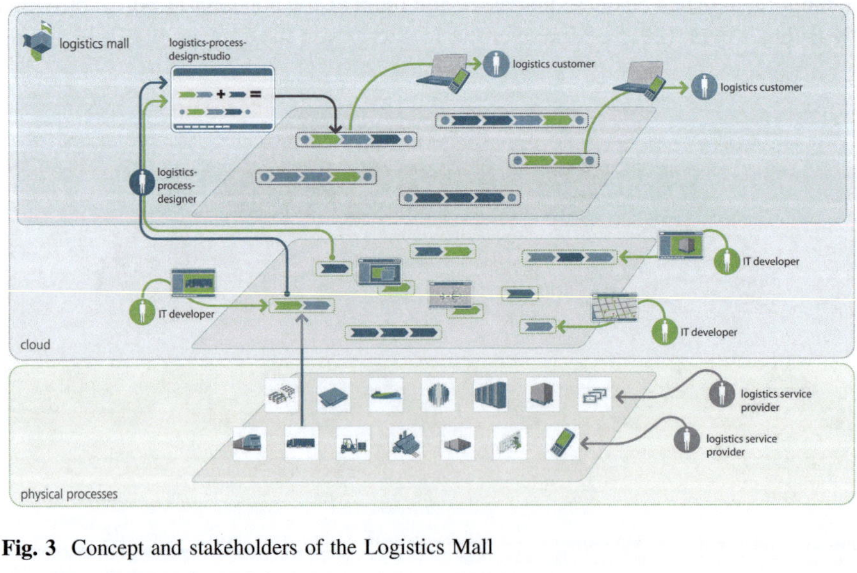

Fig. 3 Concept and stakeholders of the Logistics Mall

which are compatible to be used in the context of BPaaS, to more widespread IT solutions that support complete logistics business processes. The third group is the *logistics customers* that order and consume the IT solutions offered by the Logistics Mall.

Conceptually the Logistics Mall also supports the integration of physical services. Such serviced could be offered on the Logistics Mall platform by *logistics service providers*. An example is transport services, which encompass the transport of goods from a source to a destination.

Overall, the Logistics Mall is a comprehensive approach of an integrated platform that offers PaaS for IT developers and SaaS/BPaaS for logistics customers. The usage model behind the Logistics Mall for both groups is its operation as a public cloud or as a private cloud. Commercial operation of the Logistics Mall is available to the public at http://www.logistics-mall.com.

The design of the Logistics Mall is based on two pillars. The first pillar is the shopping frontend in the form of the mall marketplace (MMP). The MMP provides a publicly available online shop for the offered products. The second pillar of the Logistics Mall is a web portal that allows the operation and usage of the IT solutions rented on the MMP by the logistics customer. After ordering a product in the MMP it is provided on the so called Customized Access Framework (CAF). The CAF is an integrated environment that allows the user to switch between the GUIs of the rented applications and that offers a helpdesk and diverse administration functionality like user management and configuration of connections to external IT systems which are not executed on the Logistics Mall platform.

2.3 *Experiences*

The provision of BPaaS solutions using the Logistics Mall has been evaluated for the field of intralogistics. As a first step the so called *App Template* has been used. Basically the App Template is a Maven Archetype which is a Java based project templating toolkit for the build management tool Apache Maven. As part of the Logistics Mall the App Template is offered for software providers and enables them to develop IT services (Apps) that can be used in a tool that is called *Logistics Process Design Studio* (LPDS). This tool is also part of the Logistics Mall and allows to model the business process support that is needed by the logistics customer. The modeling is done by orchestration of Apps to build up a BPMN (Business Process Model Notation, a graphical representation maintained by the Object Management Group for specifying business processes) conform process model that supports the customers' individual business process. After the deployment of the process model the required IT support is provided in the CAF. In the CAF the user can access the user interface of Apps that are part of the process model and that need user interaction.

For evaluation of the Logistics Mall Apps were developed that support the intralogistics business process from incoming goods receipt to storage of these goods in a warehouse. The test scenario showed the potential of the Logistics Mall solution as the process model successfully could be modified to support changed business process. The main advantage here is that process model modification can be performed in a very short time frame. The steps required for process model modification encompass the undeployment of the active process model, the modification of the process model within the LPDS followed by the deployment of the process model. Because the deployment and undeployment of a process model is fully automated process changes can be implemented very fast. In the test scenario for example the goods in process was extended with a goods testing procedure where in specific cases incoming goods have to be inspected by warehouse personnel.

3 Empirical Results

Fraunhofer IML conducted the market analysis “Cloud Computing for Logistics” in 2011 and 2013 to identify and analyse the opinions, needs, and requirements of the users and vendors in the logistics industry in terms of both cloud computing and the Logistics Mall model. The project team conducted qualitative interviews and used the results of those interviews as the basis for examining the potential as well as the barriers of this concept and for determining which requirements should be implemented in the Logistics Mall. By examining the acceptance and readiness of those surveyed, they could draw conclusions about the attitudes of the logistics experts in the industry. To perform the analysis, they compared reference values of 2011 with the new values of 2013.

The 70 study participants of the users group belonged to small and large companies from three branches of business: logistics service providers, industry, and wholesaler/retailer.

The 102 study participants of the vendors group came from small and large providers of logistics solutions.

The results showed that the spirit of innovation of the users group is split right down the middle, which validated the theoretical assumptions that were made before the project began: it is still too early for every second participant to use a technology like the Logistics Mall. When asked more concretely 56 % of the participants from the users group would be willing to use logistics software in the cloud—if the conditions were right in their own company. This value has actually decreased by 8 % compared to the study conducted in 2011.

The interviews with the study participants from the vendors group in 2013 revealed that the innovative spirit of the vendors group is slightly less than it was in 2011. Half of the participants (49 %) are willing to sell their products or services through an external online store, which is a decline of 9 % compared to 2011 (Fig. 4). A total of 67 % (70 % in 2011) would be willing to operate their applications in the cloud. In contrast to 2011, 10 % of the study participants in the vendors group are already in the concrete planning phase for using cloud architecture (Fig. 5).

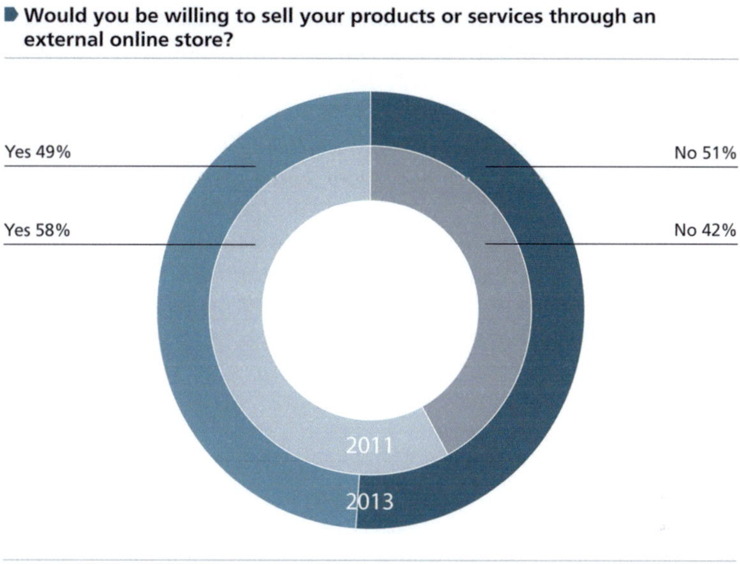

Fig. 4 The level of acceptance for selling products or services through an external online store (vendors)

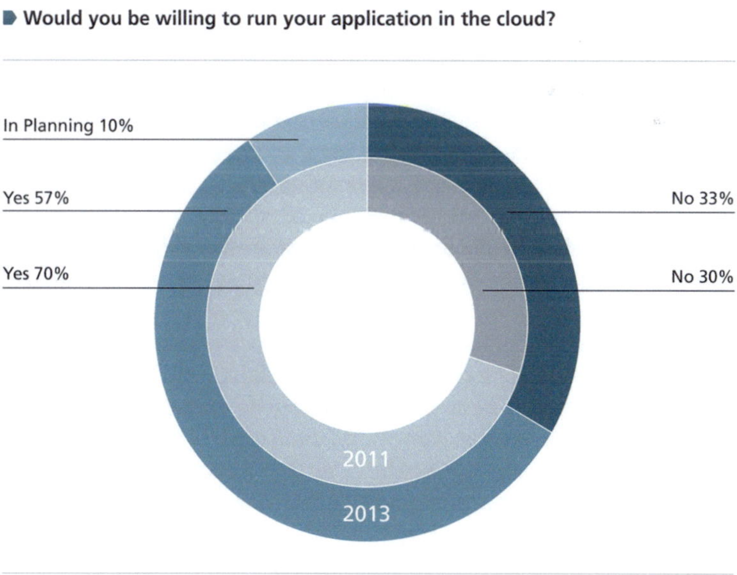

Fig. 5 The level of acceptance for operating applications in the cloud (vendors)

The agreement with these two fundamental Logistics Mall ideas diverges significantly with the size of the company. Small and medium-sized companies were more willing to change their views than larger companies. When the Logistics Mall was described in more detail, 58 % (75 % in 2011) of the vendors surveyed indicated that they would use it. In addition, 67 % (70 % in 2011) of the vendors said that they were sure that their customers would be willing to use cloud computing in logistics. A big shift in the answers to the question about the change in the relationship with the customer was observed when the 2013 data was compared to the 2011. About 56 % (39 % in 2011) accepted the fact that as a vendor they were no longer the main contact point for the user but that the operator of the Logistics Mall held that role instead.

4 Conclusions

The Logistics Mall platform aims at providing a market place for logistics services and business processes together with a cloud-based access and execution environment. It is foreseeable that the business culture will change over time. The focus will be on customized solutions. Monolithic software solutions as they are used in today's systems will be replaced by smaller dedicated subsystems that can be coupled and thereby can be assembled into higher-value services. The network concept that the Logistics Mall represents leaves a lot of creative leeway. Each company—no matter how big or small—has the chance to be a part of this development and help to shape it.

Cloud-based solutions such as the Logistics Mall presented in this paper are an important part of future scenarios like *Industry 4.0*. From the economic point of view this future scenario describes a new level of organization and control of the entire value chain through the life cycle of products. From the technological perspective there is required the availability of all relevant information in real time through integration of all entities involved in the value chain and on this basis the ability to derive the optimal value chain flow (Plattform "Industrie 4.0" 2014).

An important research field in the context of Industry 4.0 and cloud platforms is the involvement of physical processes on the shop floor level. Here, a main research question is how an interface between the applications running in the cloud and the machines on shop floor level consisting of different sensors and actors can look like. Given such an interface the applications in the cloud can provide shop floor monitoring and control functionality, which opens new possibilities in the design of business process supporting software.

References

Bayer M (2013) Die cloud wird gesellschaftsfähig Computerwoche. http://www.computerwoche.de/a/die-cloud-wird-gesellschaftsfaehig,2532192. Accessed 18 Feb 2014

Bentounsi M, Benbernou S, Atallah MJ (2012) Privacy-preserving business process outsourcing. In: IEEE 19th international conference on web services

Böhmer M, Schmidt M, Weißenberg N (2015) Seamless interoperability in logistics: narrowing the business-IT gap by logistics business objects. In: Ten Hompel M, Rehof J, Wolf O (eds) Cloud computing for logistics. Lecture Notes in Logistics (Series Editors: Clausen U, Ten Hompel M, de Souza R), Springer, Berlin

Fraunhofer Innovation Cluster (2013). Cloud computing for logistics. http://www.ccl.fraunhofer.de/en.html. Accessed 08 Apr 2013

Goyal P, Mikkilineni R (2011) Policy-based event-driven services-oriented architecture for cloud services operation and management. In: IEEE international conference on cloud computing

Hackmann J (2013) IT-outsourcing unter Druck, Computerwoche. http://www.computerwoche.de/a/it-outsourcing-unter-druck,2532536. Accessed 18 Feb 2014

Kaisler S, Money WH (2011) Service migration in a cloud architecture. In: 44th Hawaii international conference on system sciences

KPMG AG, BITKOM (2013) Cloud-monitor 2013. http://www.kpmg.com/DE/de/Documents/Cloud-Monitor-2013-KPMG-version-2.pdf. Accessed 18 Feb 2014

Mahmood Z, Hill R (2011) Cloud computing for enterprise architectures. Springer, Berlin, pp 96 ff

Mell P, Grance T (2009) The NIST definition of cloud computing, working paper national institute of standards and technology. http://www.nist.gov/itl/cloud/upload/cloud-def-v15.pdf. Accessed 08 Apr 2013

Plattform "Industrie 4.0" (2014) Forschungs- und Entwicklungsaktivitäten auf dem Weg zu Industrie 4.0. Whitepaper. http://www.plattform-i40.de/sites/default/files/Whitepaper_Forschung%20Stand%203.%20April%202014_0.pdf. Accessed 18 May 2015

Scholz-Reiter B, Rippel D, Sowade S (2011) Limitations in modeling autonomous logistic processes—challenges and solutions in business process modeling. In: IEEE international symposium on assembly and manufacturing (ISAM), pp 1–6, 25–27

Schuldt A, Hribernik KA, Gehrke JD, Thoben K-D, Herzog O (2010) Cloud computing for autonomous control in logistics. In: Fähnrich K-P, Franczyk B (eds) 40th Annual conference of the german society for computer science (GI 2010), Leipzig, Germany, September 27–October 1 2010, LNI 175, Gesellschaft für Informatik, pp 305–310

Velten C, Janata S (2013) Aktuelle Marktzahlen zum deutschen Cloud-Computing-Markt—5-Prozent-Hürde erreicht. http://www.experton-group.de/research/ict-news-dach/news/article/aktuelle-marktzahlen-zum-deutschen-cloud-computing-markt-5-prozent-huerde-erreicht.html. Accessed 18 Feb 2014

Part V
Logistics Training and Knowledge Management

Problem-Oriented Knowledge Management for Efficient Logistics Processes

Natalia Straub, Christoph Besenfelder and Sandra Kaczmarek

Abstract Due to increased market dynamics and a progressive trend towards globalization, logistics increasingly becomes the center of attention for enterprises. Today, an efficient system of enterprise logistics is mandatory in order to build up and safeguard competitive strategic advantages. In seeking to meet the demands of a dynamic and global market environment, enterprises have to develop and strengthen their employees' competences with a lasting effect, thus optimising their value chain. In this chapter, the results of the project "eQuaL 2.0—e-qualification for efficient logistics processes" are introduced. The project dealt with the concept of a new type of problem-oriented knowledge management for enterprise logistics processes' qualification. The eQuaL 2.0 knowledge base contains vital information concerning relevant issues and problems in enterprise logistics which are linked with the methods to deal with due to the semantic structure of the knowledge base.

Keywords Logistics · Vocational training · Knowledge management · Web 2.0 · Blended learning

1 Problem and Objective

Today, production enterprises operate in a global and highly competitive market environment that is characterized by rising customer demands as well as rapid and dynamic changes in processes, technologies and work tasks (Westkämper 2009). In this process, logistics performance features such as delivery time and delivery reliability are, in addition to a high quality level and low product costs, important ways of market differentiation for enterprises (Pawellek 2007; Nyhuis 2008). Against this background, logistics efficiency and especially the creation of a Lean

N. Straub (✉) · C. Besenfelder · S. Kaczmarek
Chair of Enterprise Logistics, Technische Universität Dortmund,
Leonhard-Euler-Straße 5, 44227 Dortmund, Germany
e-mail: straub@lfo.tu-dortmund.de

H. Zijm et al. (eds.), *Logistics and Supply Chain Innovation*,
Lecture Notes in Logistics, DOI 10.1007/978-3-319-22288-2_23

Production System (LPS) as well as the responsiveness and flexibility of production systems have considerably gained importance over the last few years (Spath 2003; Uygun et al. 2009). In order to be able to cope with these demands, competent and qualified employees are required who take part in these constantly processes changes and help to mold them by continuously questioning existing structures and processes as well as solving problems in a creative way (Rother 2010). Consequently, both cooperate competence and knowledge management are confronted with new and extended tasks, which leads to new requirements regarding a demand-oriented, decentralized and self-regulated qualification that has to be fulfilled on the organizational level.

Furthermore, the global economy is approaching the fourth industrial revolution and the effects of the demographic change concerning employee structure are further increasing. As a result of ongoing technological change and the use of new internet technologies in terms of industry 4.0, logistics and production systems will be prospectively understood as highly interactive sociotechnical systems (Spath et al. 2012).

In addition, the employees' focus shifts from task performance to controlling, programming, monitoring and troubleshooting (Scheer 2014; Spath et al. 2013). In this context, logistics systems require their employees to understand the value chain and its up- and downstream process steps as well as a basic knowledge of the interdependencies of logistics objectives within the process chain (Wiendahl and Fischmann 2007; Dietzen 2004).

The changing criteria imposed on employees result in an increasing demand of communication skills and a high degree of self-control and organization. High knowledge intensification, complexity and continuous changes characterize these new tasks; simultaneously companies have to develop solutions for the shortage of skilled professionals and the demographic shift in the personnel structure (Kurz 2012; Kagermann et al. 2013; Walter et al. 2013).

In view of the new challenges, the requirements for employees and their competences continue to rise. Once acquired knowledge is no longer sufficient for an entire professional life; consequently continuous adaptions of competences and qualifications are necessary (Bundesministerium des Innern 2013). As a result, companies are facing challenges in both preserving their employees' knowledge and competences in the long term, i.e. during their entire working life, and adapting them to current demands and developments.

2 Competence and Knowledge Management as a Task Within Enterprises

Since professional contexts are changing steadily in recent years, the idea of 'competence' or 'competence development' experiences a significant upturn with the result that competence development is now considered as a key objective of company training and development. Associated with this objective, the claim is to qualify employees for acting competently in their daily work environment (Scheer 2014; Kurz 2012). Therefore, independence and self-organisation are key criteria for proficient actions. This includes employees being capable of assessing the effects of their own activities, thus anticipating the consequences of their decisions. If an employee faces given working tasks autonomously while using his competence to foresee consequences, this in turn will lead to a deliberate sense of responsibility for his actions.

In this context, the main question is which competences already exist in the company and what qualifications will be needed prospectively to meet the stated long-term goals of qualification.

2.1 Knowledge Management in the Context of Enterprise 2.0: Focus on Employees and Tasks

In the context of supply chains, knowledge intensity is currently on a rise and thus knowledge management processes are increasingly important. As a result of the above described change dynamics, the half-life period of existing knowledge is decreasing (Weiß 2010). The existing knowledge in a company is understood as an enabler for the company's competitive and innovative capabilities and is thus a strategically important resource. Consequently, companies and organisations need to shape, control and support their knowledge developments. It is key to retain already existing knowledge in the company for the long term, to continually expand the knowledge base and to grant company-wide access. Knowledge management is understood as the "entirety of the personnel, organizational, cultural and technical practices that aim at an efficient usage of 'knowledge' as a resource in the context of an organisation or a network" (DIN PAS 1063 2006). Knowledge management is thus an essential element in the process of realizing the company's goals. The key processes supported by knowledge management are knowledge generation, storage, distribution and application (Orth et al. 2009).

Technological change has a profound impact on knowledge work. Demands arising from the integrative cooperation in distributed teams and in various projects in combination with the developments of Web 2.0 have led to the concept of Enterprise 2.0 solutions. Enterprise 2.0 solutions differ from typical corporate applications by a stronger user influence on the system structure (Beccera-Fernandez and Sabherval 2010). The focus is shifting from pre-structured document storage to a heavily linked collaboration platform where the user is in the centre of attention. For an overview and a comparison of different Enterprise 2.0 and Web 2.0 solutions in the application context of knowledge management we refer to the papers of Fuchs-Kittowski and Voigt (2010) and Besenfelder and Döring (2013). Knowledge provision becomes problem-oriented and individually customized to fit specific tasks. The close interconnection of knowledge management and corporate training is explained by the employee's role as the central knowledge carrier in the company. Knowledge is tied to the single employee and thus might leave the organisation once the employee is leaving. Consequently, knowledge management has the main task of passing on knowledge in order to prevent knowledge loss.

A Wiki is a specific example of an Enterprise 2.0 solution. It is a web-based and thus platform independent piece of software allowing the viewer of a hypertext webpage to change its contents (Ebersbach et al. 2008). A Wiki is a network of hypertext webpages containing formatted text, images and hyperlinks to other content. Hyperlinks can address other Wiki webpages as well as a variety of other media formats. Furthermore, a Wiki is equipped with a version management that is listing all changes performed on the page and allowing an easy reversion if necessary (Back and Gronau 2008). An integrated title and full text search engine ensures the page's accessibility. The Wiki pages are written in so-called Wiki syntax, a simplified description language. Resulting from this simplification, it is not necessary to apply programming or HTML skills when creating or editing webpages (Back and Gronau 2008). Social interaction, which is playing an important role in the generation of new knowledge, is enabled and supported by discussion pages and references to the authors provided by the Wiki (Fuchs-Kittowski and Hüttemann 2009). Contrasting these positive features, Wikis require a complex content management as well as an increasingly elaborate quality management when dealing with a growing system. These negative aspects can be ascribed to a lack of structure when establishing a Wiki (Fuchs-Kittowski and Hüttemann 2009; Happel and Treitz 2008; Orth et al. 2009). Key functions of Wikis in the context of knowledge management and corporate training are user appropriate and accessible knowledge provision, generation and storage. The combination of knowledge provision with self-organized knowledge input by the employees leads to a new learning situation in which the user has to deal with content more intensively than before. On the one hand, this leads to an internalisation of new knowledge, on the other hand, the employee's knowledge is externalised and put into the Wiki database. Thus, knowledge becomes stronger connected. Moskaliuk refers to this effect as a "co-evolution of knowledge and information" (Moskaliuk 2008).

2.2 *Problem-Orientation Enables Synergies*

Problem-orientation in knowledge management in the context of Enterprise 2.0 and employee qualification in the field of logistics processes synergistically connects these two task fields (see Fig. 1). The user-centered storage and distribution of knowledge, which is required by the work task, enables a support of the core processes of knowledge management.

Concurrently the problem-oriented provision of information, which allows self-regulated and demand-oriented employee qualification, is the basis for efficient logistics processes and a continuous improvement cycle. The knowledge domain of

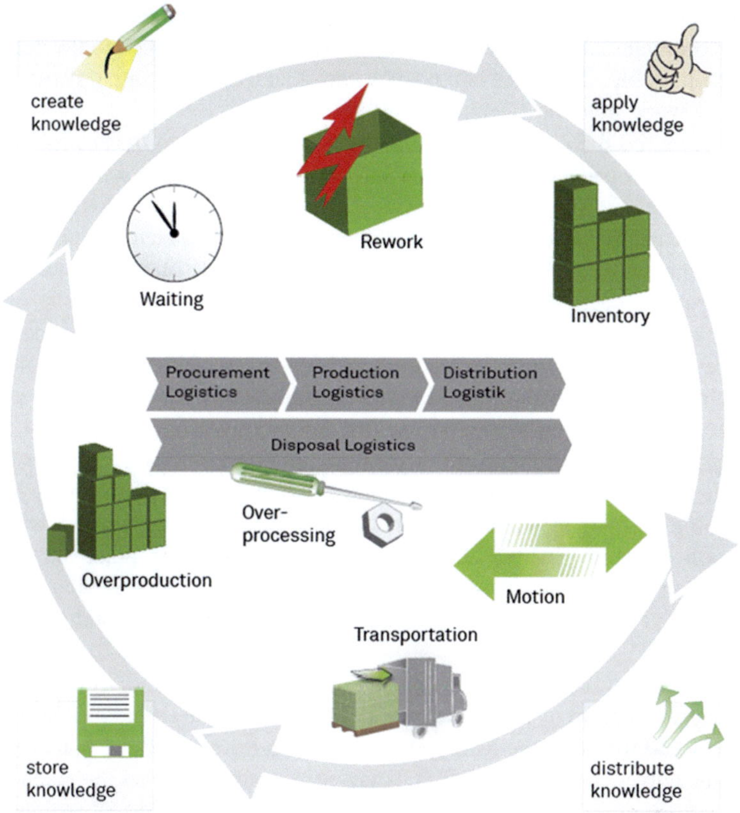

Fig. 1 Problem-orientation in logistics as the connection between knowledge management and employee qualification

the logistics processes is defined by the process orientation and the focus of waste avoidance of the holistic production systems.

The required knowledge base has to provide an optimal support of the core processes of knowledge management and of employees' competence development to generate substantial synergies.

3 The Concept of Problem-Oriented Knowledge Management

Due to the increasing demand of enterprise qualification resulting from the introduction of a new production system, e.g. Lean Production Systems or Industry 4.0-technologies, specific demands regarding a demand-oriented, decentralised and self-regulated qualification have to be fulfilled on the organizational level (Adami and Houben 2008; Spath et al. 2013). In doing so, the central decisions concerning planning, content, realisation and learning evaluation are assigned to the employees as far as possible. Thus, the employees can adapt the learning process to their individual work situation (Dehnbostel 2010).

Consequently, the company's competence and knowledge management is confronted with new and extended tasks, which lead to new requirements regarding the didactical methodology, the organisation as well as the methodological and technical infrastructure, also in terms of content of the competence and knowledge management, and which pushes ahead the development of innovative and new qualification concepts (Falk 2000; Arnold 2006).

From a **didactic-methodical** point of view, the competence development refers to a necessary change, from the learning and repetition of knowledge to an active, solution-oriented and exploratory learning (Arnold 2006). As exploratory learning, the approaches of the self-regulated construction of knowledge on the basis of "trial and error" are described that have their origins in constructivism. Here, learning is considered as a profoundly interactive, almost playful process of discovery in which the learning individual solves a problem that is directly embedded within the application context (Steiner 2000). Learning always takes place on the axes between "perceiving" and "understanding" or "thinking". That is why a learning process always starts with a concrete experience of the learning individual (Kolb 1984).

On the **methodical** level, competence development deals with the task of increasing the horizon of experience and competence of the employees by means of other than traditional methodical patterns and with developing the potential of the individual human being (Bohn 2007). By using effective working methods and forms of cooperation, new levels of performance can be reached by mobilising the creativity, the problem consciousness and the performance of the employees (Rother 2010; Korge and Lentes 2009).

The **organisation** of the qualification in terms of content has to be transfer and work process-oriented in order to intensify the logistics understanding of the employees and to sensitise them proactively with respect to typical production logistics problems and possibly conflicting objectives within the whole value chain as well as to provide them with the necessary methodical knowledge (Wiendahl and Fischmann 2007).

The **technical** infrastructure of the learning processes has to support the independent information search, the self-contained learning process irrespective of time and place as well as the allowance of heterogeneous target groups. Further important requirements contain the collaborative generation of knowledge as well as the flexible company-specific adaptation and further development of contents (Bundesministerium für Bildung und Forschung (BMBF) 2011).

Conceptual Design: eQuaL 2.0 Knowledge Base

Considering the challenging and comprehensive duties and responsibilities connected with qualification in logistics processes a technical solution alone is insufficient. In the following, the conceptual design of the knowledge base according to the aforementioned design fields is presented.

Methodological Concept As stated above, the original motivation for creating a qualification platform for logistics processes was the emergence and dissemination of lean manufacturing systems worldwide. According to this philosophy, employees are regarded as sources of innovative ideas; their commitment supports continuous improvement, solving multiple problems every day.

In order to enable a target-group-specific knowledge provision within the eQuaL 2.0 knowledge base, the overall system of the enterprise logistics has to be divided into functional subsystems by integrating the following knowledge domains: procurement logistics, production logistics, distribution logistics and disposal logistics. (Pfohl 2010)

The content structure of the eQuaL 2.0-knowledge base serves a problem-solving and demand-oriented knowledge transmission and contains the following sections: Basics, methods, problems, symptoms and practice examples (see Fig. 2):

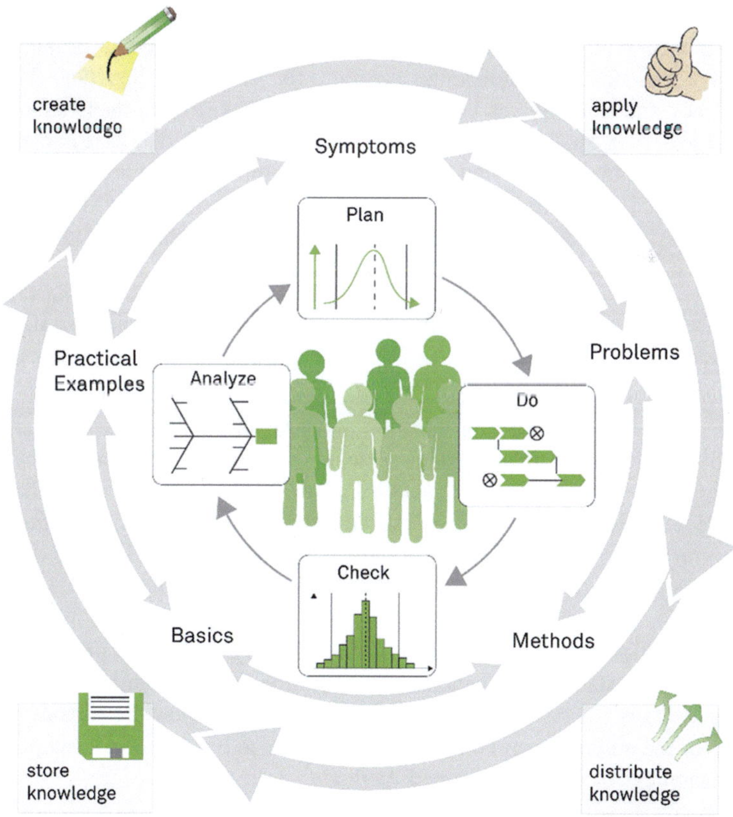

Fig. 2 Problem-solving cycle as methodical concept for eQuaL 2.0 knowledge base

- Basics: here the basic logistics terms are explained
- Logistics problems: this category outlines typical logistics problems
- Symptoms: are used to describe and diagnose a problem
- Methods: here methodical instruments for analysis, evaluation and optimization of logistics systems are explained
- Practical examples: contains internal experience reports about method application and offers applied problem-solving concepts. In this way the active experience exchange of employees of different departments is encouraged

The basic starting point for the content development within the framework eQuaL knowledge base lies in the typical problems that employees observe over the course of their working routines. The typical logistics problems were consolidated and—through symptoms visible for employees—described and semantically interlinked with analysis and optimization. Problems illustrated by reference to types of waste act as a basis for the employee's qualification in logistics processes. As a consequence of the high complexity of the logistics processes and in order to enable a professional and demand-oriented problem-solving knowledge transfer, a specification of the knowledge base's content is needed. With the help of knowledge domains and semantic correlations reflecting the logic of the problem-solving cycle, a new generic Wiki-model is developed. It essentially provides a problem-based knowledge transfer and knowledge creation for enterprise logistics. Thus, the transparency and usability can be deeply improved. Assisting in many submittals, a semantic pattern is established, which hyperlinks the knowledge-base content to the knowledge domains. In this way a consistent assignment of the content is realized. The developed semantics thus represents the norm of this assignment. Consequently, the different learning-scenarios plus the knowledge-base's maintenance and its consistent further development are guaranteed.

Therefore, eQuaL 2.0 is meant to be problem-centered. The general targets of lean manufacturing systems constitute the target system of qualification. Lean manufacturing systems already bring along a number of systematically arranged and successfully implemented methods of employee orientation and teaching. For this, the methods of process optimization and problem solving are summarized in one holistic methodical management system. In doing so, a linkage between the typical logistics problems as well as the logistics objectives and methodical tools has to be established. This enables the employees to judge diverse situations or states competently in order to track down weaknesses and obstacles on the way to perfection and to create appropriate and smart solutions (Rother 2010; Korge and Lentes 2009). The existing production logistics processes and methods are continuously questioned, exhausted and organized to a higher level of performance and a more efficient state of the whole value creation chain is aimed at. A method management system on the one hand has to be built upon the already existing methods of the enterprise and on the other hand has to be used as a basic pattern for the management of newly developed methods.

Organizational Concept For any company to exploit maximum potential, the knowledge has to be embedded within the company structures and processes. Here, structures refer to personnel responsibilities, technical systems and locations that are concerned with qualification. The project eQuaL 2.0 claims that modern training in logistics synchronizes working with learning in terms of time, site and staff. Also, the general conduct of business has to alter where it may hamper learning to develop. Top-down management, functional organization with hierarchical structures and punctual teaching need to yield. It should give way for bottom-up leadership, process organization with network organization and continuous learning.

The paramount structural coordination within the work process and thus the institutional and organizational frame is set in advance from the outside (Dehnbostel 2010). For this, the implication of the concept Blended Learning is purposeful. Blended Learning is a combination of the modern possibilities of e-learning and the traditional methods of attendance learning (Bonk and Graham 2006). The objective is to combine the advantages of single methods in such a way that educational objectives as well as efficiency criteria from an economic point of view can be met as far as possible (Kerres 2001).

Technical Concept To meet the requirements mentioned above, the eQuaL 2.0 knowledge base is realized by Web 2.0 technology and specifically by a pre-structured Wiki-system. Thereby a structured content is created with the help of templates, knowledge domains or categories (Moskaliuk 2008; Fuchs-Kittowski and Hüttemann 2009). Thus, the user is given clear procedures to place new content (creation of Wiki pages) as well as introductions concerning optional and required semantic links, to integrate the new content into the Wiki-system in a structured and standardized way. The knowledge domains describe specific knowledge fields as well as the essential content of a specific topic (Moskaliuk 2008). Consequently a thematic specification of the content is possible (Mertins and Orth 2009). The "category" as a special page type is used to connect similar pages or page groups and to realize a certain hierarchy level of the Wiki. Furthermore, the category pages allow to create an overview of the category content automatically. Thus the transparency and usability can be significantly improved. By linking the appointed content of the knowledge base to the knowledge domains it has to be ensured that a default semantic pattern is used. In this way a uniform classification of content is guaranteed. Hence, the semantics represents a norm for the allocation (Bodendorf 2006) (see Fig. 3).

The earlier mentioned sub-areas procurement logistics, production logistics, outbound logistics and disposal logistics have been realized as individual landing pages, as so called "Logistics Portals" which can be found on the Wiki-systems main page. Due to the semantic link of the Wiki pages to the knowledge domains as well as to the main and subcategories, a subject-specific assignment and contents consolidation is achieved. Thus, a fast access to the relevant logistics fields is provided for the employees.

Besides pre-structuring contents in the knowledge domains and categories, the single Wiki articles are pre-structured as well. Therefore the following article types:

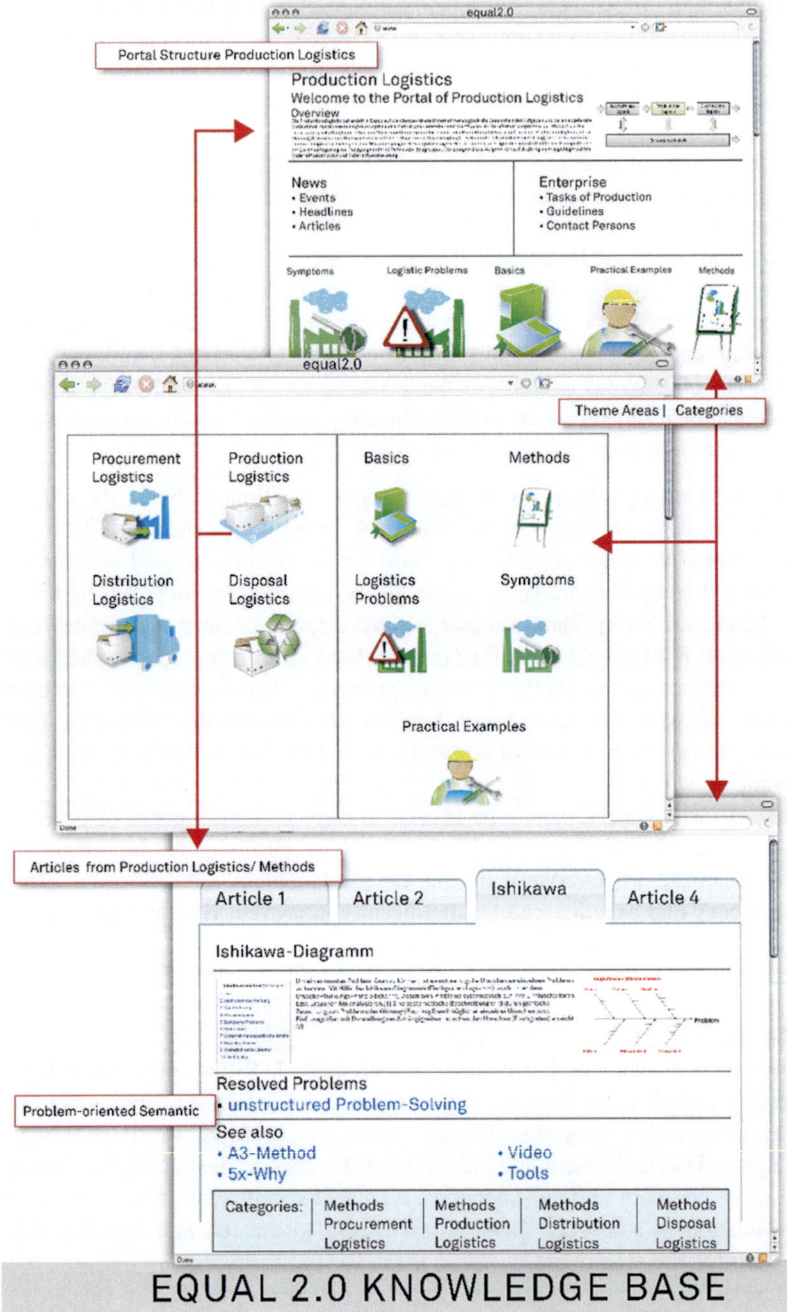

Fig. 3 Semantics of eQuaL 2.0 knowledge base

"essentials", "logistics problems", "methods", "symptoms", "practice examples", "main page" and "portal pages" are defined and supported by standardized templates. The templates contain a structure in order to offer an orientation guide for the user. Here, all value adding processes are to be addressed. The progressive technical development makes the use of new digital media with the computer as technical basis and Web 2.0 applications possible, as a consequent application of existing internet-technologies (Kerres 2001). The web 2.0-technologies reveal completely new possibilities of networking, group formation, participation, cooperative production of contents, development and spreading of information as well as quality assurance of offers. Network-based learning forms facilitate the cooperative learning, which are independent from place and time and support the active organisation of the individual learning process (Bundesministerium für Bildung und Forschung (BMBF) 2011; Schmidt 2009).

To act within the company's further training, the contents quality is of essential importance. Therefore a controlled release of contents is recommended to guarantee this quality.

3.1 Learning Scenario

The structure of contents and the semantics of the eQuaL 2.0—knowledge base enable the following three usage scenarios:

Problem and demand-oriented learning (concrete problem analysis and method acquisition). In this scenario the use of the knowledge base is initiated through the perception of a problem's symptoms, e.g. a work process. The employee can search within the knowledge base for one or multiple symptoms. In case of the availability of corresponding content, the symptom articles will offer information about the potential logistics problem. After problem description and analysis, the employee finds methods for problem-solving in the article "problem", which the employee can study and apply immediately. Afterwards, the successful application of the method and the realized problem solution can be added in the knowledge base as a practical example by means of an experience report. Thus, easily accessible and useful new knowledge is generated for the entire personnel.

Look up, inform or educate yourself In this scenario a self-organized explorative learning takes place. Thereby the content of the eQuaL 2.0-knowledge base is used by employees to extend the personal knowledge frontier or to repeat and brush up on content without a concrete educational objective. The navigation works through a classical keyword search. Another way of accessing is to select knowledge domains or categories.

Studying assigned contents or searching for necessary utility tools (e.g. method cards). In this scenario an employee receives the instruction from the human resource developer or his superior to study contents of the knowledge base or to add new content, respectively. The navigation occurs by the classical keyword search or through the selection of particular knowledge domains or categories.

4 Conclusions

As a result of risen market dynamics and technological growth in modern production and logistics, the employees' focus has shifted from performing tasks to controlling, monitoring and troubleshooting. Strong knowledge intensification, complexity and continuous changes characterize these new tasks while at the same time companies have to deal with increasing skill shortages and an employee structure facing demographic change. Extra occupational, work integrated and professionally supported competence development allows for high employees' performances throughout their entire professional life and for educating various working scenarios.

In knowledge-based economies, life-long learning is therefore essential for economic success. It is obvious that in companies operating in such economies, sophisticated equipment is applied which has to be deployed effectively and efficiently especially from a logistics point of view. In order to improve logistics processes in these companies, employees have to be trained and developed.

The qualification system of eQuaL 2.0 presented in this article is a promising approach that considers the dynamics of logistics processes thanks to the problem-oriented structured knowledge base. This kind of solution is suitable for companies which are using a lean manufacturing or lean logistics system, but it can be adapted for other domains without major changes of the logic structure.

Acknowledgment The collaborative project "eQuaL 2.0—e-qualification for efficient logistics processes" is part of the EffizienzCluster LogistikRuhr and is supported by the Federal Ministry of Education and Research as part of the "Spitzencluster"-competition (FK 01IC10L25).

References

Adami W, Houben J (2008) Mitarbeiterqualifikation für moderne Produktionsoranisationen. WT Werkstattstechnik Online 98(5):428–433

Arnold R (2006) Neue Methoden betrieblicher Bildungsarbeit. In: Arnold R, Lipsmeier A (eds) Handbuch der Berufsbildung. VS Verlag, Wiesbaden, pp 355–369

Back A, Gronau R (2008) Web 2.0 in der Unternehmenspraxis. Oldenburg, München

Beccera-Fernandez I, Sabherval R (2010) Knowledge management: systems and processes. ME Sharpe, New York

Besenfelder C, Döring H (2013) Anforderungen an das Wissensmanagement in virtuellen Netzwerken. In: Supply chain management—automotive, pp 15–20

Bodendorf F (2006) Daten- und Wissensmanagement, 2nd edn. Springer, Berlin

Bohn F (2007) Erfolgsquellen in deutschen Unternehmen. Warum ein Produktionssystem alleine nicht ausreicht. REFA-Nachrichten 1:20–25

Bonk CJ, Graham CR (2006) The handbook of blended learning. Global perspectives, local designs. Pfeiffer, San Francisco

Bundesministerium des Innern (2013) Demografiebericht. Bericht der Bundesregierung zur demografischen Lage und künftigen Entwicklung des Landes. (Online) Available at: http://www.bmi.bund.de/SharedDocs/Downloads/DE/Broschueren/2012/demografiebericht.pdf;jsessionid=C7C802222C438C99C37A6BCD6A51768E.2_cid373?__blob=publicationFile. Accessed 18 Sept 2014

Bundesministerium für Bildung und Forschung (BMBF) 2011. *eQualification. Mit digitalen Medien zu neuen Wegen der Qualifizierung.* (Online) Available at: http://www.equalification.info/_media/equalification_2011.pdf. Accessed 18 Sept 2014

Dehnbostel P (2010) Betriebliche Bildungsarbeit. Schneider Verlag, Kompetenzbasierte Aus- und Weiterbildung im Betrieb. Baltmannsweiler

Dietzen A (2004) Neue Qualifizierungsanforderungen und ihre Realisierungsbedingungen in organisatorischen Veränderungsprozessen. In: Dietzen A, Latniak E (eds) Betriebliche Qualifikationsentwicklung in organisatorischen Gestaltungsprozessen. Bertelsmann Verlag, Bielefeld, pp 17–39

DIN PAS 1063 (2006) Implementing knowledge management in networks of small to medium-sized enterprises. DIN Deutsches Institut für Normung e. V., Abteilung Normung und Standardisierung, Berlin

Ebersbach A, Glaser M, Heigl R, Warta A (2008) WIKI—kooperation im web. Springer Verlag, Heidelberg

Falk R (2000) Betriebliches Bildungsmanagement. Wirtschaftsverlag Bachem, Abeitsbuch für Studium und Praxis. Köln

Fuchs-Kittowski F, Hüttemann D (2009) Towards an intergrated collaboration and knowledge environment for SME based on web 2.0 technologies—quality assurance in enterprise wikis. In: 5th conference professional knowledge management. Experiences and visions, Köllen Druck + Verlag GmbH, Solothurn, pp 532–541, 25–27 Mar 2009

Fuchs-Kittowski F, Voigt S (2010) WEB 2.0 in produzierenden kleinen und mittelständischen Unternehmen. Fraunhofer Verlag, Stuttgart

Happel HJ, Treitz M (2008) Proliferation in enterprise wikis. In: 8. I. C. o. t. D. C. Systems, ed. Proceedings of the 8th international conference on the design cooperative systems. s.n., France, pp 123–129

Kagermann H, Wahlster W, Helbig J (2013) Umsetzungsempfehlungen für das Zukunftsprojekt Industrie 4.0. Abschlussbericht des Arbeitskreises Industrie 4.0. (Online) Available at: http://www.plattform-i40.de/sites/default/files/Abschlussbericht_Industrie4%200_barrierefrei.pdf. Accessed 18 Sept 2014

Kerres M (2001) Multimediale und telemediale Lernumgebungen. Konzeption und Entwicklung, 2nd edn. Oldenburg Verlag, München

Kolb DA (1984) Experiential learning. Experience as the source of learning and development. Prentice-Hall, Engelwood Cliffs

Korge A, Lentes HP (2009) Ganzheitliche Produktionssysteme. Konzepte, Methoden, Erfolgsfaktoren. In: Bullinger HJ (ed) Handbuch Unternehmensorganisation. Strategien, Planung, Umsetzung. Springer, Berlin, pp 569–574

Kurz C (2012) Arbeit in der Industrie 4.0. *die Fachzeitschrift für Information Management*, März, IM, pp 55–59

Mertins K, Orth R (2009) Wissenorientierte analyse und Gestaltung von Geschäftsprozessen. In Mertins K, Seidel H (eds) *Wissensmanagement im Mittelstand. Grundlagen – Lösungen Praxisbeispiele.* Springer, Berlin, pp 41–48

Moskaliuk J (2008) Konstruktion und Kommunikation von Wissen mit Wikis. Theorie und Praxis. Velag Werner Hülsbusch, Boizenburg

Nyhuis P (2008) Produktionskennlinien-Grundlagen und Anwendungsmöglichkeiten. In: Nyhuis P (ed) Beiträge zu einer Theorie der Logistik. Springer, Berlin, pp 185–218

Orth R, Finke I, Mertins K (2009) Ein Referenzmodell für Wissensmanagement. In: Mertins K, Seidel H (eds) Wissensmanagement im Mittelstand. Springer, Berlin, pp 15–22

Pawellek G (2007) Produktionslogistik. Planung-Steuerung-controlling. Hanser Verlag, München
Pfohl H-C (2010) Logistiksysteme. Springer, Berlin
Rother M (2010) Toyota Kata. Managing people for improvement, adaptivness, superior results. McGraw-Hill, s.l.
Scheer AW (2014) Industrie 4.0. Wie sehen die Produktionsprozesse im Jahr 2020 aus ? IMC Verlag, s.l.
Schmidt S (2009) Die Instrumentalisierung von web 2.0-Technologien in betrieblichen Aus- und Weiterbildung. In: Hinkelmann K, Wache H (eds) 5th conference professional knowledge management. Experiences and visions. Gesellschaft für Informatik, Bonn, pp 108–112, 25–27 Mar 2009
Spath D (2003) Revolution durch evolution. In Spath D (ed) Ganzheitlich produzieren—innovative organisation und Führung. LOG_X Verlag, Stuttgart, pp 11–45
Spath D et al (2013) Produktionsarbeit der Zukunft - Industrie 4.0. Fraunhofer Verlag, Stuttgart
Spath D et al (2012) Produktionsprozesse im Jahr 2030. Die Fachzeitschrift für Information Management, März, IM, pp 50–55
Steiner V (2000) Exploratives Lernen. Der persönliche Weg zum Erfolg. Pendo Verlag, Zürich
Uygun Y, Keßler S, Staußberg J (2009) Verbreitung Ganzheitlicher Produktionssysteme. Erkenntnisse aus einer deutschlandweiten Studie. wt Werkstatttechnik online, H. 3 99, pp 136–140
Walter N et al (2013) Die Zukunft der Arbeitswelt. Auf dem Weg ins Jahr 2030. Robert Bosch Stiftung, s.l.
Weiß K (2010) Lernen in jungen, innovativen Unternehmen. Gabler Verlag/GWV Fachverlage GmbH, Wiesbaden
Westkämper E (2009) Turbulentes Umfeld von Unternehmen. In: Westkämper E, Zahn E (eds) Wandlungsfähige Produktionsunternehmen. Springer, Berlin, pp 7–23
Wiendahl H-P, Fischmann C (2007) Mitarbeiterqualifizierung für die Produktionslogistik. Logistische Zielkonflikte spielerisch erfahren. Ind Manage 23(2):25–28

Logistics Qualification: Best-Practice for a Knowledge-Intensive Service Industry

Matthias Klumpp

Abstract The logistics industry has undergone many significant changes in the last two decades—one of these being increasing knowledge requirements necessitated by technology implementation as well as global co-operation. Whereas in the past century many blue-collar occupations in logistics like e.g. truck drivers merely required a basic school education and rudimentary qualification levels, today due to improved technology interaction with e.g. barcode and RFID systems, fleet management or toll and truck steering concepts, competence requirements for such jobs have significantly increased. The same is true for many white-collar jobs in logistics, exemplified by the increasing number of university graduates, especially in specific fields like logistics information technology, contract logistics and innovative supply chain concepts ("supply chain design"). Accordingly, the first sector-wide evaluation of competences with 1.068 logistics employees in 2013 in the German ECLR project "WiWeLo" showed competence structures and also gaps according to the Berufswertigkeit measurement concept. In the light of expected changes due to demographic change as well as further technological implementation ("industry 4.0"), there are risks as well as opportunities embedded in such quantitative analyses of competences. These are outlined in this article and will lead to a new logistics qualification paradigm: whereas past education and training in human resource management was very much driven by formal qualifications and therefore "personnel clusters" (like white- and blue-collar), especially in logistics with "mixed entry" people (from other industries as well as countries), future HRM concepts may focus on an individual analysis of gaps and potentials based on quantitative evaluations as with the Berufswertigkeit concept.

Keywords Knowledge-based services · Logistics qualification · Berufswertigkeit · ESCO project

M. Klumpp (✉)
University of Duisburg-Essen (ZLV, PIM) and FOM University of Applied Sciences Institute for Logistics and Service Management (ild), Leimkugelstrasse 6, 45141 Essen, Germany
e-mail: matthias.klumpp@fom-ild.de

H. Zijm et al. (eds.), *Logistics and Supply Chain Innovation*,
Lecture Notes in Logistics, DOI 10.1007/978-3-319-22288-2_24

1 Introduction: Logistics Personnel and Logistics Qualification

In the last decades the logistics industry has become a major knowledge-intensive service industry, similar to e.g. the financial and the health-care service industries (Klumpp et al. 2007). While about 30–40 years ago mainly manual labor and very rudimentary qualifications were commonplace both in the blue-collar as well as the white-collar side of employment in logistics companies and processes, today the qualification requirements are not only different und significantly higher but also crucial for business success (Rao et al. 1998; Lancioni et al. 2001; Mangan and Christopher 2005; Wu 2007; Esper et al. 2007). This is brought about by technology changes and implementations as for example barcode, RFID, GPS and automated systems in the physical side of the logistics "coin". On the other side—and intertwined with this—the *overall complexity* of logistics processes has risen manifold, thereby making the task for the planning, management and control side of the logistics function even more demanding. This is due to global supply chains with a new multitude of actors and stakeholders as well as to increasing demands of customers regarding quality, sustainability, price and cost-efficiency, next to flexibility and speed in delivery (Klumpp et al. 2013, pp. 2–3). Therefore, the *qualification and competence requirements* have grown steadily regarding these complex tasks in logistics management—until the point where innovation and growth may be hampered due to missing competences and regulations regarding process standards, as demonstrated e.g. in the two cases of *electric mobility* (Klumpp et al. 2014, pp. 265–266) and *3D printing* (Rideout 2011; Self 2011).

In order to provide a *quantitative overview* some statistical data may serve as a starting point for the discussion on logistic qualification. In 2010, 2.65 million people were employed in the logistic industry within Germany. The functional distribution of this logistics personnel pool is shown in Table 1. Of these, 775,803 employees (compared to 805,688 in the previous year) are employed in the field of transport, 1,212,519 persons (previous year 1,277,021) are working in the fields of warehousing and inventory management, 199,532 employees (199,624 in the previous year) are responsible for administration and 455,074 employees (previous year 474,725) are employed in a sub-sector which doesn't directly concern logistic activities but regards support services such as financing and information technology.

As column 3 of Table 1 shows the number of full time employees, in column 4 the percentage of logistic activities is presented which is practiced in the professional group (column 2), e.g. line 5 indicates that 80 % of 766,378 full time employed truck drivers works within the logistics industry (613,102 full-time employees). Column 7 displays the total number of employees in the logistics industry, including full-time employees and entrepreneurs. The logistics industry compromises in fact many segments. In general it can be classified in two main groups. i.e. *blue-collar workers* and office clerks (*white-collar*). The information

Table 1 Personnel in the logistics industry (Klaus et al. 2010, p. 57)

	No.	Occupation group	Number SVP employees 2009	Fraction logistics in percent of column 3 (%)	Number SVP employees in logistics (column 3 * column 4)	Fraction in % of all SVP employees (%)	Extrapolation to all employees (1, 2 * SVP employees)	Factor for extrapolation of indirect employees in logistics	Entire logistic employees	Service provider (%)	Industry and trade (%)	Other economic sectors (%)	Service provider	Industry and trade	Other economic sectors
	1	2	3	4	5	6	7	8	9	10	11	12	13	14	15
1	*Direct logistic occupations*														
2	711	Engine driver	35,372	20	7074	0.03	8489		8489	89	4	7	7581	329	579
3	712	Traffic controller (rail)	5359	20	10718	0.04	12,862		12,862	75	3	22	9612	400	2850
4	713	Other traffic controller	15,584	20	3117	0.01	3740		3740	42	17	42	1560	621	1560
5	714	Motor vehicle drivers	766,378	80	613,102	2.25	735,723		735,723	60	20	19	443,468	150,240	142,015
6	721	Nautical	7395	20	1479	0.01	1775		1775	73	5	22	1292	92	391
7	722	Technical ship's officer	5914	20	1183	0.00	1419		1419	51	20	29	725	289	405
8	723	Sailor	8599	20	1720	0.01	2064		2064	75	2	23	1546	38	480
9	724	Inland sailor	5939	70	4157	0.02	4989		4989	65	4	30	3251	217	1521
10	726	Aviation occupations	26,349	15	3952	0.01	4743		4743	85	3	12	4011	147	585
11	*Subtotal transport and traffic*								*775,803*	*61*	*20*	*19*	*473,045*	*152,373*	*150,386*
12	521	Quality inspector	120,647	20	24,129	0.09	28,955		28,955	8	77	15	2285	22,263	4408
13	522	Dispatcher	228,255	80	182,604	0.67	219,125		219,125	7	77	16	14,961	168,262	35,902
14	741	Storage managers	254,634	100	254,634	0.93	305,561		305,561	12	72	16	36,725	219,968	48,868
15	742	Forklift truck and other equipment	55,076	80	44,061	0.16	52,873		52,873	16	59	24	8570	31,437	12,866
16	743	Furniture remover	11,241	100	11,241	0.04	13,489		13,489	73	13	14	9858	1796	1835
17	744	Warehouse and transportworker	493,763	100	493,763	1.81	592,516		592,516	27	49	24	157,177	291,250	144,089

(continued)

Table 1 (continued)

	No.	Occupation group	Number SVP employees 2009	Fraction logistics in percent of column 3 (%)	Number SVP employees in logistics (column 3 * column 4)	Fraction in % of all SVP employees (%)	Extrapolation to all employees (1, 2 * SVP employees)	Factor for extrapolation of indirect employees in logistics	Entire logistic employees	Service provider (%)	Industry and trade (%)	Other economic sectors (%)	Service provider	Industry and trade	Other economic sectors
	1	2	3	4	5	6	7	8	9	10	11	12	13	14	15
18	*Subtotal storage and turnover*								*1212,519*	*19*	*61*	*20*	*229,576*	*734,974*	*247,968*
19	681	Wholesale and retail	512,790	10	51,279	0.19	61,535		61,535	1	85	14	613	52530	8392
20	701	Forwarding agent	101,236	100	101,236	0.37	121,483		121,483	78	14	8	94,842	17,337	9304
21	704	Estate agent	15,102	5	755	0.00	906		906	5	6	90	41	51	814
22	705	Renters, mediators, auctioneers	28,541	5	1427	0.01	1712		1712	5	11	84	83	189	1411
23	732	Postman	115,793	10	11,579	0.04	13,895		13,895	95	2	3	13,242	221	432
24	*Subtotal admin. functions*								*199,532*	*55*	*35*	*10*	*108,821*	*70,328*	*20,383*
25	*Subtotal "direct activities"*								*2,187,864*	*37*	*44*	*19*	*811,442*	*957,675*	*418,737*
26	*Indirect logistic occupations*														
27	75	Enterpriser, auditor			27,348	0.10		0.015	32,818						
28	77	Purchase accounting people			45,580	0.17		0.025	54,696				168,780	199,196	87,097
29	78	Office specialists and assistants			306,300	1.12		0.168	367,559						
30	*Subtotal "indirect activities"*				*379,228*				*455,074*						
31	*Total amount*								*2,642,927*				*980,222*	*1,156,871*	*505,834*

that can be extracted from the table is that in the logistics industry in Germany about 1,988,322 blue-collar workers (truck drivers, forklift drivers, warehouse personnel etc.) and 654,606 office clerks (administration, sales, transport planning, logistic management etc.) are employed.

The logistics industry is usually expecting a sector growth of about 1–2 % *above* the overall average economic growth per annum. This development is a good reason to pay special attention to education in the logistics industry: the enterprises need qualified personnel in the logistics field not only due to the described technology and supply chain organizational changes, but also in order to cope with the above-average growth while keeping the transport chains cost-efficient as well as sustainable for the customers and society at large. Therefore, besides the initial training in vocational as well as academic institutions also continuing education on all levels becomes very important: the logistic sector needs for example about 14,000 executives per year with an academic education (Hildebrand and Roth 2010). Table 2 highlights the importance of continuing education and *competence measurement* for personnel in the logistics industry, where 30.03 % of all full-time employees (324,299 out of 1,079,759) have an "unknown" education, 13.72 % (148,217) are totally *without* any vocational education and only 2.78 % (29,992) of all full-time employees possess an academic degree.

These figures result among other factors from the former bad image of the logistics industry (low wages, unpleasant working times and uncertain seasonal variations, unfavorable career chances, scarce qualified personnel and high workloads). In contrast, in Germany there are 43 universities, 71 universities of applied sciences and 14 universities of cooperative education that offer education programs to employees in the logistics sector (Roth 2008; Roth and Klaus 2008; Hildebrand and Roth 2010). Furthermore, there are continuing education facilities that also offer academic degree programs in logistics parallel to working employments (Roth 2010). These requirements as well as the very different education backgrounds and biographies of logistics employees are favorable for a competence measurement instrument like e.g. the "*Berufswertigkeit*" *concept which compares individual qualification profiles of persons with work requirements of business practice* (Klumpp 2007; Klumpp and Schaumann 2007). Knowledge management is one research field of the German national excellence research cluster LogistikRuhr, especially within the project "Wissenschaftliche Weiterbildung in der Logistik (WiWeLo)" conducted at the FOM Institute for Logistics and Service Management (ILD) and the University of Duisburg-Essen in Essen/Germany.

This chapter is structured as follows: Sect. 2 outlines the basic concepts and terminologies used regarding competence measurement and qualification analysis, including outlines regarding the German qualification systems. Subsequently Sect. 3 adds the challenges and requirements in today's logistics qualification and Sect. 4 provides the details of the empirical survey according to the Berufswertigkeit analysis concept in Germany (2012/2013). Section 5 presents research results as well as options for further analysis from the survey results. Additionally, Sect. 6 provides an insight into the international standardization efforts by the European Commission (ESCO project) regarding qualifications in logistics and transportation. Section 7

Table 2 Personnel qualification in the logistics industry

Economic sectors 2008		Number of employees covered by social insurance entire federal territory					
		A total of	Thereunder				
			With vocational training	Without vocational training	University of applied science degree	University degree	Vocational training unknown/no allocation possible./n.s.
		1	2	3	4	5	6
All occupations (DE)		*27,710,487*	*16,042,187*	*3,856,768*	*1,075,093*	*1,865,276*	*4,871,163*
Sum logistics occupations, thereof		1,079,759	577,251	148,217	16,088	13,904	324,299
Transport of goods in railway traffic	492	17,995	12,598	3972	491	423	511
Transport of goods in road traffic, moving transport	494	199,431	91,694	19,649	706	609	86,773
Transport through pipelines	495	157	932	79	172	280	107
Transport of goods in ocean and coastal shipping	502	1961	9986	1138	2569	858	5059
Transport of goods in inland water shipping	504	2991	168	300	50	40	921
Transport of goods in aviation and astronautics	512	459	206	105	9	12	127
Warehousing	521	71,792	38,104	16,407	1218	1142	14,921
Provision of other services in traffic	522	558,503	308,254	804	9902	9134	150,813

(continued)

Table 2 (continued)

Economic sectors 2008		Number of employees covered by social insurance entire federal territory					
		A total of	Thereunder				
			With vocational training	Without vocational training	University of applied science degree	University degree	Vocational training unknown/no allocation possible./n.s.
		1	2	3	4	5	6
Postal services of universal service providers	531	154,523	97,978	19,285	668	1077	35,515
Other postal, courier and express services	532	52,885	15,819	6882	303	329	29,552

Federal Labour Office (Bundesagentur für Arbeit) (2011)

outlines further application venues in business practices and finally Sect. 8 provides an outlook regarding the field of logistics knowledge management.

2 Qualification Terminology and Education Systems

Definitions regarding qualification and competence measurement as well as continuing education are hard to come by. The term *continuing education* is also used as further education (Hanft and Knust 2009). In 1970 the German Education Council (Deutscher Bildungsrat) acknowledged and determined continuing education in the German education structure. It can be defined as *continuation or resumption of learning after a first degree*; continuing education therefore usually might begin after entering the workforce (Bildungsrat 1970). Furthermore, continuing education includes formal, informal and non-formal learning. *Formal learning* means a regulated and structured continuing training which is organized by institutions and where students have the chance to gain acknowledged degrees and certificates. *Informal learning* indicates continuing education in project groups, networks and coaching without acknowledged degrees or certificates. *Non-formal* learning is learning by doing or learning on the job without even standardized or organized learning environments and processes. The German Ministry of Education and Research (BMBF) distinguishes continuing education into *general* (not practice- and profession-oriented), *vocational* (practice- and profession-oriented by deepening practical experience) and *higher* (education at universities and universities of applied sciences). In general, through continuing education an advantage for all involved stakeholders can be achieved. These advantages can be clustered in economic and social categories, with three tiers each (European Centre for the Development of Vocational Training 2011):

- *Macro*: *Profit for a whole society*
  - Economic profit: Economic growth and labor-market outcomes
  - Social profit: Crime reduction, social cohesion, health and intergenerational benefits
- *Meso*: *Profit for enterprises and groups*
  - Economic profit: Firms performance and employees productivity
  - Social profit: Inclusion disadvantaged groups
- *Micro*: *Profit for individuals and oneself*
  - Economic profit: Employment opportunities, earning and career development
  - Social profit: Life satisfaction and individual motivation.

In Germany, there exist 16 different *vocational education* programs for logistics such as professions in business administration, truck or train drivers, warehouse oriented professions and professions in CEP (courier, express and parcel delivery services) and moving freight service in the logistics sector (Roth and Klaus 2008). Generally, there are three ways to achieve competences in relevant fields for logistics: After a school qualification students have the option to begin directly in programs of *vocational education training* in the mentioned logistic professions and after achieving the qualification continue the education for specific competences or to begin an academic training at the university or university of applied sciences.

Furthermore, the logistic sector is also characterized by many people changing career tracks and industries and for these types of newcomers there are also ample possibilities within *continuing education* facilitating the acquisition of specific logistics skills (Berufswelt Logistik 2014). To continue education with many years of business practice experience (minimum one year) in the logistic sector there are two different ways to extend individual competences in Germany: through *academic* continuing education programs at universities (part-time or full-time) that offer an academic degree, or through practice-oriented continuing *professional* education.

In order to support success in continuing education and to motivate the current generation suitable tools of learning such as *e-learning* platforms developed within the 21st century: Because information and communication technologies find their way into everyday life, like e.g. smartphones and notebooks, e-learning is nowadays a serious concept, especially for lifelong learning scenarios. The main advantage of e-learning is the possibility of receiving information anytime and anywhere. Besides e-learning scenarios without physical presence in a classroom, blended learning concepts have been realized in which traditional face-to-face learning situations are combined with e-learning elements (Wache 2003).

Table 3 Differences between e-learning 1.0 and e-learning 2.0

E-learning	1	2
Learning environment	A closed area **in** the internet supporting content and tools	An open platform **to** the internet supporting tools for generating content
Teachers	Transfer all known resources into this closed area	Define boundaries and offer resources
Students	Consume the given content	Configure their personal learning environment (PLE) to generate own content

Most educational institutions offer their students e-learning platforms to support the face-to-face learning sessions with additional information. Two main software concepts are Moodle, which was developed at the University of Cologne. According to a definition of Web 2.0 in 2005 e-learning technologies were also developed to stimulate an active participation of the learner. Social software has been conveyed into learning environments, like wikis, podcasts or blogs. The boundary between teacher and learner disappears and collaborative learning scenarios gain in importance so that new technologies evolved (Blees and Rittberger 2009). Table 3 displays an overview of the differences between e-learning 1.0 and 2.0.

3 Competence Requirements in Logistics

Competence is a major asset when measuring logistics industries competitiveness—as also demonstrated by the *Logistics Performance Index* (LPI) published by the Worldbank, where it is one out of six indicators regarding a country-specific competence level in logistics and transportation (Worldbank 2014). Especially from the perspective of logistics practice there are several *major trends* to be recognized and reviewed for future management concepts in relationship with competence and knowledge management (Wimmer 2011, p. 16):

- Globalization: Open Supply Chains and scattered production plants demand for reliable logistic processes.
- Digitalization: High integration of information systems such as telematics, mobile handhelds, tracking and tracing, etc.
- Knowledge management: The success of logistic service providers are often intimately connected with the employees' knowledge.
- Volatility of economic development: The logistics industry experiences a more severe impact of economic fluctuations than other industry sectors.
- Security: For example attacks by pirates on ship as well as disruptions due to civil wars and natural disasters.
- Sensitivity in ecological and sustainability questions: Carbon and other emissions as well as energy and resource consumption of logistics.

Additional to these identified main subjects most high-wage countries (i.e. in Europe and America) are affected by the *demographic change* in their population. Especially the baby-boomer generation with age cohorts from the 1950s and 1960s implicate that after 2020 many employees will withdraw from economic activity because of reaching their retirement age (SBGE 2011). By regarding the identified trends and the development of the aging structure in logistics it is obvious that innovative logistics learning solutions have to be designed by offering employees *possibilities for lifelong learning*. The offer of logistics study as well as qualification and training programs increased in the last 20 years: today, in many European countries universities provide specific logistics studies as well as economic or technical programs with significant logistics content (Hildebrand and Roth 2008; Keuschen and Klumpp 2011). But the main challenge is to provide learning possibilities for employees without leaving their job. Employees have to increase their knowledge to tackle the tasks of logistics management in a global high-velocity supply chain environment. Therefore, flexible e-learning scenarios offer the possibility of knowledge acquisition on the job and account for the above mentioned dependency on current trends. The integration of technical solutions underlines the employees' capability to acquire knowledge within an e-learning scenario.

The capability of logistics learning mechanisms depends on *four components*: temporal components, cultural components, structural components and relational components. The consideration of these four components is a major requirement for a successful learning process. The cultural component can be seen as a basis of learning because the entire logistics sector and supply chains are internationally oriented. The structural component regards the specifications of the employee's organization to realize learning activities on-the-job: flexible in time and position. Relational components assist the collaboration and communication within a strongly cross-linked company structure and the temporal component supports the velocity of changes within the logistics sector and synchronizes them with the learning process (Esper et al. 2007). The goal of learning is to match an employee's knowledge with the needs of the logistics industry. The needs can be displayed as the shape of a 'T' (Fig. 1): The horizontal level displays specific logistics know-how and the vertical level displays the understanding of other company departments with connection to logistics, for example process management, engineering (R&D), production, sales or controlling. In a best case scenario, operations personnel with practical experience has been equipped with management tools and competence and therefore develop into "logistics managers of the future".

For designing education courses in logistics for working persons not only requirements on content have to be regarded. Also, *structural requirements* gain in importance because of the development of the sector. These aspects are relevant because the logistics industry is characterized by several special peculiarities: high speed and flexibility of services, significant shares of small and medium-sized companies and a typically high level of personal *tacit knowledge* leads to the structural requirements listed in Fig. 2.

Competence is thereby defined as "the ability to successfully meet complex demands in a particular context. Its manifestation, competent performance (which

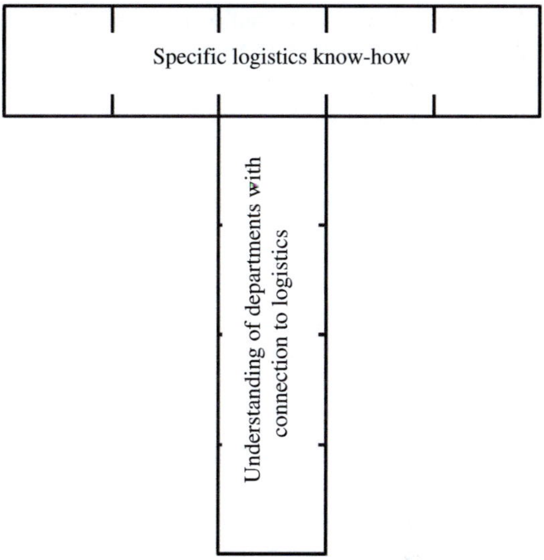

Fig. 1 Employee capabilities for logistics (Mangan and Christopher 2005)

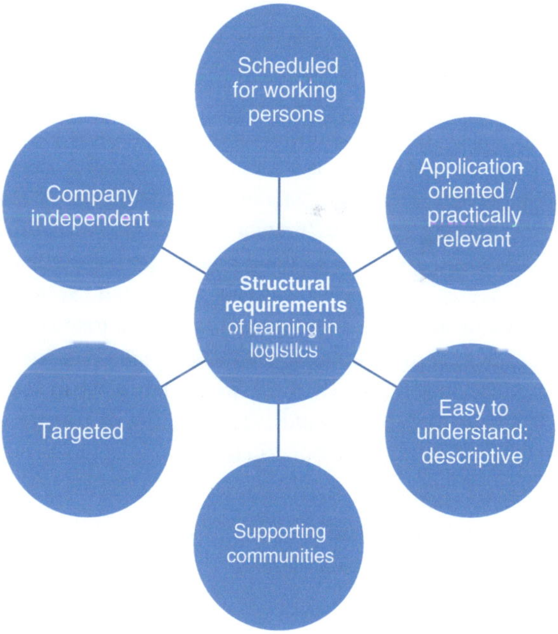

Fig. 2 Structural education requirements for the logistics industry

one may equate to effective action), depends on the mobilization of knowledge, cognitive and practical skills, as well as social and behavioral components such as attitudes, emotions, values and motivations" (Hakkarainen et al. 2004). Competence demonstrates also the level of student achievement in a science education context; competence is therefore not only skill, qualification or knowledge but all these factors together constitute the basis for the competence of an individual person (Liu 2009).

4 Research Question and the Berufswertigkeit Method

In the logistics industry, the access for everyone—and in particular the career changers—for continuing education should be improved. Furthermore, the logistics industry requires specialists, meaning that the access to specialize personal skills also has to be simplified. One major aspect is the demographic change in Germany, which has to be counteracted with more flexible continuing education offers in which the practical on-the-job experiences of employees have to be acknowledged. All these reasons call for a *competence measurement concept* which is precise, practicable and compatible in every economic sector and in particular the logistics sector. This leads directly to the research question of the results reported here:

> *RQ*: How can a quantitative analysis of logistics competences be implemented in order to support the offer and conduct of qualification measures in logistics?

One measurement concept which was developed since 2007 is the German "Berufswertigkeit" concept (Klumpp 2007; Klumpp et al. 2011). It fulfills the needs of a general competence measurement instrument as it is connected to the concept of *employability*. The main idea of "Berufswertigkeit" is a concept of competence measurement of persons with different education degrees. The criteria for an effective competence measurement regarding demands of real-world companies and work processes are *empirically evaluated and selected* from business practice (Klumpp and Schaumann 2007). With these criteria, persons with different education backgrounds and degrees can be objectively compared while the results are output-oriented (no input and curriculum analysis and comparison but competence outputs of different qualification measures). It includes *36 qualification requirement criteria* that represent the modern daily work which are used to individually measure (on a scale of 1-best to 5-worst) and calculates the aggregate "Berufswertigkeitsindex". The qualification requirement criteria are:

- Efficiency
- Independence and own initiative
- Flexibility and adaptability
- Work virtues
- Stress resistance
- Motivation and ability to lifelong learning and maintain to own competence profile
- Coordinate the work- and lifetimes
- Creativity
- Loyalty

- Risk-taking
- Charisma
- Ability to write and speak in German
- Knowledge of foreign language
- Ability to apply modern information- and communication technologies
- Communication and rhetoric
- Assertiveness
- International and intercultural competence
- Costumer focus
- Skills in mathematics and statistics
- Preparation of cost estimates and quotations
- Planning, implementation and documentation of orders and projects
- Negotiations capacity
- Analytical problem-oriented work
- Quality management (optimization of processes and products/service quality)
- Conceptual and strategic implementation of industry-specific knowledge
- Identification with the company
- Strategic orientation
- Understanding solutions of complex technical problems
- Basic knowledge of business administration
- Perception of functions of management and organization
- Conceptual analysis and work
- Planning and control procurement and logistics processes
- Staff requirements and staff mission planning/staff development
- Team, staff and leadership
- Improving responsible care
- Legal knowledge

The Berufswertigkeitsindex (BWI) value is calculated by a summed and unweighted index of an individual personal evaluation of all the 36 qualification requirement criteria. The value range of the index begins by 0 and ends at 100 (0 = evaluation of all criteria with poor and 100 = evaluation of all criteria with very good). In this way, the output-oriented measuring concept "Berufswertigkeit" serves as a basic field-evaluation concept for the development of e.g. an European Qualification Framework for the logistics industry and integrates the required investigation of competences.

A field survey with 1068 persons from the German logistics industry to be evaluated by this concept was conducted in 2012. It was executed as a telephone survey in the German states of North Rhine-Westphalia and Hesse (743 persons in North Rhine-Westphalia and 257 persons in Hesse). Both states have a significant logistics industry cluster environment, i.e. around the inland port of Duisburg (North Rhine-Westphalia) and around the airport of Frankfurt (Hesse). In that survey existing skills and competences of persons in the logistic industry are described. Additionally, traditional formal degrees in vocational and academic education can be classified according to evaluated practical competence levels.

5 Research Findings and Potential Further Analysis

The described information regarding the qualification profiles and levels of logistics personnel can be analyzed and used in different approaches, two exemplary ones are outlined here:

(a) As a *benchmarking approach*, the overall data can be used to compare individual profiles as well as unit and company average values with the total industry average or also specific averages of age cohorts and regions (Alstete 2008). One such benchmarking analysis detail is depicted in the following figure describing the comparison of different age groups within the survey population. The use of the axes in the Berufswertigkeit Index value (BWI) graph is explained as follows for different competence distributions of comparative groups. Taking the first group of up to 25 year-olds, the first value to be shown on the left side of the graph indicates that about 6 % (y axis) of the total group are featuring a BWI value of up to 65 % (out of 100 % maximum)—therefore on the left side of the graph the lower levels of qualification are to be found. Then next about 11 % of the indicated group are evaluated with a BWI level of 65 to up to 70 % and so on. The last section of the graph on the right side indicates that only about 2.5 % of this group has a BWI competence level of 95–100 %. As can be seen, in most cases (given enough individual persons embodied in the analysis groups) a normal distribution can be recognized as may be expected from a general perspective. It should be emphasized that especially not the "maximum" value of the individual graph (in the example group a value of about 28 % of the group featuring a BWI level of between 75 and 80 %) is most relevant and interesting, but the relative position of each graph to other groups regarding the distribution between the left-hand side (lower qualification levels) and the right-hand side (higher qualification levels). For example the comparative group of the "56 years and older" group displays a distribution graph which has a distinctively lower "maximum" value (only 21 % of this group show a BWI level of between 75 and 80 %) but is shifted to the right side of the figure compared to the younger age group (especially regarding the higher qualification levels). This indicates that this group has *higher competence levels* on average (Fig. 3).
Three benchmarking approaches are now feasible in this example: (i) First individuals can be compared to their specific age cohort: If we analyze and counsel for example a 24-year-old employee in a logistics company, we can compare his individual BWI value (say for example 81 %) to the age group distribution and conclude that his individual value is above the group average. (ii) Second the age profile of a complete company can be compared to the overall averages shown here as aggregate values for the company. (iii) Third in an overall industry analysis the group cohorts can be compared as indicated below.

(b) A further approach would be a *gap analysis*, searching for competence potential in the direction of further development of individuals as well as groups in

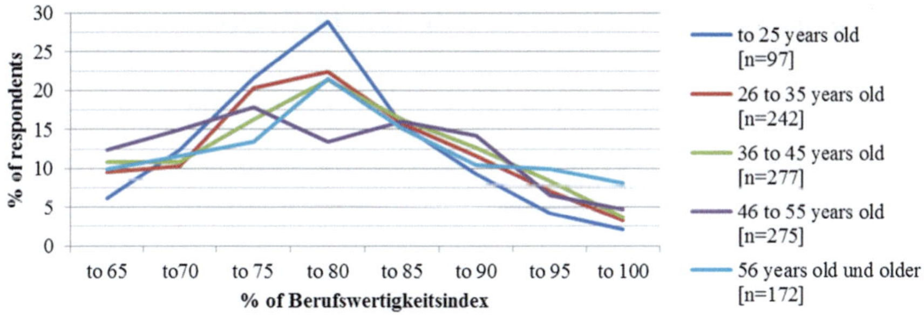

Fig. 3 Berufswertigkeit results based on the age of respondents

logistics. In the following figure different employment groups within logistics are compared to each other. Due to now significant lower numbers per groups, the picture is a little bit more "fuzzy" (not the "smooth" normal distribution). But it can be observed that for example for the interesting group of "blue collar" workers (yellow), there is an unexpectedly high value in the right part of the graph, indicating for example that about 8 % of this group features a BWI competence level of 90–95 % ("hidden talent"). This can be explained by quite a high share of "unusual" entrants into this group, e.g. from people with an even academic education especially when entering the workforce in Germany from abroad (migration background). This would make it interesting to identify these individuals and probably to train them for further purposes and tasks within logistics processes as they provide a high competence level to start with—which is currently most likely not used to its full advantage, neither for the individual employee nor for the company employer (Fig. 4).

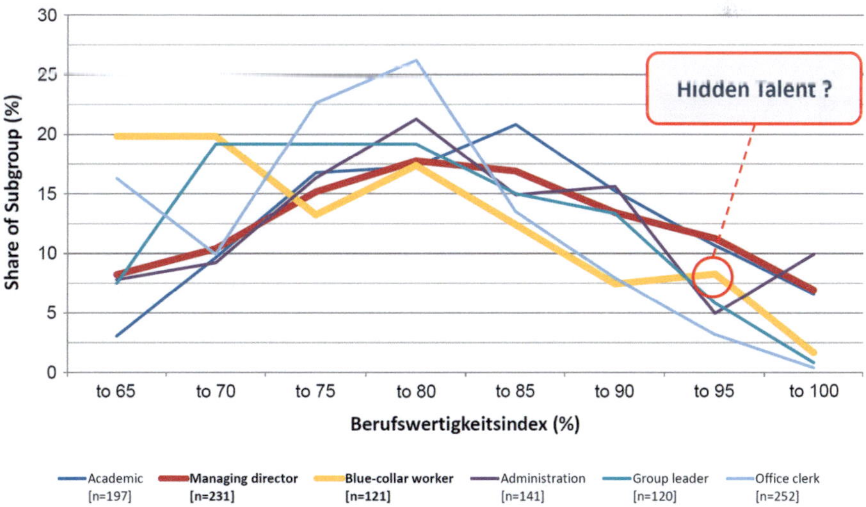

Fig. 4 Berusfwertigkeit results based on the occupational position of respondents

6 European Standardization: The ESCO Project

The ESCO initiative of the European Commission stems from the observation that many institutions—namely public and private agencies with the task of labor market matching—are missing a *common European description and system of occupations and qualifications* across industry sectors. Therefore, in 2011 the project "European Skills/Competences, Qualifications and Occupations" (ESCO) was launched and expert reference groups for different industries were established. The following figure provides the basic structure and mission of ESCO (Fig. 5).

In 2013, the ESCO reference group '*transportation and storage*' was founded and a first expert meeting took place in April in Brussels. The basic definition from this European expert group regarding transportation is as follows:

> The ESCO reference group 'transportation' deals with all occupations and required skills, competences and qualifications **addressing in a major part** the efficient, transparent, sustainable and effective **transport of persons and cargo** via the transport modes air, water, rail, road and pipelines as well as the related logistics services of transport planning and forwarding, transshipment, storage, logistics projects, reverse logistics and value added services. (ESCO Transportation 2013, p. 4)

Early results from the ESCO meetings in 2013 are interesting because a first level structure for all occupations and *qualifications* in the transportation and logistics sector was discussed and agreed upon. This structure is segmented along the transport modes air, water, rail and road as well as additional logistics services (transshipment, warehousing, project and contract logistics services etc.). Further prototyping regarding specific occupations within these top-level groups were

Fig. 5 ESCO Mission and Structure (Le Vrang 2013)

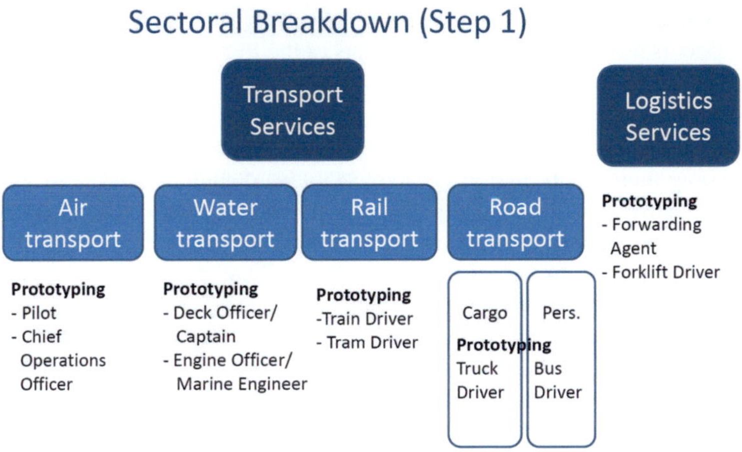

Fig. 6 ESCO reference group transportation—first level structure

decided (e.g. pilot, truck driver, forklift driver; according to ESCO Transportation 2013, p. 5) (Fig. 6).

Such elaborate *prototyping* will show whether the general ESCO approach to link occupations, skills/competences and qualifications is feasible in general and specifically for the logistics industry. In 2016 a final recommendation of the reference group regarding the whole transportation sector in Europe shall be commenced. This will also be based on a further structural outline for a second and third level of the ESCO structure (e.g. within air transportation).

It should be recognized that this development and high-level project will (i) change many structures and processes in logistics personnel management (HR) regarding personnel search, evaluation and identification as well as knowledge management and further qualification and (ii) is basically output-oriented and attempting to standardize as well as to quantify the core asset of knowledge and competences in the logistics industry (as well as in other industries across Europe). This can be linked to the further establishment of a common Industry Qualification Framework (IQF) for the Logistics industry in Europe based on the European Qualifications Framework (EQF) (Abidi et al. 2011; European Parliament 2008).

7 Applying the Findings in Logistics Practice

The following items can be outlined in order to support the further transfer and implementation of quantitative competence analyses such as the Berufswertigkeit index into logistics practice:

(a) First of all, it has to be mentioned that in general, traditional resentments towards quantitative analysis regarding education and competence in logistics have to be renounced. Though it is important to take criticism into account, a complete neglect is no longer feasible as concepts as the Berufswertigkeit concept prove to be more and more useful and the logistics and transportation sector increasingly requires professional competence management as explained.

(b) As the traditional borders between blue- and white-collar logistics employees become more permeable in the long-run, also quantitative analysis regarding competences may help: identification of potentials (blue-collar) of individuals rather than groups in order to apply educational resources has to be given priority in order to enhance the overall qualification and competence of the logistics workforce. Besides the overall competence level, such a measurement as for example with the Berufswertigkeit index scale can also identify specific competence fields with strengths and weaknesses. Therefore, especially for blue-collar workers as shown above, interesting potentials and possibly also untraditional educational measures—such as a management part-time study program—may be applicable as the individual tests provide such detailed information. In the described empirical research for example we found PhD graduates driving trucks in German logistics companies: they had been immigrating from Eastern European countries and the formal academic qualifications were not acknowledged in the German labor market—but still these people ascertain high skill and competence levels, justifying a totally different approach in further training and development than standard stereotypes regarding blue-collar workers in logistics.

(c) Finally, especially in logistics also quantitative competence information as shown with the results from the Berufswertigkeit survey in logistics can be used to flexibly source "ad hoc"-teams as e.g. needed for logistics projects in contract logistics, ramp-up-management or other large project setups. In these cases, structured qualification information may provide an easy option in sampling the necessary competences and therefore personnel inside one company and possibly throughout supply chains if shared among supply chain partners. This would enable logistics and supply chain performance to increase due to shorter lead times in preparing and setting up projects and processes in global value chains.

8 Conclusion and Outlook

This contribution has shown that especially in the logistics industry as internationally oriented, knowledge-intensive service industry *professional knowledge and qualification management tools are required*, in particular in the face of current trends and future challenges from technology as well as societal and organizational directions. This includes state of the art methods for competence measurement (using an empirical and quantitative approach), information sharing regarding competences in logistics as well as learning in modern social and communicative learning environments. These fields have for long been neglected by logistics research and are now in high demand especially after the economic crisis and with companies looking out for high-qualified personnel. The overall objective in these management fields is to identify, analyze and enhance logistics and practical management knowledge individually for each and every person employed in this sector. It is not limited to a specific group (formal qualification level or age group) but addresses all people within logistics and even those who want to enter future occupations in the industry (ESCO project). Altogether, these endeavors may be able to increase the attraction of the logistics industry as employer as well as supporting its innovation capacity for other industries and society as a whole—as innovation requires advanced knowledge levels.

Further research as for example in the German national research excellence cluster LogistikRuhr has to show how feasible toolkits and methods can be used in logistics by a broad share of employees and companies—keeping in mind that the majority of logistics companies are small and medium-size enterprises. Therefore, especially *individualized online processes* in qualification measurement as well as training will play an important role in the future in this industry. It can be safely assumed that those companies who are going to embrace this development and will implement analysis and training tools based upon these quantitative analysis and individual matching instruments will have a *competitive advantage* in the long run, especially due to impacts of demographic change in the sector.

References

Abidi H, Klumpp M, Keuschen T (2011) Industry qualifications framework logistics: explicit and tacit knowledge and qualifications in the logistics industry. In: Blecker T, Jahn C, Kersten W (eds) Maritime logistics in the global economy—current trends and approaches. Köln, Eul, pp 327–341

Alstete JW (2008) Measurement benchmarks or "real" benchmarking? An examination of current perspectives. Benchmarking Int J 15(2):178–186

Berufswelt Logistik (2014) Berufe in der Logistik. Available from internet sources: http://www.berufswelt-logistik.de/?weiterbildung_logistik. Accessed on 08 July 2014

Blees I, Rittberger M (2009) Web 2.0 learning environment: concept, implementation, evaluation. eLearning Papers 3(15)

Bildungsrat Deutscher (ed) (1970) Empfehlungen der Bildungskommission, Strukturplan für das Bildungswesen. Eigenverlag, Bonn

ESCO Transportation (2013) Meeting minutes 29.04.2013. ESCO Transportation, Brussels

Esper T, Fugate B, Davis-Sramek B (2007) Logistics learning capability: sustaining the competitive advantage gained through logistics leverage. J Bus Logistics 28(2):57–82

European Parliament (2008) Recommendation of the European parliament and the council of 23 April 2008 on the establishment of the European qualifications framework for lifelong learning. Official J European Union, 2008/C 111/01, Brussels

European Centre for the Development of Vocational Training (ed) (2011) The benefits of vocational education and training. Publications Office of the European Union, Luxemburg

Hakkarainen K, Palonen T, Paavola S, Lehtinen E (2004) Communities of networked expertise: professional and educational perspectives. European Association for Research on Learning and Instruction, Amsterdam, Oxford

Hanft A, Knust M (2009) Continuing higher education and lifelong learning—an international comparative study on stuctures, organization and provisions. Oldenburg, München
Hildebrand W, Roth A (2008) Führungskräfte für die Logistik—Akademische Ausbildung in Deutschland. In: Baumgarten H (ed) Das Beste der Logistik. Innovationen, Strategien, Umsetzungen. Springer, Berlin
Keuschen T, Klumpp M (2011) Logistikstudienangebote und Logistiktrends. FOM ild Schriftenreihe Logistikforschung Band 11. FOM, Essen
Klaus P, Hartmann E, Kille C (2010) Die Top 100 der Logistik. Ausgabe 2010/2011, Hamburg
Klumpp M (2007) Begriff und Konzept Berufswertigkeit. Arbeitspapiere der FOM Hochschule für Oekonomie & Management No. 5, 07/2007, Essen
Klumpp M, Abidi H, Marner T (2014) Qualification hurdles and chances for e-mobility. Int J Transport Traffic Eng 4(3):253–268
Klumpp M, Clausen U, ten Hompel M (2013a) Logistics research and the logistics world of 2050. In: Clausen U, ten Hompel M, Klumpp M (eds) Efficiency in logistics., Lecture Notes in LogisticsSpringer, Heidelberg, pp 1–6
Klumpp M, Dilger B, Kriebel K, Diart M (2011) Studie "Berufswertigkeit konkret"—Vergleich der Berufswertigkeit spezifischer beruflicher Weiterbildungsabschlüsse und hochschulischer Abschlüsse, Abschlussbericht. Paderborn, Eul
Klumpp M, Kandel C, Wirsing E (2013b) Cargo telematics for operational transport excellence and strategic knowledge management. In: Kreowski H-J, Scholz-Reiter B, Thoben K-D (eds) Dynamics in logistics. Lecture Notes in Logistics Series, Springer, Heidelberg, pp 71–84
Klumpp M, Krol B, Zelewski S (2007) Knowledge intensive service production analysis (DEA). In: Conference proceedings 14th international working seminar on production economics. Innsbruck, vol 1, pp 193–206
Klumpp M, Schaumann U (2007) Requirements for leadership personnel and the concept Berufswertigkeit. Kölner Blätter für Wirtschaftspädagogik 12:3–50
Lancioni R, Forman H, Smith MF (2001) Logistics and supply chain education: roadblocks and challenges. Int J Phys Distrib Logistics Manage 31(10):733–745
Le Vrang M (2013) ESCO vision and business cases. Presentation 29.04.2013, Brussels. Available from internet sources: http://prezi.com/vgvyappjudll/esco-vision-business-cases. Accessed 13 July 2014
Liu X (2009) Linking competence to opportunities to learn: models of competence and data mining. Springer, New York
Mangan J, Christopher M (2005) Management development and the supply chain manager of the future. Int J Logistics Manage 16(2):178–191
Rao K, Stenger AJ, Wu H-J (1998) Integrating the use of computers in logistics education. Int J Phys Distrib Logistics Manage 28(4):302–319
Rideout B (2011) Printing the impossible triangle: the copyright implications of three-dimensional printing. J Bus Entrepreneurship Law 5(1):6
Roth A (2008) Aus- und Weiterbildung in der Logistik. In: Klaus P, Krieger W (eds) Gabler Logistik Lexikon. Gabler, Wiesbaden
Roth A (2010) Der Logistikbildungsmarkt 2020—Bedarfsgerechte Bildungsangebote für die Logistikwirtschaft in der Logistik. Hamburg, DVV
Roth A, Klaus P (2008) Bildungsmarkt Logistik. Hamburg, DVV
Sachverständigenrat zur Begutachtung der gesamtwirtschaftlichen Entwicklung (SBGE) (2011) Herausforderungen des demografischen Wandels, Expertise im Auftrag der Bundesregierung. Statistisches Bundesamt, Wiesbaden
Self JA (2011) The use of design tools in industrial design practice. PhD thesis, Kingston University, London
Wache M (2003) E-learning—Bildung im digitalen Zeitalter. Bundeszentrale für politische Bildung, Bonn

Wimmer T (2011) Trends in der Logistik. In: Schenk M, Zadek H, Müller G, Richter K, Seidel H (eds) Proceedings of the 16th Magdeburger Logistiktage 2011, pp 15–18

Worldbank (2014) Logistics performance index global ranking. Available from internet sources: http://lpi.worldbank.org/. Accessed on 23 July 2014

Wu YCJ (2007) Contemporary logistics education: an international perspective. Int J Phys Distrib Logistics Manage 37(7):504–528

Serious Games for Improving Situational Awareness in Container Terminals

Alexander Verbraeck, Shalini Kurapati and Heide Lukosch

Abstract Container terminals decouple the long-distance, high volume deep-sea transport from the short-distance, highly fragmented hinterland transport. As around 65 % of general cargo is shipped in containers, these terminals play a very important role in global trade. Planning and aligning the activities in and around container terminals is quite difficult, even more because of frequent disturbances in the seaside and the landside operations within the transportation network. The theory of Situational Awareness (SA) might help to improve the alignment between planning activities and to increase resilience. SA asks, however, for a different way of working and communicating. Awareness and practice sessions are needed to train managers and planners to use the SA concepts. The Dinalog project SALOMO has developed tools and serious games to serve as a training, learning, and "try-out" tool for students and practitioners to gain experience in Situational Awareness, alignment of planning activities, and in dealing with disturbances in and around container terminals. This chapter discusses the Situational Awareness framework, on which the tools and games have been built, as well as first experiences in using these tools and games.

1 Container Transportation

Global container trade by deep-sea vessels is responsible for 65 % of the general cargo that is transported internationally (UNCTAD 2014), and for some routes even for 100 % (Steenken et al. 2004). Over 160 million TEU (Twenty foot Equivalent

A. Verbraeck (✉) · S. Kurapati · H. Lukosch
TU Delft, PO Box 5015, 2600 GA Delft, The Netherlands
e-mail: a.verbraeck@tudelft.nl

S. Kurapati
e-mail: s.kurapati@tudelft.nl

H. Lukosch
e-mail: h.k.lukosch@tudelft.nl

H. Zijm et al. (eds.), *Logistics and Supply Chain Innovation*,
Lecture Notes in Logistics, DOI 10.1007/978-3-319-22288-2_25

Units, the size of a 20 ft container) were transported in 2013 (UNCTAD 2014). The value of the transported containerized goods is more than 50 % of all global seaborne trade (WTO 2013, Table 3), and amounts to more than US$4 trillion annually (World Shipping Council 2014). Container terminals play a key role in the transhipment of containers between deep-sea vessels and hinterland transportation modes. More than 200 countries have container terminals to handle container ships (World Shipping Council 2014). Several functions are carried out by the container terminal: loading and unloading of deep-sea vessels, transportation of containers between the quay and the stack, storage and retrieval of containers from the stack, loading and unloading of hinterland transportation modes, and organizational activities and paperwork such as customs inspection (Steenken et al. 2004). The container terminal is organized into a number of areas (see Fig. 1): the deep-sea quay where deep sea vessels are loaded and unloaded, the horizontal transportation area to transport containers between the quay and the stack, the stack where containers are stored, loading areas for trucks and trains, and the gate where trucks are cleared for picking up or delivering a container (Brinkmann 2011). In case an inland waterway system is in use for the terminal, there is also a barge quay. Empty containers can be stored in a separate area, or in the general stack. A separate area on the container terminal or empty depot is sometimes in use for empty containers because they can be stacked higher than full containers. Containers carrying dangerous goods, and containers that are refrigerated (so-called reefers) are also stored

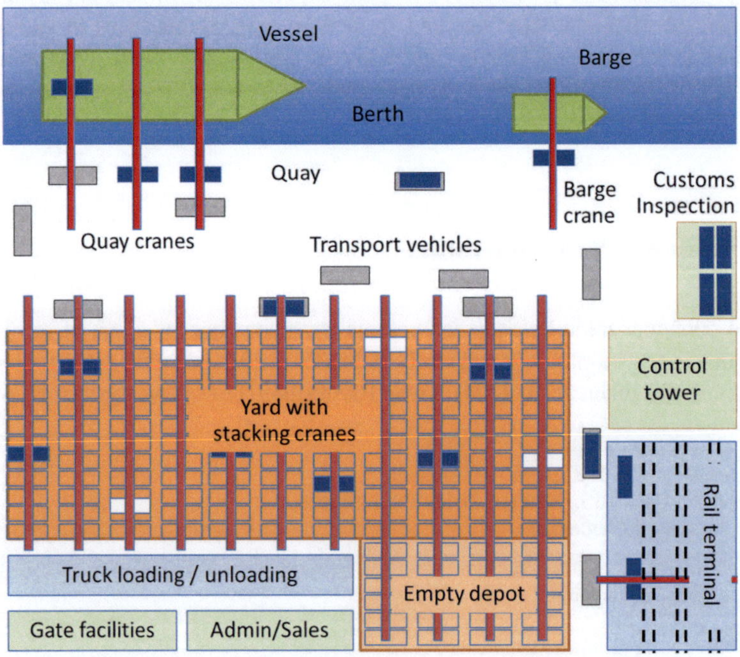

Fig. 1 General layout of a container terminal

in special areas of the terminals (Wiese et al. 2011). Reefers need to be hooked-up to a power source, which is typically a manual process.

Container ships have tremendously increased in size over the past decades, where the largest ships can now transport 18,000 TEU (Tran and Haasis 2015). This asks for a higher berth productivity in the container terminals than what is currently being offered (Port Technology International 2011), because the shipping lines want to have the ships handled as efficiently as possible in the terminals. Therefore, large vessels are served by multiple quay cranes that load and unload the containers (Brinkmann 2011). As a result, crane planning is an important planning task (Böse 2011). When unloading the ship, the quay crane places the container on the quay to be picked up by a vehicle, or containers can be directly placed on a vehicle. Vehicles can for instance be Automated Guided Vehicles (AGVs), Automatic Lifting Vehicles (ALVs), or manned straddle carriers (Kemme 2011). These vehicles transport the containers between the quay and the yard. In the yard, containers are stacked and retrieved by Rubber-Tyred Gantry cranes (RTGs) or Rail-Mounted Gantry cranes (RMGs) (Saanen 2004). There can be one or two cranes per stacking lane. In case of two cranes, one serves the sea-side containers, and the other the land-side containers. When no import or export containers have to be transferred to or from the stack, the cranes are used for so-called housekeeping moves to optimize the location of the containers in the stack. e.g., when a container that will have to be taken from the yard an hour from now is at the bottom of a stack of four containers, housekeeping moves will 'dig' for the container by first moving the containers that are on top of it (Legato et al. 2012).

One of the main functions of a container terminal is to decouple the deep-sea transport from local and hinterland transport (Steenken et al. 2004). Containers that arrive or depart by deep sea vessel can for instance be transported between the customer and the terminal by a combination of transport moves by truck, train, barge, or short-sea vessels (IDVV 2012; Notteboom 2008). Separate dedicated areas in the container terminal are available for (un)loading of trucks (Guan and Liu 2009), and (un)loading of trains (Newman and Yano 2000) (see also Fig. 1). In case of barge transportation, barges either share the quay with deep-sea vessels, or are served at a barge quay by separate cranes (Douma 2008; Melis et al. 2003).

2 Planning, Variability, and Disturbances

Planning and aligning all functions in the terminal is a difficult task (Brinkmann 2011). It involves, amongst others, planning of the berth location and timing of deep-sea vessels, feeders and barges, stowage planning of the ships (which determines the sequence of containers to be (un)loaded), planning of the number and location of quay cranes to use for each ship, planning of the horizontal transport vehicle operations, planning of the location of the container in the stack, allocation of stack resources for sea-side or land-side operations, and planning of priorities for (un)loading containers from the stack for different modalities (Böse 2011; Steenken et al. 2004). All these planning activities are interrelated, and changes in one plan have a big influence on other plans (Böse 2011). When a ship arrives late and has to use a different quay position, the location of the containers that have to be loaded onto the ship is often sub-optimal, leading to longer horizontal transportation times (Pani et al. 2014). When import containers are picked up by a different modality than planned, containers were, in hindsight, stored in the wrong part of the stack. When a horizontal transport vehicle that carries a container for a deep-sea vessel breaks down, other containers that were planned on top of that container in the ship cannot be loaded until the container of the broken vehicle has arrived to comply with the stowage plan. In addition, the broken vehicle can block other vehicles with containers, leading to a propagation of the disturbance. When a truck arrives at the terminal without a pre-notification to pick up a container, the container can be at the bottom of the stack and housekeeping moves are necessary to retrieve the container. This will take time causing waiting time for the truck, and it keeps the stacking crane busy which may cause delays for sea-side operations (Phan and Kim 2015). Again, a delay in retrieving one export container for a deep-sea vessel may cause delays for other containers because of the strict sequence of loading in the stowage plan. Usually different planners, sometimes even from different departments are responsible for maintaining the plans on the terminal. While terminals operate on a 24/7 basis, not all planners are available during the night. The so-called Terminal Operating System (TOS) helps to store and align several of the plans, but aligning all plans remains a difficult task (Lau and Zhao 2008), and human intervention is often needed to solve problems.

Many of the processes in a container terminal show variability (Han et al. 2010; Vis and de Koster 2003). An example is the number of moves per hour of quay cranes, which may differ considerably between container terminals, but also between crane operators. Weather (wind, fog) can also have a severe influence on the speed by which containers can be loaded and unloaded from vessels. Driving times in the horizontal transport system can be influenced by congestion. Stacking crane operations are impacted by the unpredictability of the arrival of trucks to pick up containers and/or bring containers. Train and barge arrivals can deviate from plan, due to delays at the previous (un)loading location or congestion. Because of the tight interrelations between plans, other processes are affected as well by the variability.

Supply Chain and Transportation executives are worried about the effect of external disruptions on their operations. On top of their list are natural disasters, conflicts and political unrest, sudden demand shocks, export or import restrictions, weather conditions, and terrorism.[1] Each of the larger disturbances or risks, described for instance in Behdani (2013), Harrington et al. (2011), Jüttner et al. (2003), Manuj and Mentzer (2008), can have a ripple effect in the terminal (Gurning

[1]World Economic Forum. New models for addressing supply chain and transport risk. http://www3.weforum.org/docs/WEF_SCT_RRN_NewModelsAddressingSupplyChainTransportRisk_IndustryAgenda%202012.

and Cahoon 2011). When a deep-sea vessel with 5000 containers arrives two days late due to bad weather conditions, the planned berth positions will probably not be available. Import containers will be too late to be loaded on the scheduled barges and trains. Truck drivers that want to pick up urgent containers arrive too early. Alerting thousands of truck drivers is impossible, because of the effort this would take, and because it is usually unknown to the terminal who are going to pick up or bring a certain container.

Therefore, important challenges are: to better align plans and make them more robust against disturbances; to be able to detect deviations and disturbances quickly; to communicate them in an efficient manner to departments or parties who need the disturbance information to adapt their processes; to design ways to effectively deal with deviations and disturbances; and to train employees in detection, communication, and problem solving. These challenges are addressed in a project called SALOMO, which is funded by Dinalog, the Dutch Institute for Advanced Logistics. SALOMO stands for Situational Awareness for LOgistic Multimodal Operations, and focuses on improved detection and communication of disturbances based on a framework for (Shared) Situational Awareness (SSA) (Endsley 1995; Endsley and Jones 1997) and Distributed Situational Awareness (DSA) (Boy 2013), and on serious gaming methods to research and train problem solving capabilities in relation to SA (Lo and Meijer 2014; Lukosch et al. 2014; Perla et al. 2000). The next section will introduce the framework for Situational Awareness, and the subsequent sections will briefly introduce a number of the developed tools and games in the SALOMO project that help to train planners and managers to use the Situational Awareness concepts.

3 Situational Awareness Framework

In spite of the fact that container terminals become more and more automated (Saanen 2004) and information technologies such as a Terminal Operating System are in use at all large terminals, dealing with variance and disturbances remains difficult. Larger terminals and larger container ships make it difficult to assess the situation at hand, choose with whom to communicate and collaborate, and jointly decide on the best course of action. According to Endsley (1995), Situation(al) Awareness or SA is ‘the perception of the elements in the environment within a volume of time and space, the comprehension of their meaning, and the projection of their status in the near future’. Endsley defines three phases for Situational Awareness (Endsley and Jones 1997), which look to be particularly applicable to handle variance and disturbances on the container terminal and in the entire transportation chain:

1. **Perception** of the elements in the environment. The first step in achieving SA is to perceive the status, attributes, and dynamics of relevant elements in the environment. In case of disturbances, this can already be hard. Finding out that

several of the thousands of trucks that arrive at a container terminal are severely delayed is outside the typical information gathering capabilities of a terminal. Assessing the effects of a crane breakdown on a terminal on the abilities to load or unload a barge is typically not possible for a barge operator. Of course it would be possible to 'push' the information to the other party rather that to try to 'pull' it, but it is hardly ever known on beforehand which stakeholders would need certain pieces of information.

2. **Comprehension** of the current situation. By combining elements from the first step into patterns, the decision maker forms a holistic picture of the environment, comprehending the significance of objects and events. In case of several trucks being late, the planner on the terminal can look for a common cause, such as road works, a strike, or an accident. If an AGV is malfunctioning and has to be repaired, the location and routes of other AGVs can be checked, as well as the status of the container on the AGV. This creates a full picture of the situation that can help to develop a course of action.
3. **Prediction** of future status. In order to prepare for action, the future status of the system has to be known. Will the situation deteriorate further? What other stakeholders will be impacted, and at what time? The future state of the system is modeled using the knowledge of the status of the elements in the system and their dynamics, and insight into the interaction patterns between different parts of the system. If a ship is estimated to arrive late, this is related to the estimated time of arrival of other ships, to the berth plan, and to the quay crane plan, to create a full picture of the future state of the berth and queue that can be used to assess whether the late arrival can be accommodated or whether replanning would be necessary.

Situational Awareness is a preparation for a design process of potential future actions, and a decision making process where the best course of action will be chosen, after which it can be implemented (Endsley and Jones 1997).

Where Endsley's model was developed for more individual settings, such as SA for fighter pilots in the Air Force, we are looking for SA in a setting that involves multiple decision makers. In such cases, different actors want to create a "common operational picture" (Allen et al. 2014; Nofi 2000) to help different decision makers to solve problems as a team, rather than individually. Instead of optimizing individual goals, which could be far from the optimal solution, a team jointly decides on the best course of action. When we want to accomplish Shared Situational Awareness (SSA), shared mental models, communication and cooperation are crucial (Nofi 2000). As Nofi (2000, p. 29) states: "*communications is the most critical issue in treating shared awareness.*" Shared awareness is often accomplished in a team setting, where actors can quickly exchange information about the evolving situation (shared mental model), and can align their planned courses of action with each other (cooperation). One can imagine that in case of a major delay of an ocean vessel, or a truckers' strike, several planners and decision makers can work together on designing the best course of action. These collaborative sessions typically take place within one organization, as there is a common frame of reference and a single authority that can decide in case of trade-offs.

When we move outside the organization into the transport network, the single authority and common frame of reference are lacking. Boy (2013) indicates that in such complex, socio-technical and distributed settings, four concepts play a major role to accomplish Distributed Situational Awareness (DSA): sharing, distribution, delegation, and trading. Let's illustrate this with the example of a deep-sea vessel arriving late. *Sharing* relates to the fact that the shipping line, which usually learns of the late arrival first, informs other stakeholders such as the terminal and freight forwarders about the disturbance that might occur. *Distribution* means that each actor can now assess the situation from their role in the system. The terminal can, for instance, make more cranes available to other ships in order to stimulate an early departure of one of the current ships to free space on the quay for the delayed ship. These changes in plan need to be shared, in turn, with other planners and outside stakeholders such as the shipping line of the currently berthed ships. Freight forwarders can inform trucking companies to reschedule deliveries and pick-ups of containers. *Delegation* is needed because not all actors have the power or ability to implement required actions. For instance, replanning of trains and barges might be needed for containers that will not arrive on time due to the late arrival of the ship. If these plans are made by external freight forwarders, the decision making and action implementation needs to be delegated to them. After delegation, sharing changes in the current situation with the parties to which decisions have been delegated becomes even more important. Finally, *trading* refers to negotiation among actors. In many cases, distribution and delegation are not sufficient. Trade-offs need to be discussed between actors. In the case of the ship arriving late, the late arriving ship could be allowed to berth when it arrives, forcing another ship to wait, or the delayed ship could wait longer for an empty spot in the plan, increasing its tardiness even further. Clearly, this is not something a single organization can decide. The effects of different options have to be discussed and potentially the decision between the options has to be escalated to higher positions in the organizations (delegation) to force a final decision.

These ideas have been implemented into a framework for Shared and Distributed Situational Awareness as depicted in Fig. 2 (Kurapati et al. 2012, 2013). For individual situational awareness, the three elements of Endsley (1995), perception, comprehension, and prediction are key. When we want to accomplish Shared Situational Awareness, shared mental models, communication and cooperation are the main concepts (Nofi 2000). On the organizational and inter-organizational level (the system's level), we have to address planning and disturbances in a distributed setting (Boy 2013). Here, sharing, distribution, delegation, and trading play a major role.

The SALOMO project uses Serious Games to study the Situational Awareness framework. The games simulate the scenarios in which situational awareness can be used to increase the resilience of planning operations and management decisions. Because the games form a controlled environment, they are ideal to observe and study the activities of the players, in their role as planners or decision makers

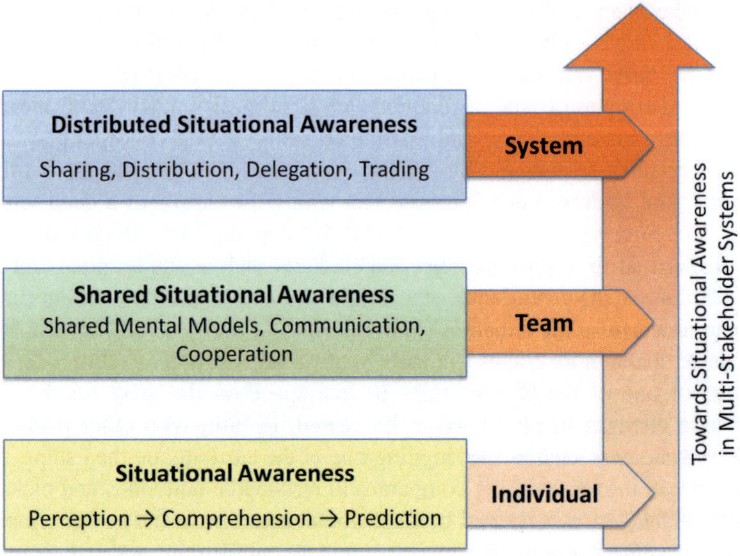

Fig. 2 Conceptual framework for shared & distributed situational awareness

(Mayer 2009). The same games can be used later as training games to replicate situations that the planners and decision makers have to be trained for (Crookall 2010). One of the main questions we had to answer was what elements to include in the game to be able to study and train the concepts of the three layers of the Situational Awareness framework. The next two sections cover two games, one game in which we studied SA and SSA for planning operations, and one in which we studied SSA and DSA for disruption mitigation. For both games we introduced a number of specific game elements that will be summarized in Sect. 6.

4 Container Terminal Planning Game: SA and SSA

To study game elements for Individual Situational Awareness and Shared Situational Awareness, an online, web-based training game was developed by SALOMO partner InThere.[2] The game is a so-called microgame (Kurapati et al. 2014) where learning takes place in short sessions, which can be repeated a few times to explore which kind of strategy works best (Lukosch et al. 2013). Microgames support situated learning, as they always start from a well-defined problem of a client, which is translated into a short simulation game (Lukosch and de Vries 2009; Overschie et al. 2013). In this particular game, the Yard Crane

[2]http://www.inthere.nl.

Scheduler game (YCS), players have to allocate resources to container terminal functions as described in Fig. 1. The main objective of the YCS game is to manage the yard and align various planning and resource allocation activities in the container terminal. The YCS game provides the top view of a container terminal with the quayside, and the yard storage areas. Deep-sea vessels that bring containers and pick-up containers arrive at the terminal. Their berth location will be determined by the scenario, but the locations where containers are stacked can be chosen by the player by making an unloading plan for the vessel. Multiple vessels have to be handled and their arrival and departure times are shown (see Fig. 3 at the right). Cranes have to be allocated to each vessel by dragging them to the right locations for unloading and loading the ship. Horizontal transportation between the quay and the stack takes place automatically. The player can move stacking cranes to the right positions to move containers between the stacking location and the horizontal transport vehicles. Several containers have to be unloaded from one ship, and loaded onto another ship that arrives later. Choosing the right stack location for those containers can be quite difficult. Other containers are picked up by trucks. Truck loading has to be handled as efficiently as possible. Several so-called missions can be played with an increasing level of difficulty. On the higher levels, equipment breaks down at random times, forcing players to adapt their plans to a continuously changing environment.

Players have to make sure that ships are unloaded and loaded in time, and equipment keeps busy. Ships that cannot be (un)loaded in time result in negative points, ships that have been fully handled and depart early in bonus points. Equipment that is idle also costs points. The player has to be aware of a number of different plans or decisions that have to be aligned to each other: the unloading schedule, the loading schedule, the quay crane allocation, and the stacking crane locations.

Fig. 3 Yard-crane-scheduler game (part of the screen)

4.1 SA Mode of the YCS-Game

The standard mode of playing the game is single-user. Players have to align several planning operations to efficiently unload and load a number of vessels during a so-called mission. This means that they will go through the SA phases perception, comprehension and prediction multiple times during a mission. By studying their scores in successive iterations of a mission, we can look whether learning takes place. To help *perception*, visualization plays an important role. Several indicators during the game help the player to perceive the state of the container terminal. Examples are pre-notification of the arrivals and departures of ships (arrival and departure information on the right in Fig. 3), and warnings for idle quay cranes and yard cranes (yellow triangles on the cranes in Fig. 3). In addition to visualization, sounds indicate loading and unloading operations and departure of ships, helping players to keep track of the status of the terminal when studying other parts of the terminal. To help *comprehension*, training missions are provided, in which players can practise patterns of activities for different operations such as making an unloading plan, a yard plan, and a loading plan, and planning quay crane and yard crane movements. These patterns come back during the more complex missions where all activities need to be combined and aligned. *Prediction* is realized by giving players detailed feed-back on several aspects of their performance and letting them play the same mission again to see whether other patterns of activities yield a better result.

Our hypothesis was that players would improve the quality of their plans in successive iterations of a mission, due to gaining more situational awareness. In order to test this, we set up an experiment in December 2014 with M.Sc. students of the R.H. Smith School of Business of the University of Maryland in the USA, who specialize in transportation and supply chain management. The students played a number of rounds of practice missions, after which they played two iterations of a more complex mission that we included in our analysis. In this way, the learning effect (getting to know container terminal operations and understanding the working of the game) should have mostly disappeared. In total, 50 students played the complex mission twice. Their average scores went up from 4160 to 5154 (with a minimum score of −772 and a maximum score of 13,658, see Fig. 4). Individual

	μ 1	μ 2	σ1	σ2	min	max	correlation	t-value	sig.	different?
Game scores	4160	5154	2913	3300	-772	13658	0.710	-2.938	0.005	yes
Playing time (s)	471	470	69	54	132	633	0.586	0.167	0.868	no
Waiting ship (s)	867	846	136	112	449	1166	0.516	1.150	0.256	no
Waiting stack (s)	434	411	192	166	0	976	0.476	0.892	0.377	no
Waiting yard (s)	323	302	195	207	6	1421	0.761	1.050	0.299	no
Waiting berth (s)	144	141	73	85	0	322	0.354	0.240	0.811	no
QC idle time (s)	142	138	70	76	9	409	0.439	0.420	0.676	no
YC idle time (s)	738	751	176	215	286	1301	0.597	-0.521	0.605	no

Fig. 4 Differences between scores for first and second iteration of mission play in YCS (n = 50)

scores between playing the mission the first time and the second time were highly correlated with a paired sample correlation of 0.71, which shows that low scoring students stayed low, and high scoring students stayed high, but both improved on average on their scores. In spite of the high standard deviation of the scores (2913 for the first iteration and 3300 for the second iteration), the difference between the first and second time score was highly significant; a paired samples t-test showed a t-value of −2.9 with a significance of 0.005 (Fig. 4). Because we challenged the students to improve their scores (the highest scorers received actual prize money), time on the second iteration of the mission was not significantly lower than on the first one (471 vs. 470 s). The hypothesis that the game time differed between the two iterations could be rejected. This also indicates that the students did not just become more "skillful" in playing the same mission, as that would certainly have led to a lower game play time. Interesting is that also here, playing time to complete the mission correlated highly between the first and second mission with a value of 0.586. Finally, we looked at a number of more detailed variables in the game to see whether students improved their scores by just getting more skilled in using the game and thereby decreasing waiting times at individual resources (see bottom 6 rows in Fig. 4). The fact that each of these *partial* indicators did not change significantly according to the paired t-tests, but the *overall* scores did, indicates that the players were better able to integrate the different patterns of activities in the second iteration of the mission, but their pure skills to handle individual aspects of tasks remained the same on average. This was confirmed by discussions with the students in the debriefing session after game play.

4.2 SSA Mode of the YCS-Game

A second version of the game was created, in which four different planning roles in the terminal are separated and played by four players in a team. The screen looks similar to the game screen of Fig. 3, but players can only influence the plans and operations they are responsible for, and cannot see all planning decisions of the other players. In order to fulfill their mission, individual SA is not enough. Players have to create a common operational picture (Nofi 2000) to efficiently unload and load the deep-sea vessels, which means that Shared Situational Awareness is needed. As indicated in Sect. 3, SSA is enabled by three concepts: shared mental models, communication, and cooperation. In the distributed YCS-game, we supported the formation of a *shared mental model* in several ways: one is the fact that all players see the same screen of the terminal. This provides the common operational picture (Allen et al. 2014; Nofi 2000) for the four planning roles. Secondly, the players carry out several missions with the single-user YCS-game to get used to the terminal and its operations. Thirdly, we have the players play all four roles in a single-user version of the multi-player game. In this version of the game, they have to explicitly switch to the right role before they can make a planning decision. This helps them to understand the information needs of the other players and the handovers between activities. *Communication* is supported by allowing the players to talk, by using a chat program (in case we want to capture their communication for research purposes), or by allowing the players to draw on a centrally placed picture of the terminal (in case we want to study the effect of exchanging location-dependent information). *Cooperation* is forced in the game because all four roles are needed to unload and load the vessels. Because not all information is shared on the screen that all players see, they will have to help each other to run the terminal in an efficient manner. Cooperation is further stimulated by giving the teams the goal to maximize their *team score*, with a team prize for the best performing team, which challenges them to align their activities even more.

In Spring 2015, we played the YCS-game in a distributed fashion for the first time. It became immediately clear that it is much more difficult to run the terminal than in the single-player version. Even in the single user version that asks for explicit role switching, scores already drop tremendously compared to individual play, and decision efficiency decreases. In the multi-player version where players were facing each other and could talk to each other, decision times increased, and players found it very hard to exchange the right information with each other, in order to help the other planners to do their work in the best possible way. As this was the whole purpose of developing the distributed version of the game, it forms an excellent starting point for further research on SSA.

4.3 Turning the YCS-Game into a Training Instrument

The game has been played by over 200 players in a large number of different sessions, involving both students and professionals from different backgrounds and countries. The sessions have been evaluated using questionnaires, focusing on the usefulness of the YCS game as a training instrument as well as the playability for potential users. The survey results show that players become very aware of the dependencies between different functions in the container terminal, and the need to align plans. The majority of the participants indicated that their understanding of the processes in container terminal operations and their interdependencies increased by playing the game. They also mentioned in the debriefing after the game that they understood the need for coordinating and aligning planning and resource allocation in container terminal operations. When playing the distributed version, it is clear that players become aware of the enormous amount of information that has to be exchanged between planners to run the terminal in an efficient manner. The microgame is able to illustrate the dynamics and interrelatedness of planning operations, and still provides a pleasant experience, as was stated by the majority of the test persons. In summary, we can conclude that the YCS game provides a well-balanced learning experience. Its conceptualization as web-based, short simulation game answers both the need for a dynamic representation of a complex problem and for flexible, situated learning approaches in complex working environments.

5 Disruption Mitigation Game: SSA and DSA

The Disruption Mitigation Game was developed by TU Delft as a table-top round-based board game that can be played by five people or five teams. In addition, SALOMO partner Open University developed a mobile version of the game that can be played using tablets or mobile phones. The game was conceived to achieve a set of key objectives (Kurapati et al. 2013):

- Understanding the impact of increased shared situational awareness on individual, group and system level performance;
- Setting a foundation to identify measures to increase SA;
- Offering a frame of reference to assess SA in the organization (SSA) and the network (DSA);
- To serve as a training tool for disruption management.

The key elements of the Situational Awareness framework have been translated into contextualized game play, using the principles of 'reality, meaning and play' of the triadic game design method (Harteveld 2011). The game focuses on how to handle disruptions in the transportation network around the container terminal, and whether and when to communicate information related to the disturbance to other players. Several roles are played in the game. The *vessel planner* has to decide on the time and location for unloading and loading the ship. One of the main goals for this player is to reduce vessel waiting times. The *yard planner* decides on the storage positions for the incoming and outbound containers in the yard. This player is responsible for ensuring sufficient stack capacity to maintain overall performance of the terminal. The *control tower* keeps track of all operations on the container terminal, and has to give permission for operations that violate existing plans. *Sales* is responsible for the booking of containers that the terminal handles and for financial transactions between the clients and the terminal. It has to arrange alternatives for clients during disruptions of the container flow in and out of the terminal. This player needs to keep the customer informed at all times, and is responsible for customer satisfaction. The *Resource planner* assigns equipment to each vessel planned for the terminal, such as quay cranes, automated guided vehicles, gantry cranes, and reach stackers. Goals for this planner are to ensure high performance and an even distribution of equipment use. Three types of disruptions can be played in the game: an equipment breakdown in the terminal with local consequences, an accident in the terminal that forces a complete shutdown while the investigation is running, and a truckers' strike with a blockage of the terminal gate that creates a major disturbance inside the terminal as well as in the hinterland network (Fig. 5).

5.1 SSA Mode of the Disruption Game

The disruption unfolds during the rounds of the game, and the players are given pieces of information (through game cards) about the disruption. The formation of

Fig. 5 Disruption mitigation game

shared mental models in the Disruption Game is supported by a shared central board with goals and decision steps of all the players that helps them to align their plans, and to consider the decisions of the other players. An important aspect of the game is that the players are evaluated on three aspects: safety in the terminal, customer satisfaction, and performance of the terminal. Evaluation is carried out both for their individual roles and as a team. Several decisions that players can make during the game affect the individual score in a negative way, but have a positive effect on the overall KPIs of the terminal, which creates a trade-off and might tempt players to not share all their ideas and plans with the other players. When playing in "SSA mode", the individual KPI scores and the team score can be viewed by all players. *Communication* is supported by a set of communication tokens that allow players to exchange information with one other player, or with all other players (a conference call). The two types of communication have different costs in terms of the number of tokens needed. Although there is a cost attached, the number of tokens is chosen in such a way that players are aware of the costs of communicating, but they don't see the costs as a burden. Mitigating each disruption needs a careful alignment of tasks between the five players. If certain steps are skipped, or intended steps are not shared with others, problems solving becomes less effective and the team score as well as the individual scores will decrease. Therefore, *cooperation* between all players is forced.

5.2 DSA Mode of the Disruption Game

According to Boy (2013), DSA is characterized by sharing, distribution, delegation, and trading. In the distributed version of the game, players cannot see the scores of others, and as a result there is more emphasis on the individual scores, which means that the players will behave more egoistic compared to the SSA version of the game. *Distribution* is implemented by giving different players different pieces of information (through game cards) about the disruption, which other players cannot see. SA is limited to awareness of individual responsibilities and goals. Players receive an individual board, with their respective KPI, goal, and decision steps based on their role description. There is a central board with the 3 overall KPIs visible to everyone, whereas individual boards are shielded from each other. Therefore, none of the players has a complete operational picture to effectively deal with the disruption. Therefore, *sharing* of information is needed to address the disruption. If certain steps are skipped, or important information is not shared with others, the problem will deteriorate and both individual and team scores will plummet. *Delegation* is implemented by allowing players to give a game card to another player at the end of a round. This enables the receiving player to start working on the issue that is mentioned on the game card. Other players do, however, not know what task was delegated. Finally, *trading* behavior in the form of trade-offs is implemented by making information exchange between players possible, but only on a one-to-one basis, and with high costs attached to communicating. They can send an email for 2 tokens (always received, but with a delay) or make a phone call for 1 token (we roll a dice to see whether the other party picks up). Players have only 10 tokens each, which makes communication much more of a burden. Trade-offs between communicating (spending tokens fast) and not communicating (hoping the information exchange was not really needed), and between individual scores (personal goal) and central scores (not really necessary to have a high score there, but on the other hand also an indicator of overall progress) can lead to interesting trading behavior, which is certainly much different from the more "open" information sharing in the SSA-mode of the game.

5.3 Turning the Disruption Game into a Training Instrument

The game has been tested with over 100 participants, both professionals and students at different universities. Players typically play on both levels, where they are able to see the effects of more or better communication and insight into each others' decisions on the individual and overall scores. Video recordings of the game play have been made, and players have answered questionnaires about playability and usability. First results indicate that players become aware of the interdependencies between different functions when addressing a disturbance in the transportation network around a container terminal, and learn about the usefulness of visibility of information and decisions of others, and the effects of (not) exchanging information. The observation of the gameplay as well as the feedback of the participants by survey show that the simulation game introduced creates a helpful learning experience in the field of disruption management and resilience of container terminal operations. The gameplay shows that players make use of communication and information sharing channels provided in the game.

6 Discussion, Conclusions and Further Research

The two games that have been described in detail in the paper address Situational Awareness in terms of the framework that was presented in Fig. 2. The YCS game looks at Individual Situational Awareness when playing in single-player mode, and at Shared Situational Awareness when playing in distributed mode. Important game elements in single player mode are visualization for perception, small training missions and detailed feedback for comprehension, and repetition for prediction. For the SSA setting of the YCS game, cooperation is forced by designing tight inter-dependencies between activities. Providing a partial shared view on the terminal for all players and supporting communication (chat, drawing or speech) helps in creating a shared mental model for the players. First results have shown, however, that it is very difficult for players to align their individual plans in the shared setting, even though they all look at the same situation of the terminal and can exchange information fully. This illustrates the importance of training for improving shared situational awareness.

The Disturbance Mitigation Game studies the differences between Shared Situational Awareness and Distributed Situational Awareness. When handling a disruption, distribution of information and not having a "common operational picture" makes it very difficult for the players to make the right decisions. Full sharing of the information and "conference calls" make it a lot easier for the players to align their decisions and increase the overall scores of the game. Players immediately understand the important role communication plays in the complex process of container terminal operation and hinterland transportation. Although the disruption game is a board game, the game elements that were used for SSA were very similar to those used in the computerized YCS game. For DSA, several new elements were introduced, such as hiding most information for other players, using game cards with information for delegation, and forcing trade-offs for the participants by having them focus on individual scores and creating a high cost and a time delay for communication.

With both games, we were able to address all three levels of the Situational Awareness framework. Our studies have provided first sets of game elements that can be used to study *and* train Individual, Shared, and Distributed Situational Awareness for planning operations and resilience in and around deep-sea container terminals. Further research will focus on studying the differences between the levels of the Situational Awareness Framework of Fig. 2 in more detail and enhancing the game elements that can help to research and train SA. Several more experiments with the different tools in the SALOMO project will be carried out to study the effectiveness of SA methods in transport networks.

The games and tools that have been developed in the SALOMO project will be made available to the wider logistics and transportation community, where they can be used for training, creating awareness, and further research on the necessary ingredients to make transportation networks more resilient.

Acknowledgments This research was funded by Dinalog, the Dutch Institute for Advanced Logistics. The SALOMO project (2011–2015) was carried out by TU Delft, APM Terminals Maasvlakte 2, InThere, Open Universiteit, Rotterdam World Gateway, TBA, TeamSupport, TRAIL Research School, and University of Maryland.

References

Allen DK, Karanasios S, Norman A (2014) Information sharing and interoperability: the case of major incident management. Eur J Inf Syst 23:418–432

Behdani B (2013) Handling disruptions in supply chains: an integrated framework and an agent-based model. PhD thesis, Delft University of Technology, Delft, The Netherlands

Böse JW (2011) General considerations on container terminal planning. In: Böse JW (ed) Handbook of terminal planning. Springer, New York, pp 3–24

Boy GA (2013) Orchestrating situation awareness and authority in complex socio-technical systems. In: Aiguier M, Caseau Y, Krob D, Rauzy A (eds) Proceedings of the third international conference on complex systems design and management CSD&M 2012, Springer, Berlin, pp 285–296

Brinkmann B (2011) Operations systems of container terminals: a compendious overview. In: Böse JW (ed) Handbook of terminal planning. Springer, New York, pp 25–39

Crookall D (2010) Serious games, debriefing, and simulation/gaming as a discipline. Simul Gaming 41(6):898–920

Douma A (2008) Aligning the operations of barges and terminals through distributed planning. PhD thesis, University of Twente, Enschede, The Netherlands

Endsley MR (1995) Toward a theory of situation awareness in dynamic systems. Hum Factors 37 (1):32–64

Endsley MR, Jones WM (1997) Situation awareness, information dominance, and information warfare. Technical report, United States Air Force Armstrong Laboratory, Belmont, MA. Report AL/CF-TR-1997-0156

Guan C, Liu R (2009) Container terminal gate appointment system optimization. Maritime Econ Logistics 11(4):378–398

Gurning S, Cahoon S (2011) Analysis of multi-mitigation scenarios on maritime disruptions. Maritime Policy Manage 38(3):251–268

Han X, Lu Z, Xi L (2010) A proactive approach for simultaneous berth and quay crane scheduling problem with stochastic arrival and handling time. Eur J Oper Res 207(3):1327–1340

Harrington LH, Boyson S, Corsi TM (2011) X-SCM—the new science of X-treme supply chain management. Routeledge, New York

Harteveld C (2011) Triadic game design: balancing reality, meaning and play. Springer, Heidelberg

IDVV (2012) Multimodal international container network (in Dutch). Deliverable 3.2. Technical report, Rijkswaterstaat, Delft, The Netherlands. Impuls Dynamisch Verkeersmanagement Vaarwegen, Spoor 3, Cluster 3

Port Technology International (2011) Berth productivity will have to keep up with shipping's super-sized revolution. Port Technol Int 50:18–20

Jüttner U, Peck H, Christopher M (2003) Supply chain risk management: outlining an agenda for future research. Int J Logistics Res Appl 6(4):197–210

Kemme N (2011) RMG crane scheduling and stacking—overview and implications on terminal planning. In: Böse JW (ed) Handbook of terminal planning. Springer, New York, pp 271–301

Kurapati S, Groen D, Lukosch HK, Verbraeck A (2014) Microgames in practice: a case study in container terminal operations. In: ISAGA 2014: 45th conference of the international simulation and gaming association. SAGSAGA, Dornbirn, Austria, pp 1–14

Kurapati S, Kolfschoten GL, Verbraeck A, Corsi TM, Brazier FMT (2013) Exploring shared situational awareness using serious games for supply chain disruptions. In: Proceedings of the 10th international ISCRAM conference, May 2013

Kurapati S, Kolfschoten GL, Verbraeck A, Drachsler H, Specht M, Brazier FMT (2012) A theoretical framework for shared situational awareness in sociotechnical systems. In: Proceedings of the 2nd workshop on awareness and reflection in technology-enhanced learning, Saarbrücken, Germany, pp 47–53

Lau HYK, Zhao Y (2008) Integrated scheduling of handling equipment at automated container terminals. Int J Prod Econ 112(2):665–682. (Special section on RFID: technology, applications, and impact on business operations)

Legato P, Mazza RM, Trunfio R (2012) Simulation for performance evaluation of the house-keeping process. In: Laroque C, Himmelspach J, Pasupathy R, Rose O, Uhrmacher AM (eds) Proceedings of the 2012 winter simulation conference (WSC). IEEE, p 1422

Lo JC, Meijer SA (2014) Gaming simulation design for individual and team situation awareness. In: Meijer SA, Smeds R (eds) Frontiers in gaming simulation—44th international simulation and gaming association conference, ISAGA 2013. Springer, Switzerland, pp 121–128

Lukosch HK, de Vries P (2009) Supporting informal learning at the workplace. Int J Adv Corp Learn (iJAC) 2(3):39–44

Lukosch HK, Littlejohn A, Margaryan A (2013) Simulation games for workplace learning. In: Littlejohn A, Margaryan A (eds) Technology-enhanced professional learning: processes, practices, and tools. Routledge, New York/London, pp 158–167

Lukosch HK, van Nuland B, van Ruijven T, van Veen L, Verbraeck A (2014) Building a virtual world for team work improvement. In: Meijer SA, Smeds R (eds) Frontiers in gaming simulation—44th international simulation and gaming association conference, ISAGA 2013. Springer, Switzerland, pp 60–68

Manuj I, Mentzer JT (2008) Global supply chain risk management. J Bus Logistics 29(1):133–155

Mayer IS (2009) The gaming of policy and the politics of gaming: a review. Simul Gaming 40 (6):825–862

Melis M, Miller I, Kentrop M, van Eck B, Leenaarts M, Schut M, Treur J (2003) Distributed rotation planning for container barges in the Port of Rotterdam. In: Verduijn T, van de Loo B (eds) Intelligent logistics concepts. Eburon Publishers, Delft, pp 101–116

Newman AM, Yano CA (2000) Scheduling direct and indirect trains and containers in an intermodal setting. Transport Sci 34(3):256–270

Nofi AA (2000) Defining and measuring shared situational awareness. Technical report, Center for Naval Analyses, Alexandria, VA. Report CRM D0002895.A1/Final

Notteboom T (2008) The relationship between seaports and the inter-modal hinterland in light of global supply chains. Technical report, OECD Publishing. Report No. 2008/10

Overschie M, Lukosch HK, Mulder K, de Vries P (2013) Micro-training to support sustainable innovations in organizations. In: Appelman JH, Osseyran A, Warnier M (eds) Green ICT and energy: from smart to wise strategies. CRC Press, Boca Raton, pp 97–106

Pani C, Fadda P, Fancello G, Frigau L, Mola F (2014) A data mining approach to forecast late arrivals in a transhipment container terminal. Transport 29(2):175–184

Perla PP, Markowitz M, Nofi A, Weuve C, Loughran J, Stahl M (2000) Gaming and shared situation awareness. Technical report, Center for Naval Analyses, Alexandria, VA. Report CRM D0002722.A2/Final

Phan M-H, Kim KH (2015) Negotiating truck arrival times among trucking companies and a container terminal. Transp Res Part E Logistics Transp Rev 75:132–144

Saanen YA (2004) An approach for designing robotized marine container terminals. PhD thesis, Delft University of Technology, Delft, The Netherlands

Steenken D, Voß S, Stahlbock R (2004) Container terminal operation and operations research—a classification and literature review. OR Spectrum 26(1):3–49

Tran NK, Haasis H-D (2015) An empirical study of fleet expansion and growth of ship size in container liner shipping. Int J Prod Econ 159:241–253

UNCTAD (2014) Review of maritime transport 2014. Technical report, UNCTAD, Geneva, Switzerland. Report UNCTAD/RMT/2014

Vis IFA, de Koster R (2003) Transshipment of containers at a container terminal: an overview. Eur J Oper Res 147(1):1–16

Wiese J, Suhl L, Kliewer N (2011) Planning container terminal layouts considering equipment types and storage block design. In: Böse JW (ed) Handbook of terminal planning. Springer, New York, pp 219–245

World Shipping Council (2014) Website on global trade 2014. http://www.worldshipping.org/about-the-industry/global-trade. Accessed 2 Oct 2014

WTO (2013) WTO press release 688, 10 April 2013, World trade 2012, prospects for 2013. http://www.wto.org/english/newse/pres13%20e/pr688%20e.htm. Accessed 2 Oct 2014

MIX
Papier aus verantwortungsvollen Quellen
Paper from responsible sources
FSC® C105338

If you have any concerns about our products,
you can contact us on
ProductSafety@springernature.com

In case Publisher is established outside the EU,
the EU authorized representative is:
Springer Nature Customer Service Center GmbH
Europaplatz 3, 69115 Heidelberg, Germany

Printed by Libri Plureos GmbH
in Hamburg, Germany